Trafficking of Intracellular Membranes:

From Molecular Sorting to Membrane Fusion

NATO ASI Series

Advanced Science Institutes Series

A series presenting the results of activities sponsored by the NATO Science Committee, which aims at the dissemination of advanced scientific and technological knowledge, with a view to strengthening links between scientific communities.

The Series is published by an international board of publishers in conjunction with the NATO Scientific Affairs Division

A Life Sciences B Physics	Plenum Publishing Corporation London and New York
C Mathematical and Physical Sciences D Behavioural and Social Sciences E Applied Sciences	Kluwer Academic Publishers Dordrecht, Boston and London
F Computer and Systems Sciences G Ecological Sciences H Cell Biology I Global Environmental Change	Springer-Verlag Berlin Heidelberg New York London Paris Tokyo Hong Kong Barcelona Budapest

NATO-PCO DATABASE

The electronic index to the NATO ASI Series provides full bibliographical references (with keywords and/or abstracts) to about 50000 contributions from international scientists published in all sections of the NATO ASI Series. Access to the NATO-PCO DATABASE compiled by the NATO Publication Coordination Office is possible in two ways:

- via online FILE 128 (NATO-PCO DATABASE) hosted by ESRIN, Via Galileo Galilei, I-00044 Frascati, Italy.
- via CD-ROM "NATO Science & Technology Disk" with user-friendly retrieval software in English, French and German (© WTV GmbH and DATAWARE Technologies Inc. 1992).

The CD-ROM can be ordered through any member of the Board of Publishers or through NATO-PCO, Overijse, Belgium.

Series H: Cell Biology, Vol. 91

Trafficking of Intracellular Membranes:

From Molecular Sorting to Membrane Fusion

Edited by

Maria C. Pedroso De Lima

Departamento de Bioquimica
Universidade de Coimbra
3049 Coimbra, Portugal

Nejat Düzgüneş

Department of Microbiology
University of the Pacific
San Francisco, CA 94115, USA

Dick Hoekstra

Department of Physiological Chemistry
University of Groningen
9712 KZ Groningen, The Netherlands

Springer

Published in cooperation with NATO Scientific Affairs Division

Proceedings of the NATO Advanced Study Institute "Trafficking of Intracellular Membranes: From Molecular Sorting to Membrane Fusion", held at Espinho, Portugal, June 19-30, 1994

ISBN-13: 978-3-642-79549-7 e-ISBN-13: 978-3-642-79547-3
DOI: 10.1007/978-3-642-79547-3

CIP data applied for

Softcover reprint of the hardcover 1st edition 1995

SPIN 10134835 31/3130 - 5 4 3 2 1 0 - Printed on acid-free paper

Preface

This volume contains the lectures presented at the NATO Advanced Study Institute (ASI) on "Trafficking of Intracellular Membranes: From Molecular Sorting to Membrane Fusion", held in Espinho, Portugal, from June 19 to June 30, 1994.

The objective of this Institute was to survey recent developments and to discuss future directions in the rapidly advancing field of membrane cell biology, with particular emphasis on the dynamical properties and intracellular flow of membranes. A wide range of interrelated topics around the central theme of intracellular trafficking of membranes was covered, including lipid flow, membrane fusion, dynamics of membrane components, protein folding and assembly, vesicular transport in membrane biogenesis, exocytosis and endocytosis. A large variety of experimental techniques and systems, including the application of viruses and model systems, to study these processes was also considered.

Membrane cell biology is a broad discipline which encompasses many scientific areas including cell biology, biochemistry, biophysics, virology, immunonology and genetics. Indeed, recent advances in the cell biology of membranes could not have been made without this multidisciplinary approach. Significant progress achieved during the last few years in understanding how newly synthesized lipids and proteins find their way to the cell organelles, how molecular sorting and the continuous flow of membranes allow each cellular membrane to maintain its own distinct molecular composition, and, thereby, the individuality of the various intracellular compartments, was discussed in considerable detail in this Institute. An understanding of the dynamics of cellular membranes is highly relevant to issues such as virus infections, but also to topics like intracellular and transmembrane signalling, cell differentiation and oncogenesis.

We hope that the availability of the papers presented at the Institute as chapters in this book, and the original data they contain, will be useful as a reference source and, in particular, will stimulate future work. In addition, we hope that the Institute itself has provided an opportunity to beginners and experienced scientists alike, for bringing forward a fertile exchange of ideas, leading to productive collaborations.

Finally, we wish to acknowledge the financial support of the NATO Scientific Affairs Division, the main sponsor of this Advanced Study Institute. The support given by the Portuguese institutions, Fundação Luso-Americana para o Desenvolvimento, and the Junta Nacional de Investigação Científica e Tecnológica (JNICT) is also much appreciated and is gratefully acknowledged. Funds provided by the Rectory of the University of Coimbra and the City Council of Espinho were also indispensable to the activities of the Institute.We are indebted to all the authors for their contributions and, last but not least, to all participants for their active interest throughout the entire meeting.

The editors

Maria C. Pedroso de Lima
Nejat Düzgüneş
Dick Hoekstra

November 1994

Contents

Protein-Lipid and Lipid-Lipid interactions in model systems and in biological membranes

A. Bienvenüe, J. Sainte-Marie and L. Maurin
"Dynamique Moléculaire des Interactions Membranaires" URA 1856 CNRS.
Université Montpellier II, cc 107
F-34095 MONTPELLIER Cedex 5
France.

A good indication for the importance of lipids is that cells need a huge amount of different lipids in strictly regulated location and composition. Essentially, the cell membranes fulfil 5 essential functions: acting as a semi permeable frontier with respect to ions, metabolites, peptides, proteins and larger assemblies; participating in cell organization and motion by bending, fusion and budding processes; solvating hydrophobic parts of integral proteins, stabilizing them in native conformation and correct lateral organization; interacting with extrinsic proteins in order to regulate their enzymatic activity or their binding and aggregation properties; stocking metabolites (fatty acids such as arachidonic acid, alkyl phospholipids, diacyl glycerol, inositol phosphates...).

This review is devoted to the physico-chemical aspect of lipid-lipid and lipid-protein interactions. The major question is how these interactions are responsible for lipid localization in membranes (lateral and transverse domains) and in whole cell (membrane lipid composition, membrane traffic and lipid sorting). Indeed, the tiny regulated localization of lipids obviously is due to more or less specific interactions either between lipids and proteins either between lipids themselves. In any case, the relative affinities and the lifetime of the interactions have to be considered in order to delineate the sorting process. For a clear answer, lipid-protein interactions should be considered from two complementary, although rarely used, points of view: lipid effect on protein properties (catalytic activity, binding, structure, movement, aggregation state ...); protein effect on lipid properties (essentially structure, movement and possibly phase separation). In fact, generally only one aspect was studied among the following: lipid lateral or transverse organization, lipid movement and order, lipid/protein boundary layer composition and dynamics, lipid effect on protein aggregation, association, conformation or functioning. In the last part, one example will be presented where lipids influence one important whole cell function (endocytosis).

NATO ASI Series, Vol. H 91
Trafficking of Intracellular Membranes
Edited by M.C. Pedroso de Lima N. Düzgüneş and D. Hoekstra

In the present review, we discuss results obtained with quite different kinds of experimental systems and interpret them by a convergent model of the structure-function relationship in biological membrane: the diversity in lipid composition, organization and distribution is the compromise solution between the need for integral and extrinsic proteins to find a correct (hydrophobic and polar head) environment and the need for the membrane components to rapidly move and react during membrane traffic, shape changes and cell activation.

Lipid organization in model and biological membranes: experimental methods and results.

Many different methods (Devaux, 1983;Seelig andSeelig, 1980) give a quite precise survey on: a) the isomerization rates between lipid conformations; b) the hydrophobic chain order relative to the bilayer normal, measured by a parameter contained within 0 (for a completely free and isotropic movement) and 1 (for a full ordered conformation)

Magnetic resonance methods give dynamic informations on the correlation time of the lipid conformational rearrangements, on lateral diffusion rates and chain order. The most convenient method is the EPR spectra of nitroxide spin labeled lipids, because its sensitivity is quite high while it detects movements on the 10^{-11}-10^{-6} second time scale (down to 10^{-3} second with some loss in sensitivity), corresponding to the major lipid movement rates. High local concentrations also may be detected thanks to the exchange broadening of the corresponding spectra. In spite of its very low sensitivity, deuterium NMR is very useful too, because the ^{2}H atoms are far less disturbing for the membrane structure that 5-atoms cyclic structures containing nitroxide. Its characteristic time range is in the order of 10^{-6}-10^{-3} second, complementary to the EPR time domain. It allows a very simple and direct measurement of order parameters of the deuterated carbon atoms. ^{31}P NMR also provides us with very interesting informations on phospholipid head groups organization and movements, mainly related to the non bilayer structures.

Other methods are currently used, generally giving informations on the average conformations and order. For example: infra-red absorption or reflection spectroscopies, as well as Raman spectroscopy, give a snapshot photograph of the chain conformation; fluorescence spectroscopy allows an easy (quite doubtful however) measurement of the so-called membrane "fluidity", and gives some insights into the lipid-lipid and lipid-protein contacts mainly by resonance energy transfer or quenching methods. Also, calorimetry is very useful to draw lipid mixture phase diagrams, while X-ray or neutron diffraction methods are essential for studying the overall lipid phase organization.

Thanks to these methods, the lipid organization in pure phospholipid bilayers is well known. Fast isomerization rates and a very low order characterize the center of the bilayer. By contrast, the relatively high bilayer viscosity is due to the interfacial part of the membrane, where the isomerization movements are slow and the order parameter high (Seelig andSeelig, 1980, Venable et al.,1993). Molecular dynamics simulations of a lipid bilayer and of hexadecane: an investigation of membrane fluidity. Broadly speaking, we safely can assert that these results are valid for biological membranes, with a very similar order parameter profile for lipids in model and in Acholesplasma laidlawii membranes (Seelig and Browning, 1978). In the same line, Fluorescence Recovery after Photobleaching (FRAP) experiments give the same lateral diffusion coefficient in the two cases.

However, diverging from this apparently common scheme, some differences appeared, then leading to a less isotropic image of biological membranes.

LIPID-LIPID INTERACTIONS

Lipid mixtures of two different types of phospholipids, for example, give generally full miscibility of the two components at a temperature higher than their transition points. In some cases however, more complicated phase diagrams appeared, showing some azeotropic behaviour (see in Bultman et al., 1991 for example). In mixtures of phospholipids containing two lipid chains of very different lengths, two phospholipid molecules associate tail-to-tail in the interdigitated state: the bilayer thickness is given by adding the shorter and the longer chain lengths (Shah et al., 1990). Phospholipid-cholesterol mixtures exhibit an unexpected phase diagram. Two different liquid phases coexist: the first one, observed when the cholesterol molar ratio is low, contains almost pure disordered phospholipids. The other one (called liquid ordered phase) is present at low as well as at high temperature for a cholesterol/phospholipid molar ratio greater than 0.20. In this phase, the lipid chains are quite ordered, while the molecules are able to diffuse almost as fast as in ordinary liquid phases.

A special case of lipid-lipid interaction is exhibited in sphingolipid-cholesterol mixtures. Detected by lateral pressure measurements at the air-water interface (Lund-Katz et al., 1988), this interaction leads to a striking cell property. For example, it was shown by Slotte and Biermann (Slotte and Biermann, 1988). that sphingomyelin degradation in the plasma membrane by sphingomyelinase produces cholesterol acylation and the storage of cholesterol esters in intracellular lipid droplets. This phenomenon was shown to be reversible, emphasizing the very special role played by cholesterol in animal cells (Bloom and Mouritsen, 1988). In the same line, cholesterol seems to be very important in the stabilization of the so-called caveolae structures, recently discovered to allow some of the GPI-linked proteins to be internalized into cells by a non endocytic pathway (Rothberg et al.,1990).

LIPID-PROTEIN INTERACTIONS

The function of many membrane-associated proteins depends on the lipid composition of the bilayers (table I and reference Bienvenüe and Sainte Marie, 1994). It is important to emphasize that all classes of proteins appear in this table: integral proteins (enzymes, receptors and transporters) as well as extrinsic ones (enzymes, cytoskeleton and proteins able to translocate between cytoplasm and membranes).

Among them, the nicotinic acetylcholine receptor (AChR) was studied particularly (see review by Barrantes, 1993). This pentamer of very homologous subunits ($\alpha_2\beta\gamma\delta$), generally purified from Torpedo electrocytes, presents many functional effects of lipids or hydrophobic molecules. The high affinity binding of acetyl choline to AChR triggers a short lasting open state (about 4ms) of a high conductance (about 40pS) sodium ion channel. As many other proteins (see later) AChR has a loose fit for 24 phospholipids. Interestingly, one face of the so-called M4 transmembrane a-helix of AChR specifically interacts with one cholesterol molecule, producing a four fold increase in the frequency of the channel opening, with no change occurring in conductance or mean open time. It was possible to inhibit by local anaesthetics that partially overlapped the sterol site. Other less specific interactions also occur with PE and with free fatty acids, corresponding to a decrease in the mean open time of the channel. Finally benzyl alcohol (a general anaesthetic) decreases both the conductance and the mean open time of the AChR, probably by interacting with another type of site than the one used by local anaesthetics.

The discussion below is devoted to the physico-chemical basis for the lipid-protein interactions. Integral then extrinsic proteins deserve special attention since the forces implied obviously are different.

TABLE I

EXAMPLES OF PROTEINS WHOSE FUNCTIONS ARE REGULATED BY LIPID HEADGROUPS

PROTEIN	LIPIDS
INTRINSIC MEMBRANE ENZYMES	
Adenylate cyclase	acidic
Cytochromec oxidase	CL
Cytochrome P450 scc	CL>PG>PC
RECEPTORS	
Nicotinic acetylcholine receptor	Acidic (?), cholesterol
Rhodopsin	PE>PC
Vibronectin receptor	Acidic
Insulin receptor	PA (inhibit)
TRANSPORTERS	
Glucose transporter	PS>PA>PG>PC
Lactose carrier	PE,PS
Calcium ATPase	Acidic
(Sodium, Potassium) ATPase	Acidic
GABA transporter	Cholesterol
PERIPHERAL MEMBRANE ENZYMES	
β Hydroxybutyrate dehydrogenase	PC
Calcineurin	PG>PI>PS
Casein kinase I	PIP2>PIP (PA inhibit)
Protein kinase C	PS
SecA	Acidic
Prothrombinase complex	PS
PERIPHERAL MEMBRANE PROTEINS	
Annexins	PS,PS+PE
Profilin	PIP2>PIP (PI inhibit)
Myosin	Acidic
Actin	non specific(?)

LIPID-INTRINSIC PROTEIN INTERACTIONS

Integral proteins spanning the membrane and possessing aqueous domains can interact with lipid head groups by electrostatic interactions as well as with fatty acid long chains or sterol moiety by hydrophobic interactions.

Specific interactions between lipid head-group and protein have been studied from early in the 80's until now, mainly by using spin labeling experiments. Broadly speaking, it is only by using extreme conditions (either quite low lipid/protein (L/P) ratio or low temperature) that spin labeled phospholipids exhibit the so-called "immobilized" EPR spectrum when they are mixed with pure proteins in reconstituted membranes. Even when a spin-labeled long chain fatty acid is covalently bound to bovine rhodopsin (Davoust and Devaux, 1982), the probe experiences a fast exchange (exchange frequency about 25MHz) between the protein interface and the bulk lipids. Recently, the same kind of results were found with different spin-labeled phospholipids (PA,CL, PS, PG, and PC) or fatty acids: the on and off frequencies were very great (about 10MHz) for the association between lipids and the mitochondrial ADP/ATP carrier(Horvath et al., 1990). However, off and on rates vary slightly (at most by a factor of 4) between the various lipids. This is sufficient to allow for some enrichment in fatty acid or PA at the direct vicinity of the protein. However it is important to underline that the lifetime of a lipid in the protein boundary layer is very short, in such a way that the bilayer composition just around an integral protein is capable of changing completely on the ms scale. The same kind of results have been found with all the proteins studied until now. Of course, it does not prove that it is the case for any other integral protein, but a longer life-time has still to be documented under normal conditions. This rule was confirmed by many deuterium NMR experiments, showing that no "immobilized" component has ever been detected by this method, even in the case where EPR spectra present some evidence for a lipid movement restriction. For example, in bovine rhodopsin perdeuterated DMPC recombinants, there is a very minute effects (if any) of protein on the lipid organization above the transition temperature of pure lipids (Bienvenüe et al., 1982) , even at a very low L/P ratio.

Hydrophobic interactions between lipids and integral proteins have been the topics of many recent papers (see Mouritsen and Bloom, 1993 for a review). An interesting thermodynamic theory has been developed, based on the matching of the hydrophobic spans of lipids and intrinsic proteins. It is clear that large differences between these two lengths must have a profound influence on many membrane properties:

-too short (long) lipids in the boundary layers of protein membrane spanning domains must be more elongated - or more ordered- (curled up - or less ordered) than the bulk lipids, changing the overall properties of the lipids (thermal transition, mean membrane thickness).

-if the lipids are not able to accommodate the difference between the two lengths, some phase exclusion or protein aggregation must occur.

A semi-quantitative theory (Mouritsen and Bloom, 1993) was able do predict or explain some experimental results concerning thermal phase diagrams and the lipid organization of protein-lipid recombinants, lipid selectivity and lateral distribution of proteins. It must be emphasized that the Monte-Carlo simulations predict that large and long lasting lipid or lipid-protein domains, as well as protein aggregates are not very likely at temperatures above the transition temperature (as it is the case in biological membranes). Probably on line with this theory, a phase exclusion of spin labeled analogues of cholesterol was shown to be produced by a small amount of glycophorin mixed with DMPC (Tampé et al., 1991)

LIPID-EXTRINSIC PROTEIN INTERACTIONS

Besides some possible weak hydrophobic contribution (Segrest et al., 1990), the major physico-chemical interactions between lipids and extrinsic proteins are electrostatic.

The activation of Protein Kinase C (PKc) by PS, in the presence of calcium and diacylglycerol (DAG), has been fully documented (see Newton, 1993, for a review) . In some occurrence, PS binds to up to 12 specific binding sites on the protein in a highly cooperative manner. This activation, very important on the biological point of view, occurs only after PKc translocates to PS-rich DAG-containing membranes. This lipid composition mainly corresponds to that of the plasma membrane internal leaflet where many proteins are known to be phosphorylated. Since PKc binds many PS molecules, a high local concentration in PS could appear. This point was hardly observed by biophysical methods (unpublished results with spin labeled lipids), except for a decrease in the lateral diffusion of fluorescently labeled PE (Bazzi and Nelsestuen,1992).

Since lipid polar head groups directly interact with extrinsic proteins, some biophysical studies were undertaken in order to detect changes in the head group conformation or movement by deuterium and ^{31}P NMR spectroscopies. Very recently Pinheiro & Watts (Pinheiro and Watts, 1994 a,b) analyzed the ^{31}P NMR spectra of various complexes between Cytochrome c and anionic lipid. Only minor changes occured in the spectra corresponding to a partial restriction of the amplitude of motion on PS head group. Indeed, the major and quite unexpected result, is that the binding of cytochrome c to the anionic lipids results in a loosening and/or a destabilization of the overall protein structure. "Magic angle" spinning solid-state ^{31}P NMR spectroscopy showed that a weak interaction between cytochrome c and CL occurred. It induces some lateral-phase separation of PE in CL/PE/PC mixtures, in which PE is segregated into isotropic structures on the NMR time-scale.

LIPID MODULATION OF CELL FUNCTIONS

In a recent review (11) we described many of the changes observed in cell functions when the lipid composition of cell membranes changed. We only discuss here some aspects of local anaesthetics on endocytosis or cell activation. For example, dibucaine and benzyl alcohol (two local anaesthetics) were shown to strongly perturbate TfR endocytosis, without any change in the number of sites or in their affinity for Tf (Sainte Marie et al., 1990, Hagiwara and Ozawa, 1990). Interestingly, inhibition by benzyl alcohol was shown to be rapid and fully reversible by simple washing (Sainte Marie et al., 1990). By spin labeling experiments with human erythrocytes, we have shown (Bassé et al., 1992) that benzyl alcohol induces a slow lipid scrambling while increasing slightly the rate of aminophospholipid translocase. These two effects produce a very high lipid excess in the inner leaflet of the erythrocyte membrane, as detected by a change of the cell shape towards the stomatocytic conformation. Benzyl alcohol thus favoured a slow membrane deformation analogous to what happens during endocytosis by vesicle invagination. Moreover, the reversibility in morphology is also slow after benzyl alcohol washing out. Thus, the hypothesis according which benzyl alcohol inhibits endocytosis by maintaining a lipid transverse distribution equilibrium is not convenient. A more convincing hypothesis is that benzyl alcohol, being a very short molecule with respect to phospholipids, disturbs the hydrophobic matching between lipids and proteins, inducing partial and reversible denaturation of many proteins, some of them being implied in the endocytosis mechanism. A model was recently proposed by Sandermann (Sandermann, 1993) taking into account the very high benzyl alcohol/lipid ratio needed for large effects. This last hypothesis seems much more likely than the one affecting to an increase in membrane fluidity the inhibitory effect of benzyl alcohol on platelet activation (Kitagawa et al., 1993). It should be stimulating to discuss in this line the effects observed in many cells (Bienvenüe and Sainte Marie, 1994) by cholesterol addition or depletion, since the role of cholesterol in the hydrophobic lipid-protein matching has already been underlined (Bloom and Mouritsen, 1988) .

CONCLUSION

Lipid-protein interactions (based on matching of hydrophobic regions of lipids and protein and on polar head groups effects) allow integral proteins to maintain their native conformation, with no uncontrolled aggregation. Lipids are also important in forming convenient large anionic surfaces on which active enzymes can assemble with their substrate, a two-dimension diffusion increasing the catalytic rate, and with which the cell structure is modulated by interacting with cytoskeleton associated proteins. The functional role of the lipid chemical structure and location is thus of major importance in the regulation of many metabolic pathways and of signal transduction mechanism, of cell shape and thus of harmonizing the cell function. The unravelled paradox is that such large effects are not caused by strictly defined interactions, in term of sites, stoichiometry, and lifetime, apart from very few cases. The notion of interface between lipid medium and other media is probably more relevant, but many more experiments are needed to fully understand the molecular functioning of the modulation of protein activity by lipids.

Indeed, lipid-intrinsic protein interactions fulfil two complementary while antagonistic functions:

- providing proteins with selective interactions which allows native conformation stabilization of non aggregated proteins. and accompanying the membrane traffic in order for the lipids to be properly sorted then addressed to their target membranes. This last property is essential for the fast regulation of the lipid content of membranes. The molecular interactions between cholesterol and other lipids could have a function in sterol sorting from its synthesis site to its final destination (essentially the plasma membrane).

- securing short lifetime for any lipid- protein contact in order to avoid a permanent binding of each intrinsic protein with many lipids. Due to the high protein/lipid ratio of most of the biological membranes, such interactions should produce large complexes, incompatible with the fast diffusion rate observed as well in biological as in model membranes for lipids and for monomeric proteins, when the latter do not interact with the cytoskeleton.

According to this point of view, the great diversity in membrane lipids is a compromise solution allowing to secure every protein with its appropriate mean environment, with no long lifetime complexes however, in order for the lipid and protein distributions to change rapidly, as needed for membrane traffic and shape flexibility.

Regarding now the lipid-extrinsic interactions, the general scheme should be of the same type: non specific electrostatic interactions (mainly with anionic lipids but also with PE, or even PC) provide the surface essential for a high local concentration in PKc, cytochrome c, PLA2, profilin, prothrombinase complex etc...In some cases a more specific interaction (as between PKc and PS) occur. Sometimes too, a lipid phase exclusion is induced by the specific interaction. Nothing is known precisely about the size and the lifetime of the domains created by this mean, but simple logical consideration lead to the idea that these domains contain a few lipids, and are very mobile, at least at the physiological temperature.

REFERENCES

Barrantes, F. J. (1993). Structural-Functional correlates of the nicotinic acetylcholine receptor and its lipid microenvironment. *FASEB J* 7:1460-1467.

Bassé, F., Sainte-Marie, J., Maurin, L. & Bienvenüe, A., (1992), Effect of Benzyl alcohol on phospholipid transverse mobility in human erythrocyte membrane. *Eur..J..Biochem.* **205**, 155-162.

Bazzi, M., and Nelsestuen, G.L. (1992) Interaction of annexin VI with membranes: highly restricted dissipation of clustered phospholipids in membranes containing phosphatidylethanolamine. *Biochemistry* **31**, 10406-10413.

Bienvenüe, A. and Sainte-Marie, J. (1994) Modulation of protein function by lipids. In "Current topics in Membranes, vol.40: Cell Lipids" (D.Hoekstra ed.), pp319-354. Academic Press, SanDiego.

Bienvenüe, A., Bloom, M., Davis, J.H. & Devaux, P.F., (1982) Evidence for protein associated lipids from deuterium NMR studies of Rhodopsin-dimyristoylphosphatidylcholine recombinants. *J.Biol.Chem.* 257, 3032-3037.)

Bloom, M. & Mouritsen, O.G. (1988) The evolution of membranes. *Can.J.Chem.* 66, 706-711

Bultman, T., Vaz, L.C.W., Melo, C.C.E., Sisk, R.B., and Thompson, T.E. (1991) Fluid-phase connectivity and translationnal diffusion in a eutectic, two-component, two-phase phosphatidylcholine bilayer, *Biochemistry*, **30**, 5573-5579.

Davoust, J. & Devaux, P.F., (1982) Simulation of electron spin resonance spectra of spin labeled fatty acids covalently attached to the boundary of an intrinsic membrane protein. A chemical exchange model. *J. Magn. Res.* 48, 475-494

Devaux, P.F. (1983) ESR and NMR studies of lipid protein interactions in membranes. In "Biological Magnetic resonance" (L.J.Berliner and J.Ruebens, eds.), pp.183-229. Plenum press, New-York.

Hagiwara,Y., & Ozawa, E. (1990) Suppression of transferrin internalization in myogenic L6 cells by dibucaïne. *Biochem. Biophys. Acta* 1051 237-241

Horvath, L.I., Drees, M., Beyer, K., Klingenberg, M., and Marsh, D. (1990). Lipid-protein interactions in ADP-TP carrier/egg phosphatidylcholine recombinants studied by spin label ESR spectroscopy. *Biochemistry* **29**,10664-10669.

J. Sainte-Marie, M. Vignes, M. Vidal, J.R. Philippot & A. Bienvenüe .(1990) Effects of benzyl alcohol on transferrin and low density lipoprotein receptor mediated endocytosis in leukemic guinea pig B lymphocytes. *FEBS Lett.* **262**, 13-16.

Kitagawa, S., M. Orinaka, & H. Hirata. 1993. Depth-Dependent change in membrane fluidity by phenolic compounds in bovine platelets and its relationship with their effects on aggregation and adenylate cyclase activity. *Biochim.Biophys Acta* 1179:277-282.

Lund-Katz, S., Laboda, H.M., McLean, L.R. & Philipps, M.C., (1988) Influence of molecular packing and phospholipid type on rates of cholesterol exchange. *Biochemistry* 27, 3416-3423)

Mouritsen,O., and Bloom, M. (1993) Models of lipid-protein interactions in membranes. *Annu.Rev.Biophys.Biomol.Struct.*, **22**,145-171.

Newton, A. C. (1993)Interaction of proteins with lipid headgroups: lessons from Protein Kinase C. *Annu.Rev.Biophys.Biomol.Struct.* 22, 1-25

Pinheiro, T. J. T., and A. Watts. 1994. Lipid specificity in the interaction of cytochrome c with anionic phospholipid bilayers revealed by Solid-State p-31 NMR. *Biochemistry* **33**,2451-2458.

Pinheiro, T. J. T., and A. Watts. 1994. Resolution of individual lipids in mixed phospholipid membranes and specific lipid-cytochromec interactions by magic angle spinning solid-state phosphorus-31 NMR. *Biochemistry* 33,2459-2467.

Rothberg,K.G., Ying, Y., Kamen,B.A. and Anderson R.G.W. (1990) Cholesterol controls the clustering of the glycophospholipids-anchored membrane receptot for 5-methyltetrahydrofolate. *J. Cell Biol.* 111, 2931-2938

Sandermann, H. (1993) Induction of lipid-protein mismatch by xenobiotics with general membrane targets. *Biochim.Biophys.Acta* 1150, 130-133

Seelig, J. & Browning, J.L. (1978) General features of phospholipid conformation in membranes, *FEBS Letters*, 92, 41-44)

Seelig, J. & Seelig, A. (1980) Lipid conformation in model membranes and biological membranes, *Quart.Rev.Biophys.* 13, 19-61;

Segrest, J.P., De Loof, H., Dohlman, J.G., Brouillette, C.G. & Anantharamaiah, G.M. (1990) Amphipatic helix motif: classes and properties. *Proteins: structure, function and genetics*, 8, 103-117),

Shah, J., Sripada, P.K. and Shipley, G.G. (1990) Structure and properties Mixed-chain phosphatidylcholine bilayers *Biochemistry* **29**, 4254-4262.

Slotte, J.P. & Biermann, E.L. (1988) Depletion of plasma membrane sphingomyelin rapidly alters the distribution of cholesterol between plasma membranes and intracellular cholesterol pools in cultures fibroblasts. *Biochem.J.* 250, 653-658

Tampé, R., von Lukas, A. & Galla, H.J., (1991) Glycophorin-induced cholesterol-phospholipid domains in dimyristoylphosphatidylcholine bilayer. *Biochemistry* 30, 4909-4916)

Venable, R.L., Zhang, Y.,Hardy, B.J. & Pastor, R.J. (1993) Molecular dynamic simulation of a lipid bilayer and of hexadecane: an investigation of membrane fluidity.*Sciences* 262, 223-226.

Dynamical Properties of Membranes: Application of Fluorescent Lipid Analogs

Dick Hoekstra, Teresa Babia, Mirjam Zegers, Kristien Zaal, Eugene G.J.M. Arts* and Jan Willem Kok
University of Groningen, Department of Physiological Chemistry, Bloemsingel 10, 9712 KZ Groningen, The Netherlands

Introduction

Recent advances in studies involving the structure and dynamics of membranes have shown that fluorescently-tagged lipid probes have become versatile and, occasionally, indispensable tools in this area of research. These probes are applied in investigations as diverse as those dealing with biophysical aspects of membranes, such as lateral mobility, lipid phase transitions and phase separations (domain formation), nonbilayer formation (hexagonal phases) and lipid translocation, but also in studies of the cell biology of membranes, including membrane flow and lipid trafficking. In this brief overview, some aspects of the properties and the application of fluorescent lipid(-like) probes will be summarized. The primary focus will be on the use of nitro-benzoxadiazole-derivatized lipids (NBD-lipids).

Fluorescent lipid(-like) probes. Some properties and applications.

A wide variety of fluorescently-tagged lipid probes have been synthesized in recent years (fig.1; references: Loew, 1988;

---

* Department of Obstetrics and Gynaecology, University of Groningen, The Netherlands

NATO ASI Series, Vol. H 91
Trafficking of Intracellular Membranes
Edited by M.C. Pedroso de Lima N. Düzgüneş and D. Hoekstra

Pagano and Martin, 1988; Haugland, 1992; Kok and Hoekstra, 1993). Obviously, the type of probe to be used very much depends on the purpose of the study. For example, a number of probes exists that structurally bear no relationship to the structure of a typical lipid molecule, but that behave like a lipid-like probe. Such probes include, for example, 1,6-diphenyl-1,3,5-hexatriene (DPH) and its cationic derivative, trimethylammonium-DPH (TMA-DPH). Upon exogenous addition, the neutral DPH becomes embedded in the hydrophobic core of a lipid bilayer, whereas the presence of the charged group (TMA) stabilizes the probe's association with the plasma membrane. The fluorescence properties are affected by the environment of the probe allowing DPH to register changes in membrane 'fluidity', as inferred from changes of the internal motions of the phospholipid acyl chains. It should be noted that such changes are not exclusively restricted to fluidity changes, an argument that often contibutes to lifely debates as to the interpretation of DPH fluorescence changes in terms of 'fluidity' (Kleinfeld et al., 1981). TMA-DPH is virtually nonfluorescent in water and binds to membranes in proportion to the available surface area. Therefore, the fluorescence intensity changes when the surface area changes, a phenomenon that is observable upon exocytosis, when an increase in available surface area arises as a result of fusion of exocytic vesicles with the plasma membrane (Bronner et al., 1986). Obviously such measurements are only valid when rapid internalization of the probe in a particular cell type can be excluded. Anchorage of the probe to the cell surface due to the polar TMA linker, appears to prevent such rapid randomization to intracellular membranes, a phenomenon often occurring in case of the neutral DPH.

Evidently, DPH nor TMA-DPH resembles the structure of a lipid. However, due to their hydrophobic amphiphilic nature they readily partition into the lipid phase and consequently are capable of monitoring the occurrence of changes in that phase, thus providing useful information concerning its dynamics.

Another set of probes that more closely matches the structure of lipids in membranes are those that consist of a (charged) polar part and a nonpolar alkyl or acyl chain, some of which are shown in fig. 1. Most of these probes are commercially available (Molecular Probes, Inc.Eugene, OR). The class of anionic probes mainly consists of fluorescent analogs of fatty acids. The fluorophores, located at various positions in the alkyl chain include anthroate, fluorescein (fig. 1), NBD, Bodipy and pyrene. Cationic probes include among others carbocyanine derivatives (DiI, fig. 1),and the probe octadecyl Rhodamine B ('R18', fig. 1). The physical behavior of these probes, such as fluidity-dependent partitioning of DiI derivatives in membranes, and the spontaneous transfer properties of cationic probes between membranes, is usually determined by the alkyl chain length, i.e., the longer the chain the more lipophilic the character of the probe and the more tightly the probe associates with the (labeled) membrane. On the other hand, the spectroscopic properties of the probes are usually not affected by the tail length. DiI derivatives have found extensive application as probes to monitor lateral diffusion processes, and to detect and define lateral domain formation in artificial and biological membranes (Loew, 1988; Wolf, 1994). As will be described below, R18 has found a wide application as a probe to register membrane fusion, in particular fusion events that involve biological membranes. Fluorescent analogs of fatty acids have been used, among others, as membrane fluidity indicators and as probes for characterizing the properties of fatty-acid binding proteins and as probes for characterizing the binding of fatty acids to proteins per se (Haugland, 1992). For studies aimed at elucidating the properties of lipids as such, fluorescent derivatives of natural lipids can be synthesized, which are used as probes in an analogous fashion as isotopically- or spin probe-labeled lipid derivatives. Different fluorescent tags can be attached to either the lipid's headgroup or to the hydrophobic tail (Kok and Hoekstra, 1993). It is evident that derivatization of the head group structure will give rise to a 'lipid molecule' of a nature that is completely different from that of the parent lipid, given that many lipid-specific properties are governed by the lipid head group. Studies in which specific questions are asked about the lipid as such, are therefore preferably carried out with acyl

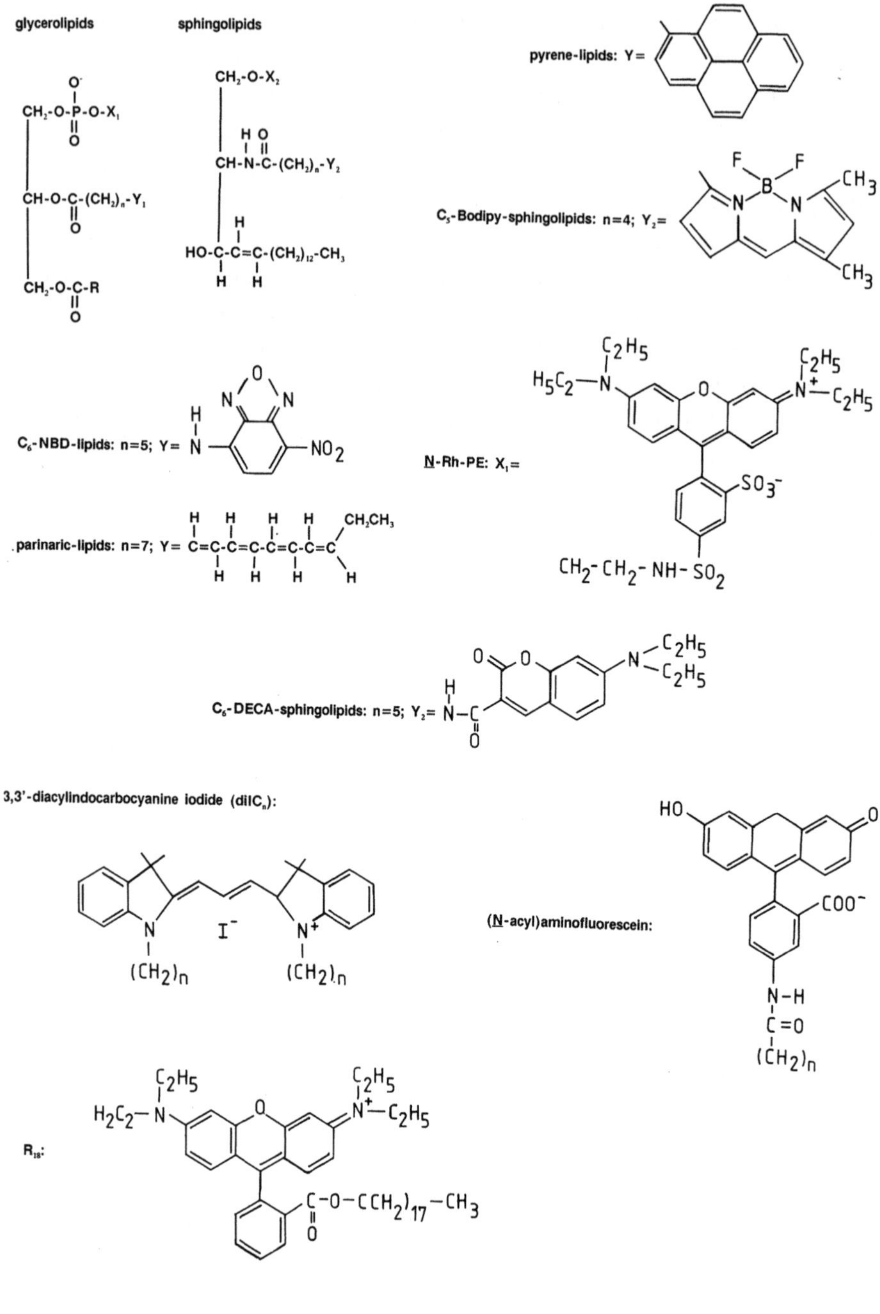

Figure 1. Some representitative fluorescent lipid analogs.

chain-derivatized lipids. C6-NBD-lipids belong to the most frequently used probes for such purposes.

Fluorescent lipid analogs, containing nitrobenzoxadiazol (NBD)

NBD-labeled lipids contain a 7-nitrobenz-2-oxa-1,3-diazol-4-yl (NBD) group. The probe can be attached via either a short (C6) or longer (C12) carbon chain to a glycerol (phospholipid) or sphingosine (sphingolipid) backbone. It is also possible to couple this fluorescent probe via an amide bond to the head-group of PE. The acyl chain-labeled lipids can be synthesized from the deacylated lyso-form of the parent lipid. The fatty acid which is removed is then replaced by C6-NBD-hexanoic acid or C12-NBD-dodecanoic acid. Although many of these derivatives are now available commercially, some of them may not, and consequently need to be synthesized in the laboratory (for synthetic procedures, see e.g. Kok and Hoekstra, 1993). The acyl chain labeled NBD-derivatives of phospholipids and (glyco-)sphingolipids have found a wide application in studies involving intracellular trafficking of lipids in animal cells. In such studies, the advantage of fluorescently-tagged lipids over isotopically labeled ones is that the fate of the former can be monitored directly by fluorescence microscopy. Thus potential artifacts that may arise from lipid scrambling and/or lipid redistribution, possibly occurring when cells are homogenized and the intracellular localization of lipids subsequently determined by gradient fractionation, can be excluded. On the other hand, fluorescence bleaching makes visualization sometimes difficult and photographic documentation somewhat tedious.

The availability of head group-labeled fluorescent lipid analogs has greatly facilitated investigations of the structure and function of membranes in general, including membrane fusion and membrane flow. In particular NBD-labeled phosphatidylethanolamine (N-NBD-PE) should be mentioned and, in conjunction with this probe, as they are often applied together, the rhodamine derivative, N-Rh-PE. The probes are attached to the headgroup's amino group and their presence changes the properties of the lipid in such a way that these properties bear no longer any resemblance to those of natural PE. For example, native PE readily translocates across membranes, which is mediated by a specific protein-translocase or 'flippase'. Translocation occurs when the lipid is inserted into the outer leaflet of the plasma membrane of eukaryotic cells, including erythrocytes (see chapter by Schroit, this volume). This specific property is no longer apparent when the head group has been modified by attaching either fluorescent probe. Interestingly however, the presence of the headgroup tags makes that these lipid probes, when incorporated into a liposomal bilayer, do not significantly transfer between the labeled liposome and unlabeled target membranes. Due to an enhanced aqueous solubility compared to natural lipids, acyl chain-labeled derivatives (C_6-NBD-lipids) rapidly randomize when the labeled membranes are incubated with unlabeled membranes.

Extensive work from numerous laboratories, published over the last decade has shown that these derivatized analogs represent valuable tools in biophysical and cell biological studies. Of major significance to an appropriate interpretion as to their application in the latter studies, is the requirement of a close resemblance to natural lipids. Arguments supporting the similarity in behavior of the probes and their natural counterparts, thus validating their reliability, have been discussed elsewhere (Pagano and Sleight, 1985; Kok and Hoekstra, 1993). In the following a few typical examples of the fruitful application of NBD- and rhodamine-labeled lipid-derivatives in membrane biophysics and cell biology, will be described.

Lipid Analogs and Membrane Fusion

Membrane fusion constitutes a crucial event in a variety of intracellular and intercellular events (Düzgünes, 1985; Hoekstra and Kok, 1989; White, 1992; Wilschut and Hoekstra, 1991). For example, the biosynthetic delivery of membrane constituents to their appropriate (intra-)cellular locations involves a series of processes in which there is a continuous pinching off and formation of small vesicles. These vesicles acting as transport carriers, and, identified among others by employing fluorescent lipid derivatives (see below), move toward distinct intracellular target sites. Upon arrival at the site of destination, fusion between vesicles and target membrane finalizes the proper insertion. Similarly, another distinct set of vesicles is involved in expulsion of intracellular contents into the extracellular environment. This process, called exocytosis, similarly involves fusion of vesicles with the cytoplasmic face of the plasma membrane. Membrane fusion also constitutes an essential intermediate step in the infectious entry of enveloped viruses into mammalian cells (Hoekstra, 1990; White, 1992; Hoekstra and Nir, 1992; de Lima et al., this volume). The viral envelope contains virus specific proteins that are intimately involved in triggering the fusion process that leads to the intracellular delivery of the nucleocapsid for viral reproduction. Fusion occurs either at the level of the plasma membrane or, following endocytosis, at the level of the endosomes. Finally, fertilization also critically depends on membrane fusion, namely between the acrosome-reacted spermatozoa, thereby exposing the fusion-expressing equatorial segment (see below), and the plasma membrane of the oocyte (Arts et al., 1993).

Much of our current knowledge regarding the physicochemical requirements for membrane fusion has been derived from investigations of fusion of artificial membrane systems (Düzgünes, this volume). Not only did these studies provide a wealth of knowledge concerning fundamental issues of membrane fusion, they particularly emphasized the significance of acquiring such data by means of kinetic measurements of the fusion reaction rather than rely on indirect parameters such as growth of vesicle size, electron microscopy, mixing of lipids by techniques such as differential scanning calorimetry or NMR (Hoekstra, 1990a; Hoekstra and Düzgünes, 1993). These measurements were done by means of monitoring changes in energy transfer between N-NBD-PE and N-Rh-PE (Struck et al., 1981). The principle of energy transfer relies on the interaction that may occur between two different fluorophores, provided that the emission band of one fluorophore, the energy donor, overlaps with the excitation band of the second fluorophore, the energy acceptor. As long as the probes are in close proximity, as is the case when the lipid analogs are initially incorporated at a certain density (0.5 - 0.9 mol % each) in a liposomal bilayer, energy transfer occurs. In practice this means that when monitoring the fluorescence of the donor (NBD) at its excitation and emission wavelengths, NBD fluorescence will be quenched: the excited state energy of the donor will be transferred to the acceptor. Such a transfer occurs less efficiently, when the distance between donor and acceptor increases. This so-called 'relief of energy transfer' takes place when the probes dilute as a result of fusion between a labeled and a nonlabeled membrane. The event can be monitored directly in a fluorometer as an increase of NBD fluorescence as a function of time. Evidently, it is crucial in such lipid mixing assays to exclude the spontaneous transfer of probes (Hoekstra, 1990a; Nir et al., 1990; Hoekstra and Düzgünes 1993). Such a transfer is usually seen for the acyl chain labeled NBD derivatives, which are less hydrophobic than their natural counterparts and head group derivatized lipid analogs. Thus, appropriate control experiments are required for each system under study in order to exclude the occurrence of membrane mixing by a process other than fusion.

The assay as described has been extensively applied because of its relative ease to use, high sensitivity and above all its ability to detect and continuously monitor the onset and kinetics, respectively, of the fusion process. (Struck et al., 1981; Hoekstra, 1990a) These conveniences provide the possibility to readily examine parameters that affect the fusion event and, in conjunction with the ability to simulate the fusion reaction by kinetic modelling which allows the determination of fusion and aggregation rate constants (Nir et al., 1990; see also Nir et al, this volume), establish whether a certain parameter (e.g. divalent cations, proteins, dehydrating agents) affects the aggregation step, the fusion step, or both.

Several structural changes in the organization of lipids in membranes have been related to the mechanism of membrane fusion. Lipid phase separations, causing formation of phase boundaries, and the propensity of lipids to undergo nonbilayer transitions have been related as crucial events in triggering fusion (see Düzgünes, this volume). Both processes can be detected by using NBD-labeled lipid analogs (Hoekstra, 1982; Hong et al., 1988). A typical feature of many fluorescent probes is their concentration-dependent selfquenching. For monitoring of lipid phase separations, the approach involves incorporation of either acyl chain or head group NBD-lipid analogs in liposomal bilayers at a concentration of approximately 5 mol %. The occurrence of phase separation leads to an increase in the apparent concentration of the fluorescent lipid analog as it is 'squeezed out' of the separating phases. The kinetics of phase separation can thus be monitored continuously as a quenching of NBD fluorescence, and its kinetics can be compared with the kinetics of fusion, measured in a separate experiment. Similarly, fluorescent probes are usually sensitive to their environment. This environment changes when part of the membrane domain that contains the probe undergoes a polymorphic transition. At relatively low concentrations (less than 1 mol % NBD-labeled lipid), the bilayer to nonbilayer transition is then reported as an increase in NBD fluorescence, correlating with a partial dehydration of the headgroups of hexagonal phase forming lipids (such as PE, containing unsaturated fatty acids). Hence, the change in quantum yield is used here to report a particular structural change in the plane of the membrane. This change in quantum yield is thus derived from dehydration, i.e., a change in polarity which in general strongly affects the quantum yield of a given fluorophore, and which is accompanied by a blue-shift in the emission maximum when the probe senses an environment of increasing hydrophobicity. Hence, the change as such is not specifically related to hexagonal phase formation. Therefore, the occurrence of such transitions per se need to be derived from, i.e., confirmed by other experiments (e.g. NMR and/or electron microscopy, in this case). Once established, fluorescent lipid analogs represent convenient tools, in providing a simple assay to monitor the change.

The described changes that occur in lipid bilayers (both hexagonal phase formation and lipid phase separation have been monitored only in artificial phospholipid bilayers) can be kinetically compared to other events such as membrane fusion, which can be monitored continuously by resonance energy transfer. In such a way, useful insight can be obtained as to the cause and consequences of a biologically relevant process such as membrane fusion.

Concentration-dependent selfquenching of fluorescence has also been exploited in the development of an assay that allows in principle the continuous monitoring of fusion of biological membranes (Hoekstra, 1991; Hoekstra and Düzgünes, 1993). In particular, this assay has been used extensively in studies in which the fusion between viruses and membranes is investigated. The procedure involves the insertion, by exogenous addition, of a fluorescent lipid-like analog, octadecyl Rhodamine B ('R18') into a membrane at a selfquenching concentration. This can be readily done by incubating the membrane preparation of interest with an ethanolic solution of the probe, followed by removal of non-inserted material. The principle of the approach relies on the surface density-dependent selfquenching properties of the dye, implying that the fluorescence intensity increases when the density of the probe decreases. This occurs when a labeled membrane fuses with an unlabeled target membrane. Obviously, a very important prerequisite is that the probe that should reflect the occurrence of fusion is nonexchangeable. Hence extensive control experiments are needed to exclude this possibility. In case of R18, it was found that once the probe is inserted into a membrane, there is very little if any spontaneous movement of the probe between the aqueous phase and the bilayer. When labeled and unlabeled target membranes aggregate, contact-mediated transfer should also be taken into account. The contribution of such a transfer to the overall process - which may vary and depend on the system under study - , can, amoung others, be determined by kinetic simulation procedures (see Nir et al., this volume). By monitoring the initial kinetics, a proper reflection of the fusion event is generally obtained. It is possible to label a wide variety of intact, isolated biological membrane preparations in this manner, because insertion is accomplished by exogenous addition. Hence, the procedure allows to register the fusion of unperturbed biological membranes. Yet, it is again appropriate to emphasize that one cannot solely rely on just the application of such an assay to claim the occurrence of membrane fusion when probe dilution is observed. Once it has been firmly established that the signal does not arise from spontaneous or contact-mediated transfer of free monomers, and when additional experimental approaches have confirmed the fusion event, application of such assays can be highly convenient to obtain valuable insight as to the mechanism of membrane fusion. Such additional experimental approaches may include the mixing of aqueous contents, which is primarily applicable when the fusion system includes artificial bilayers, electron microscopy and/or experiments that show specific inhibition of lipid mixing under conditions known to be inhibitory to the fusion event (for further reading see Hoekstra, 1991; Hoekstra and Klappe, 1993; Hoekstra and Düzgünes,1993; Kok and Hoekstra, 1993).

Lipid analogs and membrane structure

Fluorescent lipid analogs combine a number of properties which make them very attractive to work with, although one should realize that they are derivatives, i.e., probes. Yet as probes, they allow sensitive measurements and quantitation of certain events, which can be readily detected by means of a fluorometer. Such events include their mixing with other lipids resulting from fusion, as described above, and, as also mentioned, events related to membrane changes per se, revealed by an environment-sensitive change in fluorescence. Such changes thus reflect alterations in or at the membrane, either in a physical and/or in a structural sense. In addition, the fate of fluorescent analogs can be visualized directly in a fluorescence microscope. An example that may illustrate such a combined, versatile application of a particular fluorescent lipid analog, is a study in which we investigated the fusogenic properties of spermatozoa (Arts et al., 1993, 1994). The mammalian spermatozoon is a highly polarized cell type and the sperm head shows a strong regional differentiation. Before being able to fuse with the oocyte, the sperm has to undergo a socalled acrosome reaction. This reaction is essentially an exocytotic event, during which the plasma membrane vesiculates as a result of focal point fusion events between the anterior half of the plasma membrane and the underlying outer part of the acrosomal membrane. The inner acrosomal membrane, overlying the nucleus becomes subsequently exposed and the remnant left behind from the plasma membrane and the acrosomal membrane, resulting from the acrosome reaction, is called the equatorial segment. Only the acrosome reacted spermatozoon acquires the competence to fuse with the oocytal membrane. We have studied this fusion process, employing liposomes as target membranes and using the energy transfer couple N-NBD-PE/N-Rh-PE to measure fusion. By fluorometry it can be shown that lipid dilution takes place when labeled liposomes and spermatozoa are mixed, suggesting the occurrence of fusion. By fluorescence microscopy, the process can be visualized, showing that the fluorescent lipid analogs are specifically incorporated in the equatorial segment. Interestingly, the lipid probes can be seen throughout the head when the membrane structure of the sperm is perturbed, for example by prolonged storage, or by carrying the sperm through repeated freeze/thaw cycles. These results are entirely consistent with electronmicroscopic observations which suggest that the fusogenic capacity of spermatozoa is restricted to the equatorial segment. Moreover, the exclusive restriction of the fluorescent lipid analogs to the equatorial segment after their introduction via fusion between the spermatozoan and the liposome also indicated that this membrane domain harbors diffusion barriers to the lipids, as long as the membrane integrity of the sperm is not perturbed. This is analogous to barriers seen in polarized epithelial cells, where randomization of lipids and proteins in the outer leaflet of apical and basolateral membranes is prevented due to the presence of tight junctions. However, whereas in epithelial cells this barrier is restricted to the outer leaflet of the plasma membrane, in a spermatozoan the barrier in the equatorial segment apparently includes the inner leaflet as well, since fusion with symmetrically labeled liposomes will introduce the fluorescent lipids in both the inner and outer leaflet of the sperm's membrane domain. Thus a transbilayer lipid diffusion barrier must be present in the equatorial segment. The existence of a 'double' diffusion barrier was further corroborated by introducing lipid analogs in an asymmetric manner into the membrane, i.e., exclusively in the inner leaflet of the equatorial segment. This approach exploits the possibility to irreversibly destroy the fluorescence of NBD by a reduction reaction, using sodium-dithionite (McIntyre and Sleigth, 1991). This reaction eliminates under appropriate conditions only the fluorescence from NBD-labeled lipids in the outer leaflet of the (symmetrically-labeled) liposomes. Subsequent fusion of such liposomes with the spermatozoa resulted in a fluorescence band, associated with the equatorial segment, very similar as observed when symmetrically labeled vesicles were used. The fluorescence could not be eliminated upon treatment of the spermatozoan with dithionite, except after prolonged incubation when the quenching agent slowly diffuses across membranes, as measured in the fluorometer. This experiment thus confirms the restriction of the probe to the inner leaflet of the equatorial segment. Subsequent treatment with the chelator EDTA, which has been reported to perturb tight junctions in epithelial cells, results in a redistribution of the fluorescence over the entire sperm head.

Examination by fluorescence microscopy can also be quite informative in studies in which the interaction between particles and cells is examined. Such particles may include liposomes, vesicles prepared from synthetic amphiphiles, virusses and fluorescent lipid analogs per se. A first impression can be obtained as to the nature of the interaction process. For example, both liposomes and synthetic amphiphiles are used as carriers for intracellular delivery of drugs and other cell biologically relevant macromolecules (see Lee et al., this volume). The fate of the carrier, interacting with cells, is therefore of interest and by labeling with nonexchangeable lipid analogs, such as the head group labeled analogs N-Rh-PE and N-NBD-PE, one can determine whether the vesicles are primarily binding to the cell surface, are internalized by an endocytic mechanism, or show a merging of vesicular and cellular membrane. Binding is reflected by a patchy appearance of fluorescence. This picture frequently emerges when liposomes are incubated with cultured cells (Struck et al., 1981). With vesicles prepared from synthetic amphiphiles and with viruses, the interaction frequently results in a diffuse, ring-like plasma membrane staining. Presumably, this appearance results from fusion and subsequent randomization of the lipid probes in the membrane. Proper intercalation of the lipid probe within the membrane phase can be determined by measuring diffusion rate constants, using a technique based on fluorescence recovery after photobleaching (Kok et al, 1990). Simple binding of the particle of interest will result in the disappearance of fluorescence from the bleached spot, since recovery does not occur due to immobilization of the remaining probe molecules.

Finally, the ability to visualize fluorescent lipid analogs has been particularly informative in studies that are aimed at understanding the flow of membranes and the trafficking of lipids in cells. These studies are mainly carried out with acyl chain-labeled analogs so that head group specificity is maintained. These socalled C_6-NBD analogs are readily inserted into the plasma membrane of animal cells at low temperature by incubation with C_6-NBD-lipid-labeled liposomes, a micellar mixture of the fluorescent analog in ethanol, or lipid analogs

complexed to bovine serum albumin (BSA). Because of their aqueous solubility, the analogs rapidly move between donor and acceptor membranes as monomers, and intercalation in the lipid phase of the membrane can be revealed by photobleaching studies, as mentioned above. As a result, a ring-like, diffuse staining of the plasma membrane is seen as long as the cells are kept at low temperature. Above approximately 10 °C, the lipid becomes internalized by endocytic processes, as will be discussed below. However, two particular C6-NBD phospholipid analogs already become internalized as intact molecules at conditions where endocytosis is not yet active. In fact it could be shown that this process, which is very specific for phosphatidylserine (PS) and PE, is catalysed by a protein. These studies have contributed greatly to describing and characterizing the protein(s) now known as 'flippase(s)' or translocase(s) for PE and PS (see Schroit, this volume).
Apart from identifying the flippase activity, the fluorescent C_6-NBD analogs have also been imperative in studies that are undertaken to clarify the cell biology of lipids. A brief summary will be presented below. For details, the reader is referred to several recent reviews (van Meer, 1989; Koval and Pagano, 1991; Hoekstra and Kok, 1992; Hoekstra, 1994)

Lipid Analogs and Lipid Cell Biology

The ability to insert the acyl chain-labelled analogs by monomeric exchange into the outer leaflet of cells at low temperature, makes it possible to follow the flow of the lipid analog when the temperature is subsequently raised to 37 °C. By carrying out such experiments with C_6-NBD-labeled phospholipid analogs, it has been shown that these lipids do not randomize in the cell, but rather, reveal a lipid analog processing by the cells in a lipid-specific manner. As noted above, the PE and PS derivatives rapidly translocate, consistent with a preferential localization of such aminolipids in the inner leaflet of the plasma membrane. In eukaryotic cells, the localization of translocated C_6-NBD-PE or -PS is ,however, not restricted to the inner leaflet. Due to their ability to transfer as monomers, nonlabeld membranes of organelles now act as acceptor, causing a rapid labeling of intracellular membranes. However, in erythrocyte membranes where the flippase is also present, the C_6-NBD-labeled aminolipids are restricted to the inner leaflet after translocation, and the translocation rates can be monitored kinetically with an assay involving energy transfer between NBD and N-Rh-PE (see Schroit, this volume). In eukaryotic cells, the phosphatidylcholine derivative is partly endocytosed but returns to the cell surface via a recycling mechanism, whereas the phosphatidic acid derivative is metabolized at the plasma membrane to diglyceride (Pagano and Sleight, 1985; Kok et al., 1990). C_6-NBD-diglyceride is then converted to the triglyceride derivative which subsequently accumulates in lipid droplets, a process that bears strong analogy to the course of events in vivo.

Of special interest is the sphingolipid precursor C_6-NBD-ceramide. This probe also translocates across the plasma membrane at low temperature (i.e., below 10 °C) and eventually accumulates in the Golgi complex. Natural Cer, which is synthesized at the endoplasmic reticulum is also directed toward the Golgi complex, where glycosphingolipid biosynthesis is completed by carbohydrate addition to ceramide. Similarly, C_6-NBD-Cer serves as a precursor for biosynthesis of primarily C_6-NBD-glucosylceramide (GlcCer) and C_6-NBD-sphingomyelin (SM). At elevated temperature, these sphingolipids are transfered to the cell surface in transport vesicles, that have been isolated from permeabilized cells (Babia et al., 1994). In polarized (epithelial) cells, GlcCer is preferentially transported to the apical membrane surface, whereas SM distributes evenly over both (apical and basolateral) membrane surfaces (van 'tHoff and van Meer, 1990). The distribution of the fluorescent GlcCer analog, reflecting the enrichment of natural GlcCer in the apical membrane, arises as a result of lipid sorting, which is thought to occur in the trans Golgi network (TGN), the organelle identified as a sorting site for membrane proteins. A similar sorting event can also be demonstrated in HepG2 cells, a human hepatoma cell line, often used as a model for hepatocytes (Zaal et al., 1993). The plasma membrane of hepatocytes consists of distinct domains, including the sinusoidal or basolateral membrane, and the apical bile canalicular membrane via which bile acids are secreted. In HepG2 cells an intracellular compartment can be identified which is surrounded by a microvilli coated membrane. These microvilli lined vesicles (MLV) are very reminiscent of a bile canalicular space. As a model, the membrane of these vesicles may thus be seen as the bile canalicular membrane (apical), whereas the plasma membrane of the cells represents the basolateral membrane. When these cells are labeled with C_6-NBD-Cer, the analog also accumulates in the Golgi complex. After a brief incubation period at 37 °C, fluorescent lipid appears in the MLV, a process that is inhibited when an inhibitor of the biosynthesis of GlcCer and SM is included in the incubation medium. Evidence was obtained which demonstrates that the flow of the fraction of the fluorescent lipid that reaches the MLV, acquires access to the compartment by a direct transport between Golgi and MLV (Zaal et al., 1994). Thus also in this system, specific sorting seems possible, involving, similarly as noted for polarized cells, direct transport of newly-sythesized sphingolipid from Golgi to the apical membrane.

Fluorescently-tagged acyl chain lipids are readily backexchangeable, implying that they can be retrieved from membranes when the analog is exposed at the membrane surface. This socalled 'backexchange reaction' can be done by using nonlabeld liposomes or bovine serum albumin as acceptor for the NBD lipid derivatives. In this manner, internalized lipid can be readily quantified by determining the exchangeable pool. Non-exchangeability means that the lipid must be located at an intracellular site. This procedure of quantitation is very convenient when following the fate of surface-located or surface-inserted lipids. We have shown that both GlcCer and SM, after their insertion in the plasma membrane, can be internalized by cells along the pathway of receptor-mediated endocytosis. This pathway of entry can be examined and demonstrated by using appropriate markers such as transferrin, LDL, growth factors or ricin, of which the routes of entry and processing by cells have been extensively documented (Hoekstra et al., 1989, and references therein). For interpretation of the data it is also relevant to point out that following endocytic internalization, the NBD-lipids are located at the luminal site of vesicles and organelles, because the probes are initially inserted in the outer leaflet of the plasma membrane. Thus, assuming that their initial topology will be maintained, the probes will be 'trapped' in the vesicles in which they are transported or in the organelles to which they are delivered. For GlcCer, initially inserted in the plasma membrane of Baby hamster kidney (BHK) cells it could be shown that after internalization by endocytosis, efficient recycling of the lipid occurs from both early and late endosomes, with a minor fraction cycling through the Golgi area, possibly TGN (Kok et al., 1992).

Similar work in a human derived colon tumor cell line, HT29, has shown that in the endocytic pathway, GlcCer is specifically sorted from other sphingolipids such as SM, galactosylceramide and lactosylceramide (Kok et al., 1991). In contrast to the latter sphingolipids, GlcCer is preferentially transported to the Golgi apparatus, from which compartment the lipid returns to the cell surface. This sorting event was exclusively restricted to undifferentiated cells. In the differentiated counterpart all lipids were processed along the endocytic/lysosomal pathway. These results thus indicate that a specific mechanism may be operating in undifferentiated cells to maintain a relatively high level of GlcCer. Part of this mechanism apparently involves salvage of this particular lipid via a specific sorting and recycling mechanism. Possibly, the lipid plays a role in growth promotion of the undifferentiated cells, a topic which is currently under investigation. At any rate, the relevance of these studies becomes evident when taking into account the role that sphingolipids may play in signal transduction and the modulating effects they may exert in events such as oncogenesis and cell differentiation (Hakomori, 1981; Hannun, 1994). From a molecular point of view it is equally exciting that, given that sorting occurs both in the endocytic and biosynthetic pathway, mechanisms must exist that recognize, select and target specific sphingolipids. Another challenge in forthcoming years will thus be to identify these mechanisms and, for example, to determine whether or not proteins act as lipid chaparones (or vice versa) during membrane flow (Hoekstra and Kok, 1992).

References

Arts EGJM, Kuiken J, Siemen J, Hoekstra D (1993) Fusion of artificial membranes with mammalian spermatozoa. Specific involvement of the equatorial segment after acrosome reaction. Eur J Biochem 217:1001-1009

Arts EGJM, Jager S, Hoekstra D (1994) Evidence for the existence of lipid diffusion barriers in the equatorial segment of human spermatozoa. Biochem. J., in press.

Babia T, Kok JW, Van der Haar M, Kalicharan R, Hoekstra D (1994) Transport of biosynthetic sphingolipids from Golgi to plasma membrane in HT29 cells: involvement of different carrier vesicle populations. Eur J Cell Biol 63:172-181

Bronner C, Landry Y, Fonteneau P, Kuhry J-G (1986) A fluorescent hydrophobic probe used for monitoring the kinetics of exocytosis phenomena. Biochemistry 25:2149-2154

Düzgünes N (1985) Membrane fusion. in: Roodyn DB (ed) Subcellular biochemistry, vol. 11, New York, Plenum Press, pp. 195-286

Hakomori SI (1981) Glycosphingolipids in cellular interaction, differentiation and oncogenesis. Ann Rev Biochem 50:733-764

Hannun, YA (1994) The sphingomyelin cycle and the second messenger function of ceramide. J Biol Chem 269:3125-3128

Haugland RP (1992) Molecular probes; Handbook of fluorescent probes and research chemicals; pp. 235-274. Molecular Probes, Inc. Eugene OR

Hoekstra D (1982) Fluorescence method for measuring the kinetics of Ca^{2+}-induced phase separations in phosphatidylserine-containing lipid vesicles. Biochemistry 21:1055-1061

Hoekstra D (1990) Membrane fusion of enveloped viruses; especially a matter of proteins. J Bioenerg Biomembr 22:121-155

Hoekstra D (1990a) Fluorescence assays to monitor membrane fusion; Potential application in biliary lipid secretion

and vesicle interactions. Hepatology 12:61-66
Hoekstra D (1991) Membrane fusion of enveloped viruses: From microscopic observation to kinetic simulation, in: Membrane Fusion (Wilschut J, Hoekstra D (eds), Marcel Dekker Inc., New York, pp. 289-311.
Hoekstra D (ed) (1994) Cell lipids. Current topics in membranes, vol. 40. Acad Press, New York
Hoekstra D, Düzgünes N (1993) Lipid mixing assays to determine fusion in liposome systems. Meth Enzymol 220;15-31
Hoekstra D, Eskelinen S, Kok JW (1989) Transport of lipids and proteins during membrane flow in eukaryotic cells, in; Organelles in eukaryotic cells: Molecular structure and interactions, Tager JM, Azzi A, Papa S, Guerrieri F (eds) Plenum Press, NY, pp. 59-83
Hoekstra D, Klappe K (1993) Fluorescence assays to monitor fusion of enveloped viruses with target membranes. Meth Enzymol 220:261-276
Hoekstra D, Kok JW (1989) Entry mechanisms of enveloped viruses: implications for fusion of intracellular membranes. Bioscience Resp 9:273-305
Hoekstra D, Kok JW (1992) Trafficking of glycosphingolipids in eukaryotic cells; sorting and recycling of lipids. Biochim Biophys Acta 1113:277-294
Hoekstra D, Nir S (1991) Cell biology of entry and exit of enveloped viruses, in: The structure of biological membranes, Yeagle PL (ed), CRC Press Inc, Boca Raton, pp. 943-990
Hong K, Baldwin PA, Allen TM, Papahadjopoulos D (1988) Fluorometric detection of the bilayer-to-hexagonal phase transition in liposomes. Biochemistry 27:3947-3955
Kleinfeld AM, Dragsten P, Klausner RD, Pjura WJ, Matayoshi ED (1981) The lack of relationship between fluorescence polarization and lateral diffusion in biological membranes. Biochim Biophys Acta 649:471-480
Kok JW, Babia T, Hoekstra D (1991) Sorting of sphingolipids in the endocytic pathway of HT29 cells. J Cell Biol 114:231-239
Kok JW, Hoekstra D (1993) Fluorescent lipid analogues.

Applications in cell and membrane biology, Chapter 7, in: Fluorescent probes for biological cells - a practical guide, Mason WT Relf G (eds) Academic Press, London, pp. 100-119

Kok JW, Hoekstra K, Eskelinen S, Hoekstra D (1992) Recycling pathways of glucosylceramide in BKH cells: Distinct involvement of early and late endosomes. J Cell Sci, 103:1139-1152

Kok JW, Ter Beest M, Scherphof G, Hoekstra D (1990) A non-exchangeable fluorescent phospholipid analog as a membrane traffic marker of the endocytic pathway. Eur J Cell Biol 53:173-184

Koval M, Pagano RE (1991) Intracellular transport and metabolism of sphingomyelin. Biochim Biophys Acta 1082:113-125

Loew LM (ed) (1988) Spectroscopic membrane probes, vol. 1-3. CRC Press, Boca Raton, FL

McIntyre JC, Sleight RG (1991) Fluorescence assay for phospholipid membrane asymetry. Biochemistry 30:11819-11827

Nir S, Düzgünes N, Pedroso de Lima MC, Hoekstra D (1990) Fusion of enveloped viruses with cells and liposomes: Activity and inactivation. Cell Biophys. 17:181-201

Pagano RE, Martin OC (1988) A series of fluorescent N-acylsphingosines: Synthesis, physical properties, and studies in cultured cells. Biochemistry 27:4439-4445

Pagano RE, Sleight RG (1985) Defining lipid transport pathways in animal cells. Science 229:1051-1057

Struck DK, Hoekstra D, Pagano RE (1981) Use of resonance energy transfer to monitor membrane fusion. Biochemistry 20:4093-4099

Van't Hof W, Van Meer G (1990) Generation of lipid polarity in intestinal epithelial (Caco-2) cells:sphingolipid synthesis in the Golgi complex and sorting before vesicular traffic to the plasma membrane. J Cell Biol 111, 977-986

Van Meer G (1989) Lipid traffic in animal cells. Ann Rev Cell Biol 5:247-275.

White JM (1992) Membrane fusion. Science 258: 917-924

Wilschut J, Hoekstra D (eds) (1991) Membrane fusion, Marcel Dekker Inc. New York, USA

Wolf DE (1994) Microheterogeneity in biological membranes Current Topics in Membranes, vol. 40 (Hoekstra, D (ed) pp. 143-165. Acad Press, New York.

Zaal KJM, Kok JW, Kuipers F, Hoekstra D (1994) Lipid trafficking in hepatocytes. Relevance to biliary lipid secretion, in: Advances in Molecular and Cell Biology, volume 8, (Chen LB, ed) JAI Press, Inc., Greenwich, CT, pp. 133-150

Zaal KJM, Kok JW, Sormunen R, Eskelinen S, Hoekstra D (1994) Intracellular sites involved in the biogenesis of bile canaliculi in hepatic cells. Eur J Cell Biol 63:10-19

THE ERYTHROCYTE AMINOPHOSPHOLIPID TRANSLOCASE

Alan J. Schroit
Department of Cell Biology
The University of Texas M. D. Anderson Cancer Center
1515 Holcombe Blvd.
Houston, Texas 77030

A. INTRODUCTION

Our understanding of cellular membranes has rapidly progressed from the point of view of the lipid bilayer being only a simple permeability barrier and matrix for membrane proteins to one in which membrane lipids are viewed as dynamic components capable of initiating and regulating various cellular functions. Only recently, however, has it become clear that the distribution of lipids between bilayer leaflets plays an important role in many cellular processes. While most of these phenomena are directly associated with what is considered to be normal membrane lipid asymmetry, the "atypical" display of phosphatidylserine (PS) on the cell's outer leaflet has significant physiological consequences. PS participates, for example, in various cell-cell interactions (Schroit et al., 1985; Schlegel et al., 1985), cell activation and hemostasis (Bevers et al., 1983; Rosing et al., 1985; Sims et al., 1989), cell aging (Shukla and Hanahan, 1982; Herrmann and Devaux, 1990), membrane fusion events (Farooqui et al., 1987; Song et al., 1992; Schewe et al., 1992), and apoptosis (Fadok et al., 1992a, 1992b).

Recent studies have provided convincing evidence that membrane lipid asymmetry is generated and probably maintained by specific transport proteins or flippases. One of these transport proteins, the aminophospholipid transporter, is responsible for the movement of PS and phosphatidylethanolamine (PE) from the outer leaflet to the inner leaflet of the

NATO ASI Series, Vol. H 91
Trafficking of Intracellular Membranes
Edited by M.C. Pedroso de Lima N. Düzgüneş and D. Hoekstra

erythrocyte membrane (Seigneuret and Devaux, 1984; Daleke and Huestis, 1985; Connor and Schroit, 1987) and is primarily responsible for keeping PS in its inner leaflet. Because of the dramatic consequences associated with the exposure of PS in the cell's outer leaflet, the studies reviewed here focus on the spatial distribution of PS in red cell membranes with particular emphasis on its transbilayer distribution, and on the elements that control its distribution.

B. BACKGROUND

1. Erythrocyte membrane lipid asymmetry

Erythrocyte membranes contain several major classes of lipids that are composed of different fatty acyl side chains or, in the case of sphingolipids, of different long-chain bases. Studies described during the last decade have established that these lipids are not randomly distributed in the plasma membrane, but certain species and lipids of a specific molecular composition are distributed asymmetrically across the membrane bilayer. This is especially evident for the aminophospholipids, PS and PE, which reside preferentially in the plasma membrane's inner leaflet.

With the exception of diacylglycerol (Ganong and Bell, 1984) and ceramide (Lipsky and Pagano, 1985), most lipids cannot move across the lipid bilayer of artificially generated vesicles (Pagano and Sleight, 1985; Tanaka and Schroit, 1986). This is probably due to the lipid's charge and amphipathic nature, properties that make it energetically unfavorable to pass through the hydrophobic membrane core. Because of this, lipid asymmetry was, for the most part, believed to be a consequence of processes other than transbilayer lipid movement, particularly reorganization of the membrane's inner surface by the action of phospholipases and acylases. Although these processes contribute to membrane lipid asymmetry, the discovery of an aminophospholipid-specific translocase (Seigneuret and

Devaux, 1984) indicated that the nonrandom distribution of PS and PE is controlled by an active energy-requiring process. The existence of a "flippase" that shuttles PS and PE across the bilayer membrane suggested that a specific transmembrane distribution of these lipids is of major importance in cell physiology (Figure. 1)

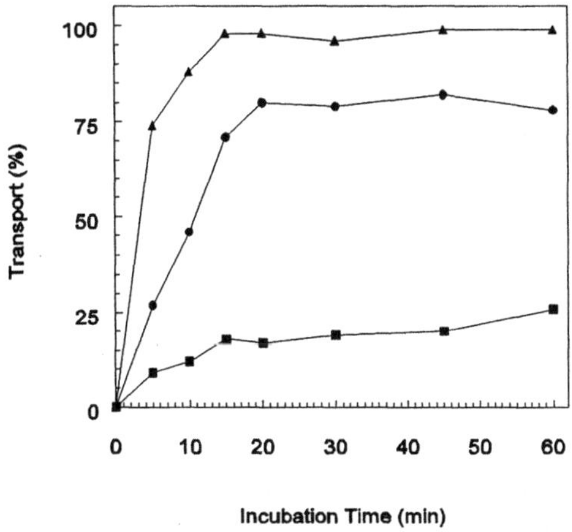

Figure 1. Uptake of fluorescent-labeled lipid analogues from the outer to inner membrane leaflet in human RBC. RBC were incubated with the indicated analogous at $0^{O}C$ for 5 min, washed and then incubated at $37^{O}C$ for the indicated amount of time. (▲) NBD-labeled-PS; (●) -PE, (■) -PC.

Considerable progress has been made in understanding the molecular requirements of aminophospholipid movement in the human red blood cell (RBC). Many experiments have shown that the transbilayer movement of aminophospholipids is ATP- and temperature-dependent (Seigneuret and Devaux, 1984), stereospecific (Martin and Pagano, 1987), and sensitive to oxidation of membrane sulfhydryls (Daleke and Huestis, 1985; Connor and Schroit, 1988; Connor and Schroit, 1990) (see Table 1). Although several candidate proteins for the transport function have been suggested (Morrot et al., 1990; Schroit et al., 1990), conclusive evidence of its identity is still lacking.

Table 1: Inhibition of lipid transport by inhibitors.

	Lipid transport (%)	
	NBD-PC	NBD-PS
Control	18	87
$4^{o}C$		17
NEM		18
Iodoacetamide		15
Vanadate		19
Azide/deoxyglucose		14

RBC were treated with the indicated inhibitors for 30 min at $37^{o}C$, except for NEM (N-ethylmaleimide) which was at $4^{o}C$. For details see Connor and Schroit (1988).

2. Characterization of the aminophospholipid translocase.

Without sufficient data to unequivocally identify the protein responsible for the transmembrane movement of aminophospholipids, designation of a specific transport protein and its mechanism of movement is still premature. Nevertheless, there is a significant amount of data that is compatible with an aminophospholipid transport role for two different red blood cell proteins. These are the 120-kDa Mg^{2+}-ATPase proposed by Zachowski and Devaux (Morrot et al., 1990), and the 32-kDa, Rh-expressing, band-7 proteins suggested by Connor and Schroit (Schroit et al., 1990; Connor et al., 1994). In the absence of structural data, models for the mechanism of these putative transport proteins must be based on several assumptions. These are (i) Specificity of aminophospholipid recognition is determined by a substrate recognition site on a specific membrane-associated transport protein. (ii) The initial interaction between the substrate and the transporter occurs at a recognition site that is at or very close to the membrane water interface. (iii) The transmembrane transport protein forms a canalicular type of structure that "protects" the lipid's polar head from the hydrophobic membrane as the lipid moves through what is probably a hydrophilic channel. (iv) Aminophospholipid movement requires hydrolyzable ATP. Although both the 120-kDa Mg^{2+}-ATPase and the 32-kDa band-7 proteins are known to have some of these characteristics, neither protein, at the present time, completely fulfills all of the assumed requirements of a specific aminophospholipid translocase.

a. The 120-kDa Mg^{2+}-ATPase

Based on corollary evidence for the biochemical requirements of aminophospholipid transport and the activity of an ATPase from RBC, Devaux and colleagues suggested that a 120-kDa Mg^{2+}-ATPase functions as the aminophospholipid transporter (Morrot et al., 1990; Zachowski et al., 1989). This hypothesis is based on observations that aminophospholipid transport, as well as a partially purified Mg^{2+}-ATPase from erythrocytes, is stimulated by PS, requires hydrolyzable ATP, and is inhibited by vanadate, fluoride, N-ethylmaleimide, and Ca^{2+} ions. That this enzyme may be the lipid transporter is supported by data showing that this Mg^{2+}-ATPase has no known function and, in contrast to the Ca^{2+}- and Na^+/K^+-ATPases, is not an ion pump (Forgac and Cantley, 1984; Morris et al., 1993).

In addition to triggering the formation of membrane invaginations (Devaux, 1991), the aminophospholipid translocase was recently proposed to be responsible for also maintaining cell shape (Devaux, 1992; Farge and Devaux, 1992). Indeed, RBC require ATP for their normal discoid shape, and they undergo dramatic shape changes (Nakao et al., 1960; Patel and Fairbanks, 1986; Xu et al., 1991) and membrane fluctuations (Levine and Korenstein, 1991) when their ATP levels are manipulated and when exogenous lipids are inserted into their membrane (Daleke and Huestis, 1985, 1989). These ATP-dependent shape changes depend on a "shape-change Mg^{2+}-ATPase" (Patel and Fairbanks, 1986) which, like the aminophospholipid transporter, is vanadate-sensitive and functional in the presence of ouabain and EGTA. Since ATP-dependent lipid transport generates ATP-dependent echinocyte-to-discocyte transitions (Daleke and Huestis, 1985, 1989) and the same inhibitors inhibit both transport and Mg^{2+}-ATPase (Morris et al., 1992), both functions may be associated with the same protein. However, the similar properties of the aminophospholipid translocase and of the shape-change Mg^{2+}-ATPase, does not unequivocally prove these proteins to be analogous. Much of the data is, in fact, consistent with the characteristics of various protein kinases, enzymes that exhibit regulatory properties similar to those of the Mg^{2+}-ATPase.

b. The 32-kDa band-7 proteins

The first indications that band-7 proteins (RBC polypeptides that migrate to the 30-kDa - 32-kDa region in SDS-polyacrylamide gels) might be involved in the transbilayer movement of aminophospholipids was obtained from photolabeling experiments using ^{125}I-labeled-N_3-analogs of PS and of phosphatidylcholine (PC) (Schroit et al., 1987). Photolysis of RBC incubated with these analogs at 37°C crosslinked the probes to membrane proteins. Analysis of SDS-PAGE gels by autoradiography revealed that, while the PC analog was randomly distributed among many membrane proteins, the PS analog preferentially labeled proteins in the band-7 region. Additional evidence for the participation of these proteins in lipid transport came from similar studies carried out under conditions where lipid movement was inhibited (Connor and Schroit, 1991). In contrast to the preferential labeling of 32-kDa proteins, photolysis of ^{125}I-N_3-PS-labeled RBC preincubated with transport inhibitors resulted in a random labeling pattern that was indistinguishable from that obtained with the non-transportable PC analog.

Further support for the involvement of 32-kDa proteins was obtained by labeling RBC with the thiol disulfide exchange reagent ^{125}I-labeled pyridyldithioethylamine (PDA), a potent inhibitor of aminophospholipid transport (Connor and Schroit 1988). Similar to the results obtained with photoactivatable PS, autoradiography of SDS-PAGE gels revealed that the probe preferentially labeled 32-kDa proteins and specifically and reversibly inhibited PS transport. The proteins labeled with photoactivatable PS and with ^{125}I-labeled PDA were later shown to be the same (Connor and Schroit, 1991).

While these findings implicated a role for 32-kDa proteins in the inward movement of PS, these proteins were also found to be involved in the general maintenance of membrane phospholipid asymmetry (Connor et al., 1992). This was concluded from data showing that the distribution of appropriately labeled PC, PE, and PS analogs localized exclusively in the cells inner leaflet were capable of (1) selectively labeling 32-kDa proteins upon photolysis, and (2) outward movement that reestablished the normal membrane asymmetry of each lipid species.

Findings that the same protein or proteins are involved in the inward movement of aminophospholipids as well as in the outward movement of aminophospholipids and PC suggests that the cell slowly, but continuously, moves lipids to the outer leaflet. Since the movement of a particular lipid species between bilayer leaflets is independent of retrograde movement of other lipid species (Connor et al., 1992), the net effect of different rates of inward and outward movement is, in the case of PE and PS, dominant localization in the cell's inner leaflet. Since PC, in contrast, is not actively transported to the inner leaflet, its equilibrium distribution favors the outside. Building a general model compatible with aminophospholipid-specific inward rates of movement and lipid species-independent outward rates of movement is difficult, but these results are compatible with the bidirectional transport model of Herrmann and Muller (Herrmann and Muller, 1986), which predicts that the translocase is a bidirectional pump with different lipid species-dependent rates of movement in each direction.

c. Is the "transporter" more than one protein?

Although the band-7 polypeptides and the 120-kDa Mg^{2+}-ATPase are distinct proteins, the assignment of "transport activity" to one does not necessarily exclude the participation of the other. As discussed above, aminophospholipid transport not only requires hydrolyzable Mg^{2+}-ATP, but it is also likely to require a structure that forms a protective environment for the lipid's polar head group to traverse the hydrophobic bilayer membrane. Some of the 32-kDa band-7 proteins were shown by recent studies to be part of a set of closely related isoforms that are associated with the Rh blood group system. Rh polypeptides may, therefore, be involved in the maintenance of membrane lipid asymmetry, especially when one considers that Rh protein is a multispanning membrane polypeptide (Avent et al., 1990; Cherif-Zahar et al., 1990), a property common to other membrane proteins associated with transport and channel functions. Since sequencing data indicated that Rh polypeptides do not belong to the ABC (ATP-binding cassette) superfamily of active transporters, it is feasible that they provide the protective pathway for transmembrane lipid movement, while the Mg^{2+}-ATPase utilizes ATP and provides the driving force that enables transport to proceed. Thus, the "*functional transporter*" could be a complex of more than one protein. Indeed, the transport of fluorescent labeled PS analogs was shown to require the participation of a 32-kDa polypeptide and a presumably distinct protein located at the cell's endofacial membrane surface (Connor and Schroit, 1990) Interestingly, Daleke has shown that protocols used to isolate the 120 kDa Mg^{2+}-ATPase also result in the purification of two major proteins with molecular weights of 32 kDa and 50 kDa (Zimmerman and Daleke 1993).

C. INVOLVEMENT OF Rh PROTEINS IN AMINOPHOSPHOLIPID MOVEMENT

Independent observations on the similarities between the biochemical properties of Rh polypeptides and the aminophospholipid transporter led to the suggestion that Rh proteins might be involved in the maintenance of lipid asymmetry (de Vetten and Agre, 1988; Connor and Schroit, 1989). Both Rh polypeptides and the putative band-7 transporter are present in nonhuman erythrocytes, have similar molecular weights and sensitivity to sulfhydryl oxidants, are associated with membrane lipids, and are apparently nonglycosylated.

A series of experiments provided more direct evidence for the involvement of Rh polypeptides in membrane lipid asymmetry (Schroit, et al., 1990). After 32-kDa band-7 red blood proteins were labeled with the transport inhibitor ^{125}I-PDA, or with the transportable substrate ^{125}I-N_3-PS, the ability of monoclonal Rh antibodies to immunoprecipitate the labeled proteins was determined. Autoradiography of SDS-PAGE gels revealed that the immunoprecipitated Rh polypeptides were labeled with the iodinated probes, indicating that the labeled proteins were Rh polypeptides. Precipitation of the probe-labeled Rh protein was specific because immunoprecipitation occurred only when monoclonal antibody was incubated with cells of the appropriate Rh phenotype. Thus, anti-c, anti-D, and anti-E precipitated labeled 32-kDa polypeptides from cDE/cDE cells, but only monoclonal anti-c precipitated polypeptides from cde/cde cells because they do not express the D or E alleles. Similarly, only anti-D precipitated polypeptides from D-- cells because this phenotype lacks both the C/c and E/e epitopes.

While photolabeling of Rh polypeptides with ^{125}I-N_3-PS implicated involvement of the Rh complex in the inward movement of aminophospholipids, other studies suggested that these proteins might also be involved in the outward movement of lipid (Connor et al., 1992). Results from experiments using photoactivatable ^{125}I-N_3-PS, -PE, and -PC, showed that when these lipids occupied the cell's inner leaflet, photolysis resulted in labeling of 32-kDa band-7 proteins in a lipid species-independent manner. Immunoprecipitation with Rh antibodies revealed that the labeled band-7 proteins were Rh polypeptides. These results raised the possibility that the Rh blood group system might be involved not only in the inward movement of aminophospholipids, but also in the general maintenance of phospholipid asymmetry.

D. CONCLUSION

Aminophospholipid asymmetry in eukaryotic cell membranes has been under intensive study since the first observations of ATP-dependent lipid movement across the red cell membrane by Seigneuret and Devaux in 1984. While real progress has been made in this field, many important details must be resolved to complete our understanding of the mechanisms and significance of active transbilayer lipid movement. Clearly, more experimentation is needed to identify and characterize all proteins that might be involved in this process. Nevertheless, based on our current understanding of the biochemical requirements for PS translocation across the plasma membrane bilayer and on predictions of the protein structure from the nucleotide sequence of the putative transport protein, one can envision a model of the aminophospholipid transporter.

Considering the low molecular weight of the aminophospholipids and the structural similarities of the Rh protein to other known transporters, an attractive hypothesis is that the polypeptide forms a pore that operates in a manner similar to that of other transporters of low molecular solutes. While the specific amino acids that comprise the transporter's active site are not known, it is likely that one or more cysteine and histidine residues participate in the process because lipid movement is abolished by sulfhydryl oxidants (Connor and Schroit, 1988) and by histidine-specific reagents (Connor et al., 1992), agents that also alter Rh immunoreactivity (Victoria et al., 1986).

Since lipid movement requires hydrolyzable ATP and Rh protein does not contain a consensus ATP-binding site, the energy needs of active transport must be fulfilled by another protein, such as an ATPase or protein kinase. This assumption implies that enzymes distinct from Rh are involved in lipid transport (Connor and Schroit, 1990; Schroit and Zwaal, 1990) and suggests that Rh polypeptides could be complexed to these enzymes, thereby fulfilling the transporter's energy requirements. This possibility is supported by unrelated studies that have shown that the transporter/Rh protein does not exist in the red cell membrane as a single polypeptide but forms a complex (Hartel-Schenk and Agre, 1992) or "Rh cluster" (Bloy et al., 1988) with other membrane components, one of which may be an ATP-utilizing enzyme.

The detailed molecular events that control the functions of the aminophospholipid transporter are poorly understood and will remain so until all proteins involved in lipid movement are identified. Reconstitution of these proteins will set the stage for a better understanding of the processes at the molecular level and provide a model system to study the precise molecular events that lead to transmembrane lipid movement.

Acknowledgments

This work was supported in part by National Institutes of Health grant DK-41714.

References

Avent ND, Ridgwell K, Tanner MJA, Anstee DJ (1990) cDNA cloning of a 30 kDa erythrocyte membrane protein associated with Rh (Rhesus)-blood-group-antigen expression. Biochem J 271:821-825

Bevers EM, Comfurius P, Zwaal RFA (1983) Changes in membrane phospholipid distribution during platelet activation. Biochim Biophys Acta 736:57-66

Bloy C, Blanchard D, Dahr W, Beyreuther K, Salmon C, Cartron JP (1988) Determination of the N-terminal sequence of human red cell Rh(D) polypeptide and demonstration that the Rh(D), (c), and (E) antigens are carried by distinct polypeptide chains. Blood 72:661-666

Cherif-Zahar B, Bloy C, Le Van Kim C, Blanchard D, Bailly P, Hermand P, Salmon C, Cartron JP, Colin Y (1990) Molecular cloning and protein structure of a human blood group Rh polypeptide. Proc Natl Acad Sci USA 87:6243-6247

Connor J, Schroit AJ (1988) Transbilayer movement of phosphatidylserine in erythrocytes. Inhibition of transport and preferential labeling of a 31,000 Dalton protein by sulfhydryl reactive reagents. Biochemistry 27:848-851

Connor J, Schroit AJ (1990) Aminophospholipid translocation in erythrocytes. Evidence for the involvement of a specific

transporter and an endofacial protein. Biochemistry 29:37-43
Connor J, Schroit AJ (1991) Transbilayer movement of phosphatidylserine in erythrocytes: Inhibitors of aminophospholipid transport block the association of photolabeled lipid to its transporter. Biochim Biophys Acta 1066:37-42
Connor J, Pak CH, Zwaal RFA, Schroit AJ (1992) Bidirectional transbilayer movement of phospholipid analogs in human red blood cells. J Biol Chem 267:19412-19417
Daleke DL, Huestis WH (1985) Incorporation and translocation of aminophospholipids in human erythrocytes. Biochemistry 24:5406-5416
Daleke DL, Huestis WH (1989) Erythrocyte morphology reflects the transbilayer distribution of incorporated phospholipids. J Cell Biol 108:375-1385
de Vetten MP, Agre P (1988) The Rh polypeptide is a major fatty acid-acylated erythrocyte membrane protein. J Biol Chem 263:18193-18196
Devaux PF (1992) Protein involvement in transmembrane lipid asymmetry. Annu Rev Biophys Biomol Struct 21:417-439
Devaux PF (1991) Static and dynamic lipid asymmetry in cell membranes. Biochemistry 30:1163-1173
Fadok VA, Voelker DR, Campbell PA, Cohen JJ, Bratton DL, Henson PM (1992a) Exposure of phosphatidylserine on the surface of apoptotic lymphocytes triggers specific recognition and removal by macrophages. J Immunol 148: 2207-2216
Fadok VA, Savill JS, Haslett C, Bratton DL, Doherty DE, Campbell PA, Henson PM (1992b) Different populations of macrophages use either the vitronectin receptor or the phosphatidylserine receptor to recognize and remove apoptotic cells. J Immunol 149:4029-4035
Farge E, Devaux PF (1992) Shape changes of giant liposomes induced by an asymmetric transmembrane distribution of phospholipids. Biophys J 61:347-357
Farooqui SM, Wali RK, Baker RF, Kalra VK (1987) Effect of cell shape, membrane deformability and phospholipid organization on phosphate-calcium-induced fusion of erythrocytes. Biochim Biophys Acta 904:239-250
Forgac M, Cantley L (1984) The plasma membrane (Mg+2)-dependent adenosine triphosphatase from the human erythrocyte is not an ion pump. J Membr Biol 80:185-190
Ganong BR, Bell RM (1984) Transmembrane movement of phosphatidylglycerol and diacylglycerol. Biochemistry 23:4977-4983
Hartel-Schenk S, Agre P (1992) Mammalian red cell membrane Rh polypeptides are selectively palmitoylated subunits of a macromolecular complex. J Biol Chem 267:5569-5574
Herrmann A, Muller P (1986) A model for the asymmetric lipid distribution in the human erythrocyte membrane. Biosci Rep 6:185-191
Herrmann A, Devaux PF (1990) Alteration of the aminophospholipid translocase activity during in vivo and artificial aging in human erythrocytes. Biochim Biophys Acta 1027:41-46

Levine S, Korenstein R (1991) Membrane fluctuations in erythrocytes are linked to MgATP-dependent dynamic assembly of the membrane skeleton. Biophys J 60:33-737
Lipsky NG, Pagano RE (1985) A vital stain for the Golgi apparatus. Science 228:745-747
Martin OC, Pagano RE (1987) Transbilayer movement of fluorescent analogs of phosphatidylserine and phosphatidylethanolamine at the plasma membrane of cultured cells. J Biol Chem 262:5890-5898
Morris MB, Monteith G, Roufogalis BD (1992) The inhibition of ATP-dependent shape change of human erythrocyte ghosts correlates with an inhibition of Mg2+-ATPase activity by fluoride and aluminofluoride complexes. J Cell Biochem 48:356-366.
Morris MB, Auland ME, Xu YH, Roufogalis BD (1993) Characterization of the Mg2+-ATPase activity of the human erythrocyte membrane. Biochem Mol Biol Int 31:823-832
Morrot G, Zachowski A, Devaux PF (1990) Partial purification and characterization of the human erythrocyte Mg+-ATPase. FEBS Lett. 266:29-32
Nakao M, Nakao T, Tatibana M, Yoshikawa H (1960) Shape transformation of erythrocyte ghosts on addition of adenosine triphosphate to the medium. J Biochem 47:694-695
Pagano RE, Sleight RG (1985) Defining lipid transport pathways in animal cells. Science 229:1051-1057
Patel VP, Fairbanks G (1986) Relationship of major phosphorylation reactions and MgATPase activities to ATP-dependent shape change of human erythrocyte membranes. J Biol Chem 261:3170-3177
Rosing J, Bevers EM, Comfurius P, Hemker HC, van-Dieijen G, Weiss HJ, Zwaal RFA (1985) Impaired factor X and prothrombin activation associated with decreased phospholipid exposure in platelets from a patient with a bleeding disorder. Blood 65:1557-1561
Schewe M, Muller P, Korte T, Herrmann A (1992) The role of phospholipid asymmetry in calcium-phosphate-induced fusion of human erythrocytes. J Biol Chem 267:5910-5915
Schlegel RA, Prendergast TW, Williamson P (1985) Membrane phospholipid symmetry as a factor in erythrocyte-endothelial cell interactions. J Cell Physiol 1123:215-218
Schroit AJ, Madsen JW, Tanaka Y (1985) In vivo recognition and clearance of red blood cells containing phosphatidylserine in their plasma membranes. J Biol Chem 260:5131-5138
Schroit AJ, Madsen J, Ruoho AE (1987) Radioiodinated, photoaffinity-labeled phosphatidylcholine and phosphatidyl-serine: Transfer properties and differential photoreactive reaction with human erythrocyte membrane proteins. Biochemistry 26:1812-1819
Schroit AJ, Bloy C, Connor J, Cartron JP (1990) Involvement of Rh blood group polypeptides in the maintenance of aminophospholipid asymmetry. Biochemistry 29:10303-10306
Schroit AJ, Zwaal RFA (1990) Transbilayer movement of phospholipids in red cell and platelet membranes. Biochim Biophys Acta 1071:313-329
Seigneuret M, Devaux PF (1984) ATP-dependent asymmetric distribution of spin-labeled phospholipids in the

erythrocyte membrane: Relation to shape changes. Proc Natl Acad Sci. USA 81:3751-3755
Shukla SD, Hanahan DJ (1982) Membrane alteration in cellular aging: Susceptibility of phospholipids in density (age)-related human erythrocytes to phospholipase A2. Arch Biochem Biophys 214:335-341
Sims PJ, Wiedmer T, Esmon CT, Weiss HJ, Shattil SJ (1989) Assembly of the platelet prothrombinase complex is linked to vesiculation of the platelet plasma membrane. Studies in Scott syndrome: An isolated defect in platelet procoagulant activity. J Biol Chem 264:17049-17057
Song LY, Baldwin JM, O'Reilly R, Lucy JA (1992) Relationship between surface exposure of acidic phospholipids and cell fusion in erythrocytes subjected to electrical breakdown. Biochim Biophys Acta 1104:1-8
Tanaka Y, Schroit AJ (1986) Calcium-phosphate-induced immobilization of fluorescent phosphatidylserine in synthetic bilayer membranes: Inhibition of lipid transfer between vesicles. Biochemistry 25:2141-2148
Victoria EJ, Branks MJ, Masouredis SP (1986) Rh antigen immunoreactivity after histidine modification. Mol Immunol 23:1039-1044
Xu YH, Lu ZY, Conigrave AD, Auland ME, Roufogalis BD (1991) Association of vanadate-sensitive Mg(2+)-ATPase and shape change in intact red blood cells. J Cell Biochem 46:284-290
Zachowski A, Henry JP, Devaux PF (1989) Control of transmembrane lipid asymmetry in chromaffin granules by an ATP-dependent protein. Nature 340:75-76
Zimmerman ML, Daleke DL (1993) Regulation of a candidate aminophospholipid transporting ATPase by lipid. Biochemistry 32:12257-12263

PURIFICATION AND SUBSTRATE SPECIFICITY OF THE HUMAN ERYTHROCYTE AMINOPHOSPHOLIPID TRANSPORTER

David L. Daleke, Jill V. Lyles, Edward Nemergut, and Michael L. Zimmerman

Department of Chemistry, Indiana University, Bloomington, IN 47405

INTRODUCTION

The asymmetric transmembrane distribution of phospholipids across biological membranes is maintained by a combination of slow transmembrane flip-flop, cytoskeletal-protein interactions and protein-mediated inward transport of aminophospholipids. These latter proteins are members of a growing class of transporters, or flippases, that catalyze the transmembrane transport of phospholipids (Devaux and Zachowski, 1993). The aminophospholipid flippase selectively transports phosphatidylserine (PS) and phosphatidylethanolamine (PE) to the cytofacial surface of the membrane. The flippase is Mg^{2+}-ATP-dependent and is inhibited by sulfhydryl reagents, vanadate and Ca^{2+} (Daleke and Huestis, 1985; Bitbol, et al., 1987; Connor and Schroit, 1988). This transporter demonstrates a high degree of specificity for its lipid substrate. The amine functionality is essential; acylation of the amine group of PS prevents transport by the flippase (Morrot, et al., 1989; Drummond and Daleke, 1994) and *N*-methylation of PE reduces transport of this lipid (Morrot, et al., 1989). Only the natural *sn*-1,2 glycerol isomer of PS is a substrate; the *sn*-2,3 isomer of PS is not transported in fibroblasts (Martin and Pagano, 1987). In contrast, the flippase is insensitive to acyl chain length (Daleke and Huestis, 1985) and is relatively insensitive to acyl chain composition.[1] These characteristics allow the inclusion of spin labeled (Seigneuret and Devaux, 1984), short chain (Daleke and Huestis, 1985), fluorescent (Connor and Schroit, 1987) or radiolabeled (Tilley, et al., 1986; Daleke and Huestis, 1989; Anzai et al., 1993) fatty acid as reporter groups for the measurement of lipid transport in intact membranes.

[1]However, transport of phospholipids containing NBD-fatty acids is dependent on the length of the fatty acid linker and the length of the fatty acid in the *sn*-1 position (Tilly et al., 1990; Colleau et al., 1991).

NATO ASI Series, Vol. H 91
Trafficking of Intracellular Membranes
Edited by M.C. Pedroso de Lima N. Düzgüneş and D. Hoekstra

Although flippase activity has been well characterized, the protein or proteins responsible for aminophospholipid transport have not been positively identified. Two proteins have been proposed as candidate flippases. A 32 kDa protein, that may be a member of the Rh family of antigenic erythrocyte proteins, has been nominated as a participant in flippase activity based on its ability to react with two flippase inhibitors: a radio-iodinated sulfhydryl reagent and photoaffinity PS (Connor and Schroit, 1988; Connor, et al., 1992). Another protein, a vanadate-sensitive Mg^{2+}-ATPase, has been suggested to be involved in flippase activity (Zachowski, et al., 1989). Attempts have been made to purify this enzyme from human erythrocytes (Morrot, et al., 1990; Daleke, et al., 1990; Zimmerman and Daleke, 1993; Morris et al., 1993; Auland et al., 1994) and chromaffin granules (Moriyama et al., 1991).

We have purified a candidate aminophospholipid transporting ATPase from human erythrocytes using a combination of gel filtration and anion exchange chromatography. The ATPase shows enzymatic characteristics similar to that of the flippase and is selectively activated by lipids that are flippase substrates (Morrot et al., 1990; Zimmerman and Daleke, 1993). Using synthetic PS analogs, we show that the interaction of the flippase and the ATPase with PS involves several recognition elements, including the headgroup and glycerol backbone, but is insensitive to PS acyl chain composition.

RESULTS

ATPase Purification

The Mg^{2+}-ATPase was purified from detergent-solubilized erythrocyte membranes using a combination of ammonium sulfate precipitation and Q Sepharose ion exchange chromatography (Zimmerman and Daleke, 1993). Vanadate-sensitive ATPase activity was measured in the presence of PS as described (Zimmerman and Daleke, 1993). Most of the ATPase activity eluted from the Q Sepharose column at 0.25 M NaCl and was pooled for the following studies. A smaller ATPase fraction, containing both vanadate-sensitive and vanadate-insensitive activities eluted at approximately 0.12 M NaCl (Daleke, et al., 1990). The bulk of the bound protein contains no ATPase activity and eluted from the column at high salt concentrations (> 0.4 M NaCl). The ATPase mixture exhibited a specific activity of 300 nmol/min/mg and contained proteins of molecular weights 165, 110, 83, 47, and 32 kDa (Daleke, et al., 1990; Zimmerman and Daleke, 1993). For some experiments, the ATPase was further purified by sequential size exclusion and high performance ion exchange chromatography, resulting in a single protein of molecular weight 80 kDa. Enzymatic characteristics of the purified enzyme were indistinguishable from those of the Q Sepharose pool. In addition, a 120 kDa protein, previously reported to be the ATPase (Morrot et al., 1990), was separated from the ATPase and identified as the human erythrocyte ATP citrate lyase by partial sequence analysis, enzymatic activity, and immunoreactivity.[2]

Effect of Flippase Inhibitors on Mg^{2+}-ATPase Activity

PS-stimulated ATPase activity was measured in the presence of ouabain or three established inhibitors of the flippase: vanadate, N-ethylmaleimide, and Ca^{2+} (Zimmerman and Daleke, 1993). With the exception of ouabain, each of these compounds inhibited ATPase activity (Figure 1).

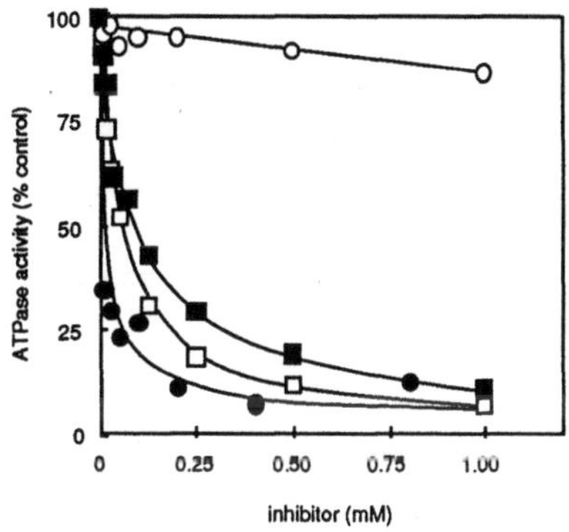

Figure 1. *Effect of flippase inhibitors and ouabain on Mg^{2+}-ATPase activity.* PS-stimulated ATPase activity was measured in the presence of ouabain (○), vanadate (●), N-ethylmaleimide (□), and Ca^{2+} (■) and normalized to activity in the absence of inhibitors (adapted from Zimmerman and Daleke, 1993, with permission).

Of the inhibitors, vanadate was the most effective (IC_{50} = 1.8 μM). N-ethylmaleimide and Ca^{2+} were equally potent (IC_{50} = 45 and 65 μM, respectively). Inhibition by N-ethylmaleimide implies that the ATPase, like the flippase, possesses an essential cysteine residue, while lack of inhibition by ouabain and potent inhibition by Ca^{2+} indicate that this ATPase is distinct from the erythrocyte Na^+,K^+-ATPase and Ca^{2+}-ATPase.

---

[2]Unpublished observations.

Phospholipid Headgroup Specificity

The ability of phospholipids to activate the ATPase was measured in detergent-phospholipid-protein mixed micelles (Zimmerman and Daleke, 1993). The detergent solubilized ATPase was inactive in the absence of added phospholipids and was not activated by the addition of zwitterionic lipids, such as phosphatidylcholine or sphingomyelin. PS provided maximal activation, while PE and negatively charged lipids, such as phosphatidylglycerol (PG), phosphatidic acid (PA), and phosphatidylinositol (PI) partially activated the ATPase (Figure 2).

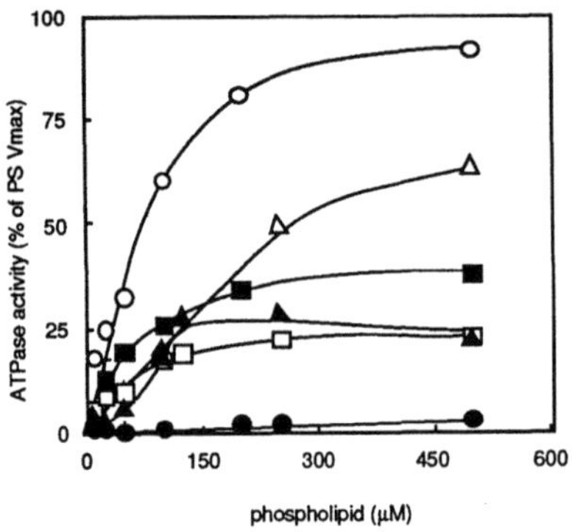

Figure 2. *Lipid specificity of the Mg^{2+}-ATPase.* ATPase activity was measured in the presence of detergent ($C_{12}E_9$) and the indicated concentration of PS (○), PI (Δ), PG (▲), PA (❑), PE (■), or PC (●). Data are expressed as percent of the maximal activity observed in the presence of PS (adapted from Zimmerman and Daleke, 1993, with permission).

The structural features of PS essential for activation of the ATPase and transport by the flippase were studied using synthetic PS analogs. Structural analogs of PS were synthesized from PC by phospholipase D-catalyzed headgroup exchange with the appropriate alcohol headgroup (Juneja, et al., 1989) and were purified as described (Comfurius and Zwaal, 1977). ATPase activity was measured in detergent mixed micelles as described (Zimmerman and Daleke, 1993). Short chain (dilauroyl) analogs of these lipids were also synthesized and introduced into intact erythrocytes by spontaneous transfer from sonicated vesicles. Transport was detected by measuring characteristic changes in cell shape (Daleke and Huestis, 1989).

The amine group is required by both enzymes; phosphatidylhydroxypropionate (POPP) did not activate the ATPase, nor was it transported by intact erythrocytes (Table I). However, *N*-methylation (POPS-*N*-Me) had no effect on transport and only partially reduced the activation of the ATPase, indicating that these enzymes tolerate both primary and secondary amines.

Table I. ATPase activation and transport of synthetic PS analogs.

lipid	ATPase activity[a]	flippase activity[b]
POPS	42.0	+++
PI	8.2	–
POPG	54.0	–
POPE	4.8	+
POPA	4.5	–
sn-2,3-DLP-L-S	7.7	–
sn-2,3-DLP-D-S	4.0	–
POP-D-S	38.0	+++
POPS-*N*-Me	16.0	+++
POPS-*O*-Me	8.8	–
POPP	1.6	–

[a]Catalytic efficiency (Vmax/Km) of the ATPase was measured in the presence of the indicated lipid, detergent and POPC as described (Zimmerman and Daleke, 1993).
[b]Transport of dilauroyl analogs of the indicated lipid was measured as described (Daleke and Huestis, 1989). Transport activity was scored as equivalent to PS (+++), 10% of PS (+), or not detectable (–).

The carboxyl group is not essential, but is required for maximal activity; PE activated the enzyme 10-fold less well than PS and was transported at a rate 10-fold slower (Seigneuret and Devaux, 1984; Daleke and Huestis, 1985). Most significant, however, is the sensitivity of these enzymes to PS stereochemistry. While the stereoconfiguration of the serine headgroup is unimportant (Hall and Huestis, 1993), both enzymes were selectively stimulated by the naturally occurring *sn*-1,2 glycerol isomer (*sn*-1,2-DLP-L-S) and not the *sn*-2,3 isomers (*sn*-2,3-DLP-L-S or *sn*-2,3-DLP-D-S) of PS.[2]

PS Acyl Chain Dependence

The acyl chain dependence of ATPase activation by PS was studied using a homologous series of saturated phosphatidylserines and a series of unsaturated phosphatidylserines with increasing number of double bonds. As shown in Figure 3A, ATPase activity was proportional to fatty acyl chain length for the saturated series of phosphatidylserines up to 14 carbons in length.

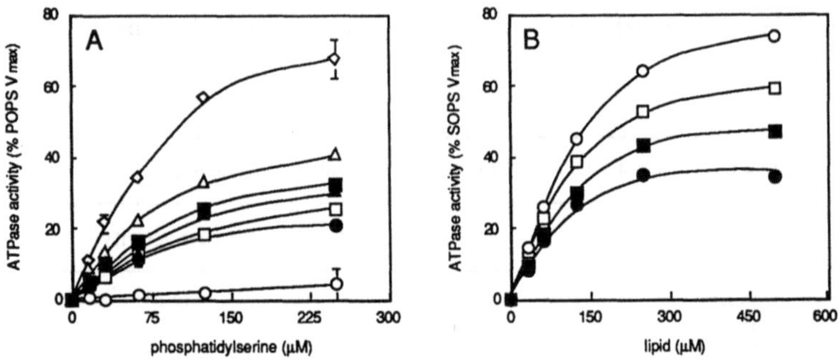

Figure 3. *The effect of PS acyl chain length and unsaturation on ATPase activity.* A. ATPase activity was measured in the presence of POPS (◊) or in the presence of symmetrical, saturated phosphatidylserines containing fatty acids of 6 (○), 8 (●), 10 (□), 12 (■), 14 (Δ), or 16 (▲) carbons. B. Effect of fatty acyl chain unsaturation on PS-stimulated ATPase activity. ATPase activity was measured in the presence of SOPS (○), DPPS (●), DOPS (□) or DLiPS (■). (from Zimmerman and Daleke, 1993, with permission).

The catalytic activity of the ATPase in the presence of these lipids was linearly dependent on fatty acyl chain length up to 14 carbons (Figure 4A), due to an increase in Vmax. The apparent Km of the enzyme for these lipids remained constant (not shown).

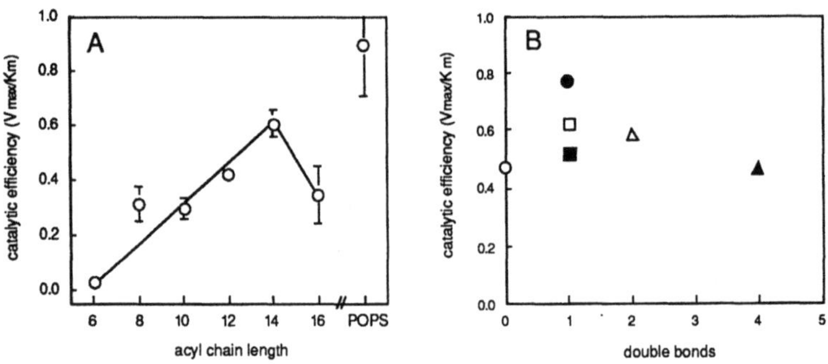

Figure 4. *Catalytic efficiency of the ATPase as a function of PS acyl chain length and unsaturation.* Kinetic parameters were determined from data shown in Figure 4 by least-squares fits of each set of data points to the Michaelis-Menten equation. A. Data from Figure 4A. B. Data from Figure 4B: DPPS (○), POPS (●), SOPS (□), brain PS (■), DOPS (Δ), and DLiPS (▲)(adapted from Zimmerman and Daleke, 1993, with permission).

Although activity varied with acyl chain length for saturated lipids, activity in the presence of unsaturated lipids was always greater (Figures 3 and 4). ATPase activity in the presence of unsaturated PS was dependent on double bond content (Figure 3B). Those lipids with a single double bond stimulated the highest degree of ATPase activity, though catalytic activity in the presence of mono- and polyunsaturated lipids varied relatively little (Figure 4B).

To further study this apparent acyl-chain specificity, ATPase activity was measured in the presence of short chain PSs and a long chain, unsaturated PC. ATPase activity in the presence of mixtures of di-8:0 PS or di-12:0 PS with POPC was increased compared with short chain PS mixed with an acyl chain matched PC (Table II).

Table II. Effect of acyl chain unsaturation on non-activating lipids in the presence of short chain PS.[a]

lipids	ATPase activity
di-8:0 PS	38.8 ± 2.5
+ di-8:0 PC	32.7 ± 1.7
+ POPC	51.6 ± 0.4
di-12:0 PS	48.6 ± 2.0
+ di-12:0 PC	55.6 ± 1.0
+ POPC	79.8 ± 2.5

[a]ATPase activity (as % POPS Vmax) was measured in the presence of PS or PS + an equimolar amount of either acyl-chain matched PC or POPC (adapted from Zimmerman and Daleke, 1993, with permission).

The apparent acyl chain length and unsaturation requirement can be met by non-activating lipids and is not a structural requirement for the activating PS. Thus, the acyl chain dependence of ATPase activity likely reflects the generation of the proper hydrophobic environment to stabilize the enzyme and not a specific substrate-enzyme interaction.

DISCUSSION

A number of enzymatic and structural characteristics of the human erythrocyte aminophospholipid flippase have been determined in intact cells or cell membranes. Of these, only two enzymatic properties of the enzyme are known: 1] transmembrane aminophospholipid transport activity and 2] vanadate-sensitive ATPase activity. Either of these activities could be exploited to purify candidate flippases. Although reconstitution is required to positively identify the flippase, ATPase assays are technically more efficient for the purpose of enzyme purification. The strategy used in the present work was to isolate a subfraction of ATPases from erythrocyte membranes that possess enzymatic characteristics in common with the flippase.

The ATPase described here shares a number of enzymatic properties with the erythrocyte flippase. Flippase inhibitors, such as N-ethylmaleimide, Ca^{2+}, and vanadate, inhibit the ATPase. Though the sensitivity of the ATPase and the flippase to these inhibitors is qualitatively similar, the absolute sensitivity of the ATPase to these inhibitors differs quantitatively from that of the flippase (Bitbol et al., 1987; Zimmerman and Daleke, 1993). These differences are most likely due to differences in the assays employed; ATPase activity was measured in detergent micelles, while flippase activity was measured in intact cells. Although these inhibitors are not specific for the flippase, their effects on the ATPase assure that the enzyme falls within the subset of ATPases likely to be associated with flippase activity.

The vanadate-sensitive ATPase is inactive in the absence of added lipid and displays a unique specificity for its lipid activator. Enzyme activity is selectively reconstituted by negatively charged lipids and PS is the optimal lipid activator. This negatively-charged lipid requirement is not dependent on the charge per molecule, but rather is dependent on surface charge density. For example, maximal ATPase activity in the presence of the polyphosphoinositides PIP or PIP_2 is lower than in the presence of PI (Zimmerman and Daleke, 1993). However, accompanying this decrease in Vmax is a greater decrease in the apparent Km, resulting in a high catalytic efficiency. This indicates that PIP and PIP_2, although present in low concentrations (< 1 mol%) in erythrocytes, may regulate this enzyme. Activation by PG is unique in that it is apparently sigmoidal, indicating a possible cooperative interaction with the protein. The nature of this cooperativity is unknown, but it may be due to a threshold effect, similar to the apparent cooperativity exhibited by other membrane enzymes in the presence of lipid (Koshland, 1987). Preferred activation of this enzyme by PS, PI and PG indicates that these lipids share a common mechanism for activation of the enzyme. Indeed, molecular modeling of these lipids indicate that they can adopt similar conformations in which hydrogen bond donors and acceptors can be identically positioned. However, activation of the ATPase by PS can be distinguished from activation by negatively charged lipids. Preliminary results have shown that, in contrast to PS, the interaction of PI, PG and PA with the enzyme is primarily electrostatic.[2] Thus, this ATPase may possess two distinct binding sites for lipids, one which can be filled with negatively charged lipids, including PS, and another which is specific for PS.

The ATPase selectively recognizes each of the structural elements of the PS headgroup. The amine group is essential for activity, yet secondary amines are tolerated by both the ATPase and the flippase. These data indicate that the amine group may have an ionic, rather than a hydrogen-bond, interaction with the protein that is essential for transport and ATPase activity. Although the carboxyl group is not essential, the presence of a negative charge enhances transport and ATPase activity, indicating that the carboxyl group may also form a specific ionic interaction with the enzyme. Finally, although serine headgroup stereoconfiguration is not important (Hall and Huestis, 1993), the absolute configuration of the glycerol backbone is an important recognition element (Martin and Pagano, 1987).

The specific activation of the Mg^{2+}-ATPase by PS is unique for erythrocyte ATPases. Other erythrocyte ATPases (Na^{+},K^{+}-ATPase, Ca^{2+}-ATPase) are activated by a broad class of negatively charged lipids and do not exhibit a strong preference for PS. The extreme specificity of the Mg^{2+}-ATPase described here for its lipid activator places this enzyme in a unique class of lipid-specific enzymes, such as protein kinase C and β-hydroxybutyrate dehydrogenase, that display an absolute specificity for their lipid cofactors (Newton, 1993).

In contrast to the absolute specificity for the glycerophosphoserine headgroup, both flippase and ATPase activity are insensitive to the length of the fatty acyl chains on the substrate lipid. Flippase activity is independent of PS acyl chain length (Daleke and Huestis, 1985; Loh and Huestis, 1993) though not absolutely independent of acyl chain composition. Both spin labeled and fluorescent labeled lipids exhibit transport activities that qualitatively differ from those observed with non-labeled lipids and that depend on acyl chain length (Tilly et al., 1990; Colleau et al., 1991). This may indicate that the flippase is sensitive to the presence of bulky or hydrophilic fatty acyl groups. These reporter groups may equilibrate at the membrane-aqueous interface (Chattopadhyay and London, 1987) and interfere with transport directly or by modifying the configuration of the glycerol backbone. Increasing the hydrophobicity of the molecule by increasing the acyl chain length of the linker or adjacent fatty acid is sufficient to restore transport activity (Colleau et al., 1991). Although the ATPase displays an apparent specificity for acyl chain length and unsaturation, this requirement can be met by non-activating lipids, such as PC. This apparent acyl chain specificity may be an artifact of the detergent-phospholipid-ATPase micelle assay used in these studies or it may reflect the requirement for a more hydrophobic environment than can be provided by detergent alone. Thus, the ATPase, like the flippase, does not discriminate between substrates based on acyl chain length.

Conclusion

Both the flippase and the ATPase described here are sensitive to Ca^{2+}, vanadate, and sulfhydryl reagents. Both enzymes show a high degree of structural specificity for their lipid activators that involves several recognition elements including the headgroup and glycerol backbone. The striking similarity of the lipid specificity of both enzymes is consistent with a role for the ATPase in flippase activity.

REFERENCES

Anzai K, Yoshioka Y, Kirino Y (1993) Novel radioactive phospholipid probes as a tool for measurement of phospholipid translocation across biomembranes. Biochim. Biophys. Acta 1151:69-75.

Auland ME, Morris MB, Roufogalis BD (1994) Separation and characterization of two Mg^{2+}-ATPase activities from the human erythrocyte membrane. Arch. Bioch. Biophys. 312:272-277.

Bitbol M, Fellmann P, Zachowski A, Devaux PF (1987) Ion regulation of phosphatidylserine and phosphatidylethanolamine outside-inside translocation in human erythrocytes. Biochim. Biophys. Acta 904:268-282.

Chattopadhyay A, London E (1987) Parallax method for direct measurement of membrane penetration depth utilizing fluorescence quenching by spin-labeled phospholipids. Biochemistry 26:39-45.

Colleau M, Hervé P, Fellmann P, Devaux PF (1991) Transmembrane diffusion of fluorescent phospholipids in human erythrocytes. Chem. Phys. Lip. 57:29-37.

Comfurius P, Zwaal RFA (1977) The enzymatic synthesis of phosphatidylserine and purification by CM-cellulose column chromatography. Biochim. Biophys. Acta 467:146-164.

Connor J, Schroit AJ (1987) Determination of lipid asymmetry in human red cells by resonance energy transfer. Biochemistry 26:5099-5105.

Connor J, Schroit AJ (1988) Transbilayer movement of phosphatidylserine in erythrocytes: Inhibition of transport and preferential labeling of a 31 000-dalton protein by sulfhydryl reactive reagents. Biochemistry 27:848-851.

Connor J, Bar-Eli M, Gillium KD, Schroit AJ (1992) Evidence for a structurally homologous Rh-like polypeptide in Rh_{null} erythrocytes. J. Biol. Chem. 267:26050-26055.

Daleke DL, Huestis WH (1985) Incorporation and translocation of aminophospholipids in human erythrocytes. Biochemistry 24:2406-2416.

Daleke DL, Huestis WH (1989) Erythrocyte morphology reflects the transbilayer distribution of incorporated phospholipids. J. Cell Biol. 108:1375-1385.

Daleke DL, Cornely-Moss K, Smith CM (1990) Partial purification of a candidate aminophospholipid transporter. J. Cell Biol. 111:321a.

Devaux PF, Zachowski A (1993) "Transmembrane lipid asymmetry in eukaryotes" in New Developments in Lipid-Protein Interactions and Receptor Function. Wirtz KWA, Packer L, Gustafsson JÅ, Evangelopolous AE, Changeux JP (eds) Plenum New York 213-226.

Drummond DC, Daleke DL (1994) Synthesis and characterization of pH-dependent "caged" aminophospholipids. Chem. Phys. Lipids, in press.

Hall MP, Huestis WH (1993) Phosphatidylserine headgroup diastereomers translocate equivalently across human erythrocyte membranes. Biochim. Biophys. Acta 1190:243-247.

Juneja LR, Kazuoka T, Goto N, Yamane T, Shimizu S (1989) Conversion of phosphatidylcholine to phosphatidylserine by various phospholipases D in the presence of L- or D-serine. Biochim. Biophys. Acta 1003:277-283.

Koshland DE Jr. (1987) Switches, thresholds and ultrasensitivity. TIBS 12:225-229.

Loh RK, Huestis WH (1993) Human erythrocyte membrane lipid asymmetry: Transbilayer distribution of rapidly diffusion phosphatidylserines. Biochemistry 32:11722-11726.

Martin O, Pagano RE (1987) Transbilayer movement of fluorescent analogs of phosphatidylserine and phosphatidylethanolamine at the plasma membrane of cultured cells. Evidence for a protein-mediated and ATP-dependent process*(es)*. J. Biol. Chem. 262:5890-5898.

Moriyama Y, Nelson N, Maeda M, Futai M (1991) Vanadate-sensitive ATPase from chromaffin granule membranes formed a phosphoenzyme intermediate and was activated by phosphatidylserine. Arch. Bioch. Biophys. 286:252-256.

Morris MB, Auland ME, Xu Y-H, Roufogalis BD (1993) Characterization of the Mg^{2+}-ATPase activity of the human erythrocyte membrane. Biochem. Mol. Biol. Int. 31:823-832.

Morrot G, Hervé P, Zachowski A, Fellman P, Devaux P (1989) Aminophospholipid translocase of human erythrocytes: Phospholipid substrate specificity and effect of cholesterol. Biochemistry 28:3456-3462.
Morrot G, Zachowski A, Devaux PF (1990) Partial purification of the human erythrocyte Mg^{2+}-ATPase. A candidate aminophospholipid translocase. FEBS Lett. 266:29-32.
Newton AC (1993) Interaction of proteins with lipid headgroups. Lessons from protein kinase C. Ann. Rev. Biophys. Biomol. Struct. 22:1-25.
Seigneuret M, Devaux PF (1984) ATP-dependent asymmetric distribution of spin-labeled phospholipids in the erythrocyte membrane: Relation to shape changes. Proc. Natl. Acad. Sci. USA 81:3751-3755.
Tilley L, Cribier S, Roelofsen B, Op den Kamp JAF, van Deenen LLM (1986) ATP-dependent translocation of aminophospholipids across the human erythrocyte membrane. FEBS Lett. 194:21-27.
Tilly RHJ, Senden JMG, Comfurius P, Bevers EM, Zwaal RFA (1990) Increased aminophospholipid translocase activity in human platelets during secretion. Biochim. Biophys. Acta 1029:188-190.
Zachowski A, Henry JP, Devaux PF (1989) Control of transmembrane lipid asymmetry in chromaffin granules by an ATP-dependent protein. Nature 340:75-76.
Zimmerman ML, Daleke DL (1993) Regulation of a candidate phosphatidylserine-transporting ATPase by lipid. Biochemistry 32:12257-12263.

PHOSPHOLIPIDS IN PLATELETS : LOCALIZATION, MOVEMENT AND PHYSIOLOGICAL FUNCTION

Alain Bienvenüe, Patrick Gaffet and Nadir Bettache
"Dynamique Moléculaire des Interactions Membranaires"
URA 1856 CNRS
Université Montpellier II, case 107
F-34095 MONTPELLIER Cedex 5

Hemostasis is the control of blood circulation in vessels by a very complex and fast cascade of proteolytic reactions. In the case of vascular lesions, a major activator complex is formed between tissue factor (TF) and factor VII. At the same time, platelets adhering to subendothelial components secrete their granule contents and activate other platelets in a chain reaction. These platelets aggregate in turn while their plasma membrane rapidly becomes able to bind coagulating factors (VIIIa-IXa in tenase complex; VA-Xa in prothrombinase complex), essentially through their phosphatidylserine (PS) outer surface content. Finally, the procoagulant power of PS-containing membranes depends on their ability to assemble tenase and prothrombinase complexes and to protect activated factors against endogenous anticoagulating factors (Mann et al., 1990).

The PS (and more generally, the whole phospholipid) transverse distribution in platelet plasma membrane is of major importance for the regulation of blood coagulation. Our results on this point are presented in this review. We used a classical experimental method based on EPR spectroscopy assay of spin labeled short chain analogues of various phospholipids (known to be good reporters for endogenous phospholipid (PL) movements (Seigneuret and Devaux, 1984;Tilley et al., 1986)) after probe back exchange between cells and bovin serum albumin and ferricyanide oxidation (Bergmann et al., 1984). The following conditions are successively studied:

- phospholipid translocation in platelets at resting state,
- then in activated platelets;
- the phospholipid movements in plasma membrane during platelet activation;
- the mechanism of vesicle shedding from platelets during activation;
- the phospholipid distribution changes during platelet ageing.

The major result consists in a very strong evidence for PS and PE (phosphatidyl-ethanolamine) outflux, without concomitant phosphatidylcholine (PC) and sphingomyelin

NATO ASI Series, Vol. H 91
Trafficking of Intracellular Membranes
Edited by M.C. Pedroso de Lima N. Düzgüneş and D. Hoekstra

(SM) influx during efficient platelet activation. Various mechanisms for this fast membrane reorganization are considered and discussed in the last part.

Phospholipid movements in platelets at resting state

PS (PE) is exclusively (mainly) localized in the cytoplasmic leaflet of the plasma membrane of platelets at resting state (Chap et al., 1977), as in all other types of cells studied to date (Suné et al., 1988, Zachowski and Morot Gaudry-Talarmain, 1990). In red blood cells, the aminophospholipid distribution is controlled by a translocase using the energy of ATP hydrolysis (Seigneuret and Devaux, 1984; Suné et al., 1988; Zachowski and Devaux, 1990). Using paramagnetic analogs of different phospholipids, the distribution and rate of transfer between the two leaflets of the plasma membrane of platelets at rest have been studied. Our results (1Bassé et al., 1992) confirm the previous ones and allow to determine a fast influx rate for PS and PE analogues (about 3 min and 10 min half time respectively).

Phospholipid movements in activated platelets

The same probes were used as above, after *in vitro* activation by two agonists: thrombin in the absence of calcium and A23187 in the presence of 1 mM calcium (Bassé et al., 1993; Sims et al., 1989). In thrombin-activated platelets, translocase activity is stimulated slightly, PC passive diffusion is enhanced slightly, but no cytoskeleton proteolysis occurs and no vesicles are shed into the external medium. In calcium ionophore-stimulated cells, a very sharp rise in the internal calcium concentration causes cytoskeleton proteolysis (mainly filamin, talin and myosin heavy chain), vesicle shedding, and very strong translocase inhibition (Bassé et al., 1992). A23187 concentrations higher than 0.5μM were necessary to activate platelets in the presence of 1mM calcium, suggesting that the final intracellular calcium concentration and the velocity of calcium invasion could be important parameters for the full platelet response. The dependence on calcium was also shown by the restoration of translocase activity by adding excess external EGTA. In vesicles shedded from platelets after A23187/calcium activation, the kinetics of relocation of PS and PC analogues were similar, and after 1-2hrs led to a nearly symmetric distribution of these PLs in the vesicle membranes, as observed in remnant platelet membranes.

To summarize, except for a common slow passive PC influx, activated platelet plasma membranes and cytoskeleton have completely different properties according to the activation procedure (depending mainly on the calcium entry).

Phospholipid transmembrane movements during platelet activation

Previously internalized aminophospholipids and externally located PC can be used as probes to follow the transverse movement of phospholipids from the internal and the external leaflets respectively, during platelet stimulation by the same agonists (Bassé et al., 1993). No outward (even transient) movement of internally located spin-labeled aminophospholipids was observed during thrombin-induced activation (Fig.1A), whereas the influx of externally located probes increased slightly and slowly: immediately after the addition of thrombin, spin labeled PC internalized faster, leading to identical distribution whether the platelets were activated before or during probe reorientation.

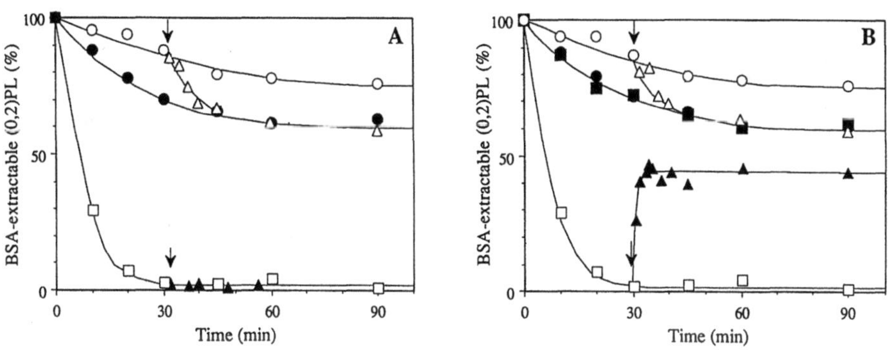

Figure 1: Modification in BSA-extractable (0,2)PC and (0,2)PS during thrombin or A23187 -induced activation at 37°C. (0,2)PC (○,●) and (0,2)PS (□,■) were added at zero time to resting (○,□) or activated platelets (●,■). 30 min after the incorporation of spin-labeled PLs, thrombin (5 units/ml in the presence of 5mM EGTA) (A) or A23187 (1µM in the presence of 1mM Ca^{2+}) (B) were added (arrows) to the suspension of resting platelets. BSA-extractable (0,2)PS (▲) and (0,2)PC (△) were then monitored.

During A23187-mediated activation (Fig.1B), a similar slightly increased influx was observed, whereas 40-50% of the initially internally located spin-labeled aminophospholipids (PS and PE) appeared on the outer leaflet. The sudden exposure on the outer face was dependent on an increase in intracellular calcium and was achieved in less than 2 min at 37°C. The probes were distributed between the vesicles and remnant cells in proportion to their phospholipid content, suggesting that no scrambling of plasma membrane leaflets occurred during vesicle blebbing, contrary to a suggestion by Sims et al. (Sims et al., 1989).

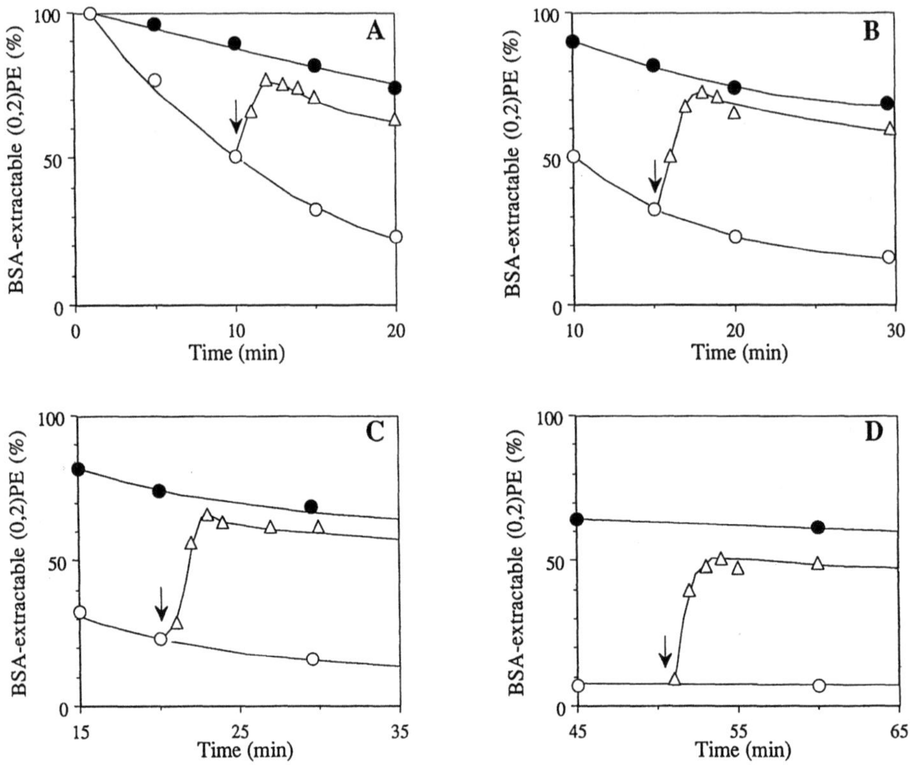

Figure 2: Modification in BSA-extractable (0,2)PE during ionophore A23187-induced activation at 37°C. (0,2)PE was added at zero time to resting (○) or A23187-activatd platelets (●). Activation of resting platelets was performed (A) 10 min, (B) 15 min, (C) 20 min, or (D) 51 min after the incorporation of (0,2)PE by adding A23187 (1μM) (arrows) to the suspension in the presence of 1mM Ca^{2+}. BSA-extractable (0,2)PE (△) was then monitored.

Another evidence for a vectorial outflux of aminoPLs during ionophore platelet activation was brought by fig 2. In these experiments, always the same ratio (about half) of internally located PE at the time of activation appeared on the external leaflet. This result is a very convincing argument in favour of a vectorial aminoPL outflux. It is also against a PL scrambling which could occur during the membrane merging accompanying the vesicle blebbing. Another important finding is that the spin-labeled aminophospholipid exposure rate and amplitude were unchanged when vesicle formation was inhibited strongly by calpeptin, a calpain-inhibitor (Fig.3).

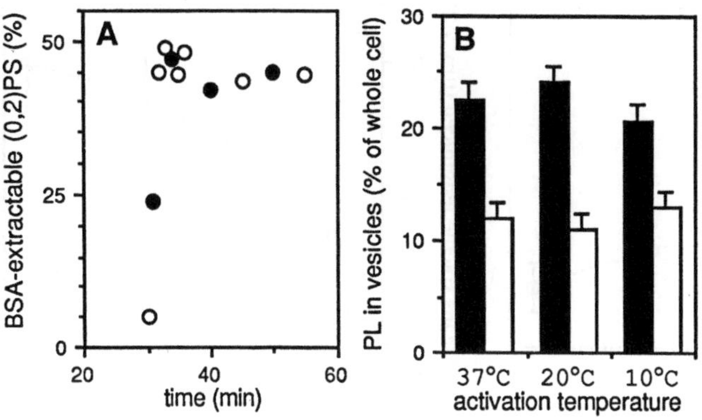

Figure 3: Effect of calpain inhibition on (A) BSA-extractable (0,2)PS during A23187-induced activation, (B) vesicle formation, and (C) calpainmediated proteolysis. (A) after platelets have been preincubated (●) or not (○) with 50µg/mL calpeptin for 30 min at 37°C, (0,2)PS was added to the cell suspension. 30 min after (0,2)PS addition, A23187 (1 µM) was added to the suspension in the presence of 1 mM Ca^{2+} (arrow). The BSA-extractable (0,2)PS was then monitored (●,○). (B) After 30 min of incubation in the presence (white column) or absence (dark column) of calpeptin (50µg/ml) at 37°C, platelets were slowly brought to the desired temperature and further activated by adding A23187 (1µM) in the presence of 1mM Ca^{2+}. PL content in vesicles was compared to the whole cell PL content. (C) SDS-PAGE of platelets activated at 37°C by 1µM A23187 in the presence of 1mM Ca^{2+}. Lane 1: resting platelets. Lane 2: activated platelets. Lane 3: platelets activated after preincubation with 50µg/mL calpeptin. Arrows indicate the greatest modifications in the protein pattern after activation.

All these results indicate that loss of asymmetry (thus inducing generation of a catalytic surface of the external face of the platelet plasma membrane) is not the consequence of vesicle formation. Conversely, we propose that vesicle shedding is an effect of phospholipid transverse redistribution and calpain-mediated cytoskeleton proteolysis during activation. According to this view, the calcium/A23187, induced platelet stimulation very rapidly causes a phospholipid excess in the external leaflet of the plasma membrane from PS and PE initially located on the internal leaflet.

Cytoskeleton proteolysis is implied in vesicle shedding during platelet activation

A large extra outer surface is caused by the PS and PE outflux without concomitant PC or SM influx. The appearing membrane transverse asymmetry can be counterbalanced only by very small bending radius shapes such as filipodia (Suné and Bienvenüe, 1988). These structures are very taut so the loss of membrane vesicles is a very likely event, provided that the cytoskeleton is not able to maintain cell integrity. In this line (14) inhibiting calpain-catalyzed proteolysis of the cytoskeleton decreases dramatically the amount of PL appearing in vesicles, while very long filipodia persist on the activated platelet surface (see figure 4).

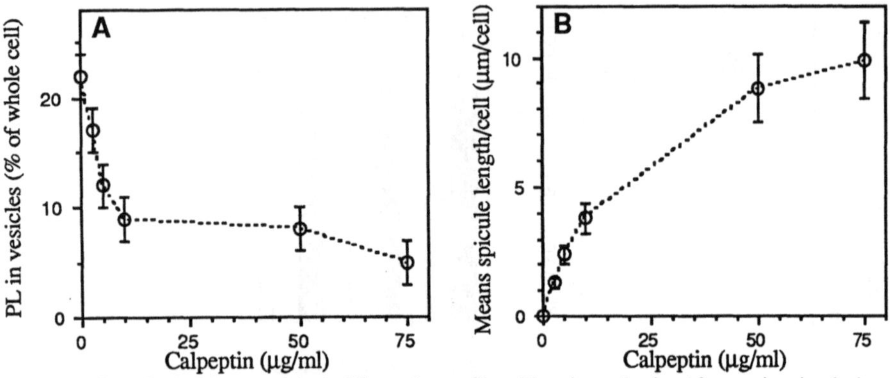

Figure 4: Effect of calpeptin on vesicle shedding and mean filopod length per platelet. After previous incubation for 30 min at 37°C in the presence of various amounts of calpeptin, platelets were activated by the addition of A23187 (1µM) in the presence of 1mM Ca^{2+}. (A) Phospholipid contents in supernatant of centrifugated suspension (12000 x g for 2 min) were expressed in percentage of whole cell phospholipid content. (B) Filopods lengths were measured on electron micrographs of activated platelets and the average length per platelet was determined by divided cumulated length by the number of observed platelets.

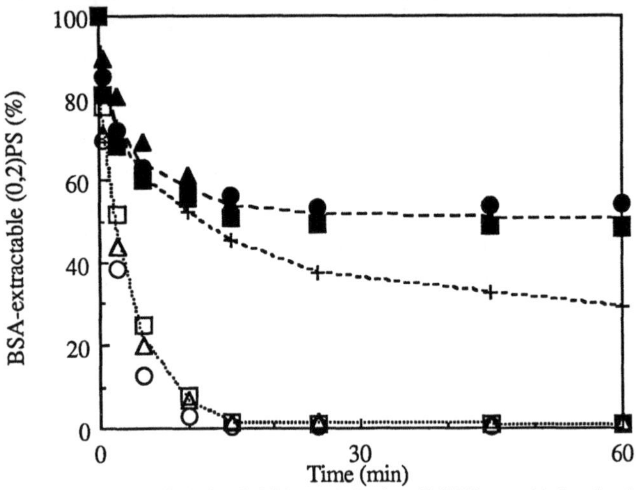

Figure 5: redistribution kinetic of (0,2)PS at 37°C in aged platelets. (0,2)PS was added to the platelet suspension at time zero at 20µM final concentration, representing less than 1% of the total amount of membrane phospholipids. BSA-extractable (0,2)PS was expressed in percent of total (0,2)PS remaining in the platelet suspension at the same time (non-hydrolysed spin-labeled PS). Redistribution kinetic at 37°C of (0,2)PS added to platelets at 1(□), 5(△), 7(○), 9(+), 10(▲), 11(●), 12(■) days of aging. Lines represent the means for 1,5,7 (··········), 9(-------) and 10,11,12 (– – – –) days of storage respectively.

Membrane PL transverse distribution during *in vitro* cell ageing

Changes similar to those occuring during a strong platelet activation also were observed when platelets were kept for 7-9 days in the blood-bank conditions (Gaffet et al., 1994). As shown above, asymmetrical distributions and translocation kinetics were very different for spin-labeled phosphatidylserine and spin-labeled phosphatidylcholine in fresh platelet plasma membranes. In freshly prepared platelets and up to day 7, spin-labeled phosphatidylserine penetrated very rapidly to the inner leaflet of the platelet plasma membrane. However, spin-labeled phosphatidylcholine was mainly retained on the external leaflet. From day 7 to 9, the two translocation kinetics became identical with symmetrical distribution of both spin-labeled phospholipids at equilibrium. Inhibition of translocase activity and modification of membrane stability accounted for these transformations.

The rapid re-exposition of spin-labeled phosphatidylserine after A23187 stimulation, measured in fresh platelet concentrates, persisted up to day 9 but disappeared between day 10 and day 12 (fig.5). From day 7 to 9, a strong cytoskeleton proteolysis and marked decrease in intracellular ATP were observed. Moreover, complete suppression of ß-N-acetyl glucosaminidase secretion and vesicle formation after A23187 stimulation of aged platelets indicated that platelets could no longer be activated beyond day 9. Taken together, these results showed that during *in vitro* aging there are metabolic and membrane modifications in platelet similar to those described for platelet activation. In addition, all of the observed events occurred simultaneously between day 7 and day 9, very likely because a sudden ATP depletion happens during cell ageing. These results highlight the importance of maintaining plasma membrane asymmetry to increase the hemostatic effectiveness of transfused platelet concentrates.

Conclusion

By generalisation of our results obtained with spin labeled PLs, a very fast PS and PE outward movement occurs in platelet plasma membrane during strong cell activation (i.e. with a strong and fast calcium influx), in agreement with previous results (Bevers et al., 1983) showing that activated platelets exhibit a high PS level at their surface. This movement is not counterbalanced by a symmetrical influx of previously externally located PC and SM molecules. This induces a very large extra outer surface which can be counterbalanced only by very small bending radius shapes such as filipod. Such structures are very taut so the loss of membrane vesicles is a very likely event, provided that the cytoskeleton is not able to maintain cell integrity as a result of its calpain-catalyzed proteolysis.

According to these results, vesiculation is caused by budding of the filopod membrane due to the large excess of PLs on its outer monolayer and to calpain-mediated proteolysis of submembranous cytoskeletal proteins. The molecular mechanism of aminoPL outflux is still unclear. Since the scrambling effect accompanying membrane merging now is excluded, we can imagine the three following other mechanisms to push previously inside located PL to translocate.

- The sea urchin model: platelet activation is accompanied by a profound polymerization of tubulin and actin filaments. Microtubules concentrate near the center of the cell, while microfilaments are distributed all along filipod. If the earlier event caused by calcium concentration rise is the microfilament lengthening, we can admit that the only way for platelet membrane to accomodate stemms longer than the cell diameter is to modify its shape

In filipodia, outer membrane surface is much higher than inner one, furnishing the driving force for PL outflux without symmetrical influx. However, some of our preliminary unpublished results indicate that neither cytoskeleton organizing nor disorganizing drugs are able to modify the PS outflux during A23187-induced platelet activation.

- The PIP2/calcium induced PL scrambling. When calcium enters into erythrocytes (via A23187 transport), previously inside located PS and PE molecules are also able to translocate to the outer leaflet (Williamson et al., 1992; Sulpice et al., 1994). However the results differ from those observed in platelets in several respects: firstly, this is definitely a scrambling mechanism in erythrocyte, since PC or SM influx accompanies the PS or PE outflux contrarily to what occurs in platelets. Secondly, the kinetics of the PL translocation is much slower in erythrocyte than in platelet (about 15 min. and 30 sec. respectively) for similar temperature and calcium or PIP2 concentrations.

- A platelet specific mechanism, likely implying special type of "translocator". Research on purified platelet plasma membrane is currently undergoing in our laboratory in this field.

REFERENCES

Bassé, F., Gaffet, P., & Bienvenüe, A. (1994) Correlation between inhibition of cytoskeleton proteolysis and antivesiculaion effect of calpeptin during A23187-induced activation of human platelets: are vesicles shed by filopod fragmentation? *Biochim. Biophys. Acta, 1190*, 217-224.

Bassé, F., Gaffet, P., Rendu, F. & Bienvenüe, A. (1992) Phospholipid transverse mobility modifications in plasma membranes of activated platelets: an ESR study. *Biochem. Biophys. Res. Commun. 189*, 465-471.

Bassé, F., Gaffet, P., Rendu, F. & Bienvenüe, A. (1993) Translocation of spin-labeled phospholipids through plasma membrane during thrombin- and ionophore A23187-induced platelet activation. *Biochemistry 32*, 2337-2344.

Bergmann, W. L., Dressler, V., Haest, C. M. W. & Deuticke, B. (1984) Reorientation rates and asymmetry of distribution of lysophospholipids between the inner and the outer leaflet of the erythrocyte membrane. *Biochim. Biophys. Acta, 772*, 328-336.

Bevers, E. M., Comfurius, P. & Zwaal, R. F. A. (1983) Changes in membrane phospholipid distribution during platelet activation. *Biochim. Biophys. Acta 736*, 57-66.

Chap, H. J., Zwall, R. F. A. & van Deenen, L. L. M. (1977) Action of highly purified phospholipases on blood platelets. Evidence for an asymmetric distribution of phophoslipids in the surface membrane. *Biochim. Biophys. Acta, 467*, 146-164.

Gaffet, P. Bassé, F., & Bienvenüe, A. (1994) Loss of phospholipid asymmetry in human platelet plasma membranes after 1 to 12 days storage: an ESR study. *Eur.J.Biochem., in press.*

Mann, K. G. , Nesheim, M. E. , Church, W. R. , Haley, P. & Krishnawamy, S. (1990) Surface dependence reactions of the vitamin-K dependent enzyme complexes. *Blood 76*, 1-16.

Seigneuret, M. & Devaux, Ph. F. (1984) ATP-dependent asymmetric distribution of spin-labeled phospholipids in the erythrocyte membrane: relation to shape changes. *Proc. Natl. Acad. Sci. U.S.A. 81*, 3751-3755.

Sims, P. J., Wiedmer, T., Esmon, C. T., Weiss, H. J. & Shattil, S. J. (1989) Assembly of the platelet prothrombinase complex is linked to vesiculation of the platelet plasma membrane. *J. Biol. Chem. 264*, 17049-17057.

Sulpice, J-C., Zachowski, A., Devaux, Ph. F. & Giraud, F. (1994) Requirement for phosphatidylinositol 4,5-bisphosphate in the Ca^{2+}-induced phospholipid redistribution in the human erythrocyte membrane. *J. Biol. Chem., 269*, 6347-6354.

Suné, A., Vidal, M., Morin, P., Sainte Marie, J. & Bienvenüe, A. (1988) Evidence for bidirectional tranverse diffusion of spin-labeled phospholipids in the plasma membrane of guinea pig blood cells. *Biochim. Biophys. Acta, 946*, 315-327.

Suné, A. & Bienvenüe, A. (1988) Relationship between the transverse distribution of phospholipids in plasma membrane and shape change of human platelets. *Biochemistry 27*, 6794-6800.

Suné, A., Bette-Bobillo, P., Bienvenüe, A., Fellmann, P. & Devaux, Ph. F. (1987) Selective outside-inside translocation of aminophospholipids in human platelets. *Biochemistry 26*, 2972-2978.

Tilley, L., Cribier, S., Roelofsen, B., Op Den Kamp, J. A. F. & van Deenen, L. L. M. (1986) ATP-dependent translocation of amino phospholipids across the human erythrocyte membrane. *FEBS Lett. 194*, 21-27.

Williamson, P., Kulick, A., Zachowski, A., Schlegel, R. A. & Devaux, P. F. (1992) Ca^{2+} induces transbilayer redistribution of all major phospholipids in human erythrocytes. *Biochemistry, 31*, 6355-6360.

Zachowski, A. & Devaux, Ph. F. (1990) Transmembrane movements of lipids. *Experientia 46*, 644-656.

Zachowski, A. & Morot Gaudry-Talarmain, Y. (1990) Phospholipid transverse diffusion in synaptosomes: evidence for the involvement of the aminophospholipid translocase. *J. Neurochem., 55*, 1352-1356.

Interactions of Peptides with Phospholipid Vesicles: Fusion, Leakage and Flip-Flop

Shlomo Nir, Elias Fattal[1,3], Roberta A. Parente[1,4], Jose L. Nieva[2,5], Jan Wilschut[2], and Francis C. Szoka, Jr.[1]

The Seagram Center for Soil and Water Sciences
Faculty of Agriculture
Hebrew University of Jerusalem
Rehovot 76100
Israel

1. Introduction

The understanding of the interrelationships among dynamics, structure, and function of membrane-interacting peptide segments has been intensely studied during the last decade.

Several studies focused on peptides corresponding to N-termini of viral glycoproteins. The homology of the amino acid sequence of the N-termini of several viral

[1]School of Pharmacy, University of California, San Francisco, CA 94143-0446, U.S.A.
[2]Department of Physiological Chemistry, University of Groningen, Bloemsingel 10 9712, KZ Groningen, The Netherlands.
[3]Universite Paris-Sud, Faculte de Pharmacie, Laboratoire de Phyhsico-Chimie, Pharmacotechnie, Biopharmacie, URA CNRS 1218, 92296 Chatenay-Malabry, Cedex, France.
[4]Ortho Diagnostics, Route 202, Raritan, N.J. 08869, U.S.A.
[5]Department of Biochemistry, University of the Basque Country, P.O. Box 644, 48080 Bilbao, Spain.

NATO ASI Series, Vol. H 91
Trafficking of Intracellular Membranes
Edited by M.C. Pedroso de Lima N. Düzgüneş and D. Hoekstra

proteins has led to the proposal that this region is intimately involved in the fusion process. The hydrophobic N-terminal peptide of the Sendai virus F-protein is normally exposed on the surface of the protein (Gething et al. 1978; Hoekstra and Kok, 1989), whereas that of the influenza virus hemagglutinin (HA2) is exposed only at low pH (White and Wilson, 1987), where the virus is fusion active (Stegmann et al., 1990; Nir et al., 1990; Duzgunes et al., 1992; Ramalho-Santos et al., 1993).

Synthetic peptides corresponding to these sequences have anti-viral effects (Richardson and Choppin, 1983). A synthetic peptide (HA2.7) corresponding to the 7 amino acids of the N-terminal of HA2 (X31) induces conductance fluctuations in planar bilayers and leakage from liposomes at neutral pH (Duzgunes and Gambale, 1988). Peptides at the N-termini of hemagglutinin mutants with impaired or no fusion activity in cell-cell fusion were found to have an impaired ability to interact with phospholipid membranes (Duzgunes and Gambale, 1988; Rafalski et al., 1991).

This presentation deals with three aspects, leakage, flip-flop and fusion induced by peptides. The aim is to explain the experimental and theoretical procedures employed, and discuss recent results.

2. Leakage and Pore Formation

The current mathematical model of pore formation was developed to explain the leakage induced by the amphipathic peptide GALA from large unilamellar PC vesicles. No fusion or aggregation of these vesicles resulted upon addition of the peptide alone. Several observations supported a pore model. First, at pH 5 and 20°C incomplete final extents of leakage were seen for peptide/lipid ratios below 1:2500. This fact could rule out leakage induced by collisions between vesicles, i.e., the leakage occurred from non-interacting vesicles. Second, as will be elaborated below the leakage occurred by an all or non mechanism; vesicles either leak or retain all their contents. Third, the leakage showed a selectivity to the size of the entrapped molecules. These results could imply that vesicles containing less than a critical number, M, of peptides do not leak at all, whereas in vesicles containing M or above peptides, surface aggregation of the bound peptides can result in the formation of a pore. Once a pore has been formed the leakage of small molecules would be rapid, whereas large molecules would not leak. The mathematical model developed (Parente et al., 1990a) could explain and predict the final extents and kinetics of leakage for a wide range of peptide/lipid ratios and provided the pore size.

2.1. Experimental Procedures

We describe here the procedures in Parente et al. (1990a). The ANTS/DPX assay (Ellens et al., 1984) was used to monitor leakage induced by peptide interaction with vesicles. The fluorescence signal resulting from the dequenching of ANTS released into the medium was observed through a Schott GG 435-nm cutoff filter (50% transmittance at 435 nm) while samples were irradiated at 360 nm; 90° light scattering was simultaneously recorded through a monochrometer at 360 nm. Data points were recorded at 0.5-s intervals over a 10-min period after which vesicles were lysed to obtain the maximal fluorescence value, which was set to 100% leakage. Fluorescence from intact vesicles in buffer was set to 0% leakage. The peptide was added to stirred vesicle suspensions (100 μm) at 20°C.

Leakage Mechanism. To differentiate if leakage of contents was an all or none event (some of the vesicles release all of their contents) as opposed to a graded event (all of the vesicles release some of their contents), we employed a modification of the fluorescence dequenching method of Weinstein et al. (1981), encapsulating the fluorophore/quencher pair ANTS/DPX in place of carboxyfluorescein in the original report. Two sets of experiments are prerequisite for this study. First, the extent of leakage of ANTS/DPX induced by GALA was determined at various lipid to peptide ratios. Second, a quench curve for vesicles containing varying concentrations of ANTS and DPX was constructed. The quencher and fluorophore were encapsulated at a 3.6/1 mole ratio.

Vesicles prepared for the fluorescence dequenching experiments contained 6.25 mM ANTS and 22.5 mM DPX. From the quench curve these vesicles are 81% quenched or have 19% associated fluorescence. This starting condition was chosen at a point on the quench curve such that a small change in ANTS concentration due to vesicle leakage would give rise to a significant change in fluorescence. Vesicles were incubated with

GALA at ratios varying from 3500/1 to 8500/1 for 15 min. The lipid concentration was kept constant at 100 μM. Samples were then applied to a Sephadex G-75 column (0.5 cm x 7 cm) and eluted with 5 mM TES-100 mM KCl, pH 7.5, to separate the vesicles fraction from ANTS and DPX which leaked from vesicles during incubation. Fluorescence of the eluted vesicle fraction was measured before and after detergent and the percent fluorescence remaining with the vesicles was calculated.

One could predict the percent fluorescence that would be vesicle associated for an all or none or graded response by making use of the results from the prerequisite experiments. For example, at 5000/1 there is approximately 80% leakage after 10 min. If this means that all vesicles leak 80% of their contents (graded release), then the ANTS/DPX concentration remaining inside would be 1.25 and 4.5, respectively. From our quench curve this corresponds to 46% quenching or 54% fluorescence remaining with the vesicles. However, if 80% of the vesicles leak all of their contents (all or none release), the ANTS and DPX concentrations inside the intact vesicles would be unchanged from the starting condition (81% quenched or 19% associated fluorescence). These predicted results were used to interpret the observed values.

2.2 Theoretical Analysis of Leakage Kinetics and Extent

The model assumes that (1) the peptides bind and become incorporated into the bilayer of the vesicles and (2) within the membranes peptide aggregation occurs. When an aggregate within a membrane has reached a critical size, a channel or a pore is created within the membrane, and leakage of encapsulated molecules can occur. The size of the pore dictates the upper bound on the size (and shape) of molecules that can leak. The size of the pore or channel depends on the number of peptides forming it.

The assumption that a critical number of peptides in a vesicle is required to cause leakage has been introduced to account for two of our findings. (i) In those instances where the final extent of leakage was less than 100%, the vesicles were divided into two sharply defined groups, those that leaked all their contents and those that leaked practically none of their contents (see results in Table 1). (ii) In systems where the final extent of leakage (e.g., after 10 min) was not complete and the rate of leakage had practically become zero, additional leakage could be induced by elevating the pH to neutral and reducing it again to pH 5. When the pH was elevated, the peptides dissociated from the membranes [see Figure 1B], and when the pH was again reduced, there was a probability that vesicles that had not leaked in the first round would incorporate a sufficient number of peptides to undergo leakage in the second round.

Final Extents. According to the model, the final extent of leakage is due to the leakage of contents from all the vesicles containing M or more peptides, where M is the critical number of peptides in an aggregate. For a completely homogeneous population of vesicles, the final extent of leakage $L \equiv L(\infty)$ is given by

$$L = \sum_{i=M}^{N} A_i \tag{1}$$

in which N is the largest number of peptides that can be bound to a single vesicle, and A_i is the fraction of vesicles containing i incorporated peptides.

In Parente et al. (1990a) we approximated the vesicle population to consist of two sizes. In more recent calculations (Fattal et al., 1994; Nieva et al., 1994) we have considered the vesicle population to have a size distribution given by measurements of dynamic light scattering. We used the vesicle number distribution for S size classes, with fractions a_j converted to lipid fractions according to

$$Q_j = a_j N_j / (\sum_{j=1}^{S} a_j N_j), \tag{2}$$

in which N_j is the number of lipid molecules per vesicle. Hence, L in Eq (1) is calculated by

$$L = \sum_{j=1,i=M}^{S,N_j} Q_j A_{ij} f_j \tag{3}$$

in which f_j is the fraction of encapsulated volume in vesicles of size J. It should be noted that vesicles of larger size contribute significantly more to the leakage than their fraction of the total population. Procedures for calculating the quantities A_i (or A_{ij}) are given in Nir et al. (1986); Bentz et al. (1988) and Parente et al. (1990a). In the latter reference A_i was calculated from the recursion equation:

$$A_i = A_{i-1}\, PK[N-(i-1)]/(Ni), \tag{4}$$

in which P is the peptide molar concentration in solution (at time t), and K, which is the binding constant was calculated from the amount of peptide bound. The binding constant K (units = M^{-1}) is given by K=C/D, in which C (units = $M^{-1}.S^{-1}$) is the forward rate constant of binding, and D (units = S^{-1}) is the rate constant of dissociation.

In the studies with GALA for a certain range of peptide and lipid concentrations the binding of peptides to the lipid satisfied a partition relation, where for a lipid concentration of 100 μM the fraction of peptides bound (B) is 0.645 irrespective of peptide concentration. This corresponds to an apparent bilayer/aqueous molar partition coefficient = 1.0×10^6 as calculated according to Pownall et al. (1984). For lipid/peptide mole ratios of 2000-30,000 this partition relation (0.645) was satisfied quite well with a 1-4% deviation from the value computed according to

$$(P_o-P)/P_o = \sum_{i=1}^{N} iA_i/P_o \tag{5}$$

By comparing the experimental value of L, the final extent fraction that has leaked (within 10 min), to the calculated fraction according to Eq 3 for each ratio of lipid/peptide, we aimed at finding a range of M values that could yield the critical number of peptides in a vesicle for which complete leakage occurred. As it turned out, we could explain (and predict) quite satisfactorily all the experimental results by assuming M = 10 for all cases.

Kinetics of Leakage. In calculating the kinetics of leakage, we assumed that the process of peptide incorporation in the vesicle membranes was rapid (Parente et al. 1990a) and that once a channel (or a pore) has been formed in a vesicle all its contents would leak rather fast, within less than 1 s. Thus, the kinetics of leakage was equated to that of formation of aggregates consisting of M or more peptides in the bilayer. Treatments of the problem of dynamical aggregation are given in Bentz and Nir (1981) and Nir et al. (1983). The scheme is

$$\begin{aligned} X_1 + X_1 &\underset{D_{11}}{\overset{C_{11}}{\rightleftharpoons}} X_2 \\ X_1 + X_i &\underset{D_{1i}}{\overset{C_{1i}}{\rightleftharpoons}} X_{i+1} \\ X_2 + X_2 &\underset{D_{22}}{\overset{C_{22}}{\rightleftharpoons}} X_4 \end{aligned} \qquad (6)$$

where X_i are molar concentrations of aggregates of order i. Here we have used the same scheme, but in our case X_i denotes surface concentrations. In these calculations we have introduced the computational simplifications that aggregate dissociation could be ignored and, furthermore, that the Smoluchowski (1917) treatment that assumed $C_{ij} = C$ could be employed. Recently, the treatment was extended to allow for reversibility of surface aggregation (R. Peled and S. Nir, unpublished), but no improvement to the fit of the experimental results was achieved, despite the addition of one parameter. The studies of Nieva et al. (1994) indicate that the process of surface aggregation of the peptide HIV_{arg} in POPG is essentially irreversible.

In calculating (for a given vesicle at time t) the fraction, $F_M(t)$, of peptides in aggregates composed of M or more molecules, we employed an equation derived from the Smoluchowski distribution (Nir et al. 1983):

$$F_M(t) = \left(\frac{M+\tau}{1+\tau}\right)\left(\frac{\tau}{1+\tau}\right)^{M-1} \qquad (7)$$

In Eq 7, $\tau = \text{Cit}$, where t is the time (s), i is proportional to the surface concentration of peptides (units = 1/area), and C is a forward rate of aggregation. If i has a unit of inverse area, then C has a unit of area/time and τ is a dimensionless quantity. We chose here i as a number, so that C has a unit of s^{-1}.

The fraction of encapsulated material that has leaked at time t is given by

$$L(t) = \sum_{j=1}^{S} \sum_{i=M}^{N_i} A_{ij} F_{Mj}(t) f_j \tag{8}$$

In our calculations, we assumed that the big and small vesicles have the same surface properties, irrespective of the number of peptides bound, which means that the same value of C was employed for all cases. However, for the same value of i the surface density of peptides is larger for the small vesicles. This was taken into account in the calculations. The value of C corresponds to the outer surface area of a 100 nm radius vesicle.

2.3. <u>Results and Discussion</u>

Fig. 1A (dashed line) shows a typical leakage curve obtained under these conditions (100 μM egg PC) when GALA was added to a vesicle suspension at pH 5 to give a 5000/1 lipid/peptide mole ratio. The solid line is the leakage pattern obtained when initial conditions were set for a 10,000/l lipid/peptide ratio and an equal aliquot of peptide was injected (after 10 min, see arrow) to give a final ratio of 5000/1. The final extent of leakage is the same at a given ratio whether GALA is added in one or more aliquots. This shows that GALA acts reproducibly.

The reversible pH dependency of GALA's activity is clearly illustrated in Figure 1B. Vesicles were placed into suspension at pH 5, and GALA was added to give a final mole ratio of 5000/1. After 0.75 min a concentrated TES solution was added to bring the pH to 7.3, and at 4.5 min an aliquot of concentrated acetate was added to return the pH to 5. Raising the pH dramatically stopped further leakage, while returning the pH to 5 caused an immediate increase in the rate of leakage (vs the initial rate). The dotted line shows the leakage curve obtained at pH 5 without interruption at the same lipid/peptide ratio. We believe that leakage begins as the peptide enters into the vesicles in its helical form at pH 5 and is halted at pH 7 as the peptide changes conformation and disassociates from the membrane. When the pH is returned to 5, GALA can redistribute among the vesicles and induce leakage of those previously left intact. The overshoot to 100% leakage that is observed when the pH is cycled back to 5 vs uninterrupted can be explained by such a mechanism.

Table 1. Release Mechanism of GALA at pH 5[a]

	% fluorescence remaining with vesicles		
	predicted[b]		
lipid/peptide	all or none release	partial release	observed[c]
3500/1	19	67	23 ± 3
5000/1	19	54	19 ± 2
6000/1	19	45	20 ± 2
8500/1	19	26	17 ± 1

[a]Details of how both the predicted and observed values were obtained are given under Materials and Methods.

[b]Predicted value is from a quench curve of ANTS/DPX fluorescence and has an error of ±2%.

[c]Observed value is an average of at least three measurements at each lipid/peptide ratio.

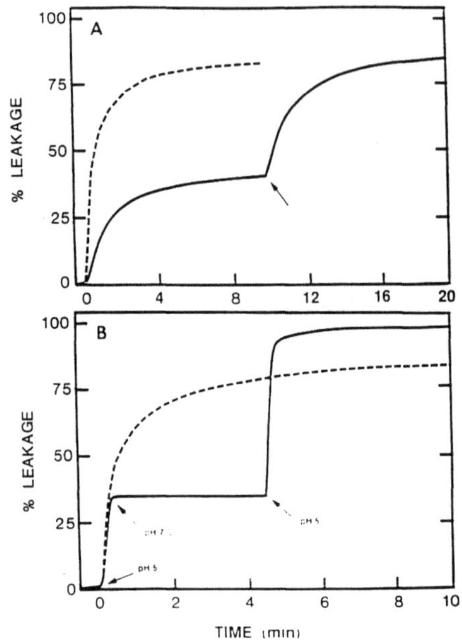

Figure 1. (A) Percent leakage as a function of time and peptide concentration at pH 5. GALA was injected into egg PC vesicles containing 12.5 mM ANTS and 45 mM DPX at a 10 000/1 lipid/peptide ratio. After 10 min, an equivalent aliquot of peptide was added to give a final lipid/peptide ratio of 5000/1. The dashed line is the leakage curve obtained when GALA is added at a 5000/1 lipid/peptide ratio at time zero. (B) Percent leakage as a function of time and pH. Leakage was arrested by the injection of a concentrated TES solution, which raised the pH to 7.3. A final injection of acetate lowered the pH back to 5, and leakage resumed. Injection points are indicated by arrows. The dashed line shows the leakage obtained for a 5000/1 lipid/peptide ratio at pH 5 without interruption. In both panels the results presented are from a single representative experiment. Replicate experiments showed no more than a ±2% variation in the final extents of leakage.

In spite of similar structural features at pH 5, the modified peptide, LAGA (Parente et al. 1990b), was unable to induce leakage of egg PC large vesicles even at lipid/peptide ratios of 50/1 (a 50-fold greater amount of peptide than needed for 100% leakage by GALA).

For several lipid/peptide ratios, Table 1 gives the percent fluorescence remaining with vesicles after being incubated and subsequently separated from peptide. The second and third columns are the predicted results (see Experimental Procedures) for the two types of release mechanisms while the last column contains the experimentally obtained values. From a comparison of the predicted and observed results, it is quite evident that the leakage induced by GALA occurs as an all or none event. This implies that the number of peptide molecules needed to induce complete leakage from a single vesicle must exceed a certain critical value. At the lipid/peptide ratios where partial leakage occurs, one can assume that on average there is a certain percentage of vesicles which do not have the critical number of peptide molecules associated and/or properly arranged in the bilayer to cause leakage.

The all or none mechanism of leakage was also operative in the case of the HIV_{arg} peptide (Nieva et al. 1994; J.L. Nieva, S. Nir and J. Wilschut, unpublished) and in the case of pardaxin and one of its analogues (D. Rapaport, S. Nir, Y. Shai, unpublished) when added to large liposomes under conditions where vesicle fusion was minimal. In contrast, Ca^{2+}-induced leakage from PS liposomes increased dramatically with lipid concentration indicating that it depended on interactions between liposomes (Nir et al., 1983). Indeed, the leakage in this case was closer to graded leakage.

Pore Formation versus Vesicle Disruption. As a result of leakage being an all or none event, a mechanism involving assembly of peptides into a pore in the bilayer seems most plausible although a detergent-like action of the peptide can also be envisioned as a means to disrupt vesicles. It was reasoned that if a pore formed, then there would be a size/molecular weight dependence on the leakage of initially encapsulated compounds, while size should not be a factor if peptides could act in a detergent-like fashion.

Compounds ranging in size from MW 445 to 5200 were encapsulated in vesicles and incubated with GALA at pH 5 or 7.5 prior to determination of how much material leaked out. At pH 7.5 none of the compounds leaked significantly in 15 min (Table 2), while at pH 5 only leakage of molecules greater than MW 800 was reduced (Table 2). The size dependence of leakage supports the idea of GALA forming a channel or pore in the vesicle bilayer.

Table 2. Size Dependence of Marker Retention in Vesicles[a]

		% contents remaining with vesicle fraction	
contents entrapped	MW	pH 5	pH 7.5
ANTS	445	0	100
NADH	660	26	95
acid blue 9	795	14	100
AP_5A	915	100	100
inulin	5200	87	100

[a]Vesicles were incubated for 15 min with GALA prior to separation of vesicle fractions from released material on a Sephadex G-75 column. Lipid to peptide ratios were 2500/1 and below.

Results of Theoretical Model. In developing a model to describe GALA-induced leakage of vesicle contents, the final extents of leakage at 10 min were first used to determine the critical number of peptides (M) needed to assemble to form a pore. The 10-min time period was selected since leakage was essentially complete by this time. When the extent of leakage was measured at 1 h, it was found to be unchanged over the value at 10 min. The value for M was required to satisfy the leakage over the range of lipid to peptide ratios studied. This determination was independent of the kinetics of the leakage. The observed and calculated extents of leakage as a function of lipid to peptide mole ratio at a constant lipid concentration are shown in Figure 2. The open squares are the calculated extents of leakage, 10 being used as the critical peptide unit. The value of M

= 10 provides a reasonable fit to the final extent leakage data over a wide range of lipid/peptide ratios. A statistical test for the goodness gave SD(M) = 1.2 and R^2 = 0.94.

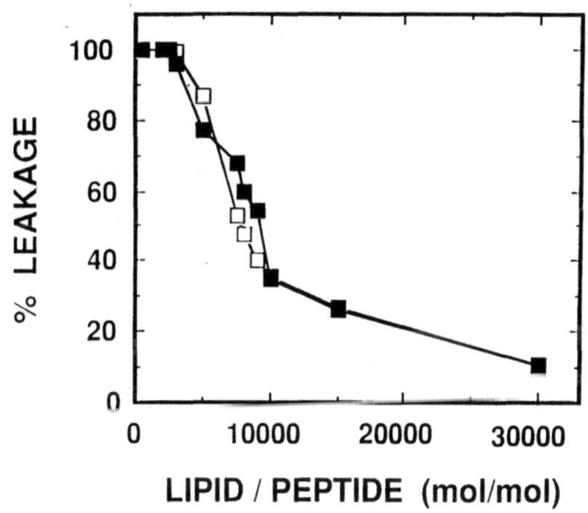

Figure 2. Comparison of experimental (■) and predicted (□) final extents of leakage as a function of lipid/peptide mole ratio at pH 5 when the lipid concentration is held constant at 100 μM. Below 2500/1 the experimental and calculated curves overlap at 100%. The curves converge again at ratios greater than 10000/1. The reproducibility in the experimental extents of leakage is excellent in replicates from the same experiment (±2%) and varies no more than ±5% between experiments. A value of M = 10 was used for predicting the final extents of leakage.

Table 3. Effect of Vesicle Size on Experimental and Calculated Extents of Leakage.

lipid/peptide (mol/mol)	extrusion limit[a] (μm)	% leakage exptl.	% leakage calcd.
5000/1	0.05	54	48
5000/1	0.1	77	86
5000/1	0.2	100	100
15000/1	0.05	11	17
15000/1	0.1	26	27
15000/1	0.2	36	33

[a]For the purpose of the calculation each vesicle preparation was assumed to be composed of two sizes in a 24/1 ratio. Vesicle radii selected to represent the resulting sample populations were as follows: for vesicles extruded through 0.05 μm, 40 and 80 nm in radius; for 0.1 μm, 50 and 100 nm; for 0.2 μm, 60 and 120 nm.

After having fixed the critical number of peptides in a vesicle to be $M = 10$, from the results of final extents of leakage, we tested the ability of the model to explain and predict the kinetics of leakage. The leakage kinetics induced by GALA calculated as described give rise to the curves represented by the closed circles in Figure 3. These are plotted with the experimental curves (solid lines) for representative lipid to peptide ratios. The calculated curves are in close agreement with the observed data. It is noteworthy that a single parameter, C, which is proportional to the forward rate constant of surface aggregation, can explain and predict the kinetics of leakage for the five curves in Figure 3. The value of R^2 for the fit of the kinetic results was 0.98, and the standard deviation of the parameter C was 0.0004 s^{-1}.

By changing the diameter of the vesicles, the peptide to vesicle ratio can be altered. A prediction of the model is that for a constant amount of lipid and a given lipid/peptide ratio the extent of leakage should decrease as the vesicle diameter decreases. To test this, a single vesicle preparation was fractionated into different size populations by sequential extrusion through 0.2-, 0.1-, and 0.05-μm polycarbonate membranes. The predicted trend was observed when leakage from these vesicles was measured 10 min after GALA addition. The experimental leakage values at a 5000/1 and 15000/1 lipid/GALA ratio are given in Table 3 along with the calculated values.

2.4. General Discussion

A detailed picture is beginning to emerge on the leakage mechanism and structure of the leaking unit formed by GALA in egg PC bilayers. The model is based upon the kinetic studies and physical measurements made on the GALA-lipid complex. Leakage of contents from phosphatidylcholine vesicles induced by GALA is a sensitive function of pH. This corresponds with the pH-dependent random coil to α-helical transition in the secondary structure of GALA observed by circular dichroism (Subbarao et al., 1987) as glutamic acid residues become protonated. We have shown that leakage is an all or none event. This finding supports the idea that a critical number of peptides must interact with a vesicle before it can become leaky. In conjunction with this, the size dependence of molecules which can be released from the vesicles supports the idea that peptides organize to form a pore.

The structural aspects of the model are quite similar to those of the model for the alamethicin pore originally put forth by Fox and Richards (1982). We propose that when the pH is reduced from 7.5 to 5, GALA rapidly partitions into the membrane and then assembles into an oligomeric complex containing between 8 and 12 GALA monomers. In the complex, individual GALA molecules have a predominantly α-helical secondary structure and the helix axis is aligned perpendicular to the bilayer surface (Figure 4). When in a helix, a 30 amino acid peptide would have an end to end length of 45 Å sufficient to span an egg PC bilayer with a phosphate to phosphate distance of 40 Å (White & King, 1985).

The notion that GALA is predominantly α-helical in the membrane is indirectly supported from X-ray diffraction studies with other synthetic peptides such as Lys_2-Gly-Leu_{24}-Lys_2-Ala-amide (Huschilt et al., 1989) in oriented multilayers. This peptide forms α-helices, which orient with the helical axis perpendicular to the lipid bilayer. The proposed orientation and extent of helical structure of GALA in the bilayer receive direct support from circular dichroism and polarized FTIR measurements on GALA in oriented multilayers (Goormaghtigh et al., 1991).

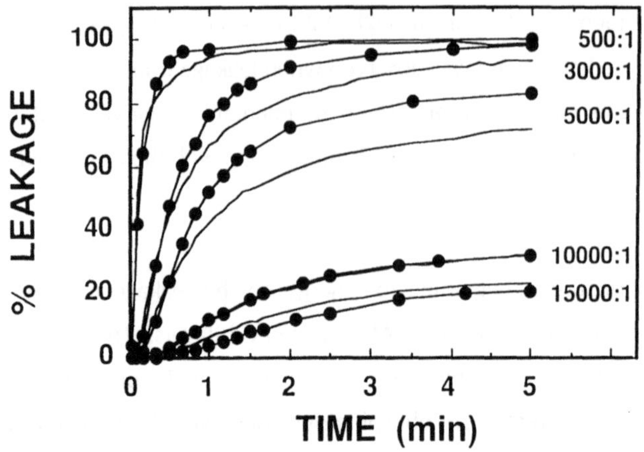

Figure 3. Extents of leakage as a function of time after GALA addition at pH 5. Lipid concentration was kept constant at 100 μM while the lipid/peptide ratio was varied. The solid lines are the experimental data recorded at the five lipid/peptide ratios indicated at the right of the figure. The reproducibility in the experimental extents of leakage is excellent in replicates from the same experiment ($\pm 2\%$) and varies no more than $\pm 5\%$ between experiments. The filled circles represent the best fit to the data at these ratios with $B = 0.645$ $C = 0.002$ s^{-1}, and $M = 10$ as described in the text.

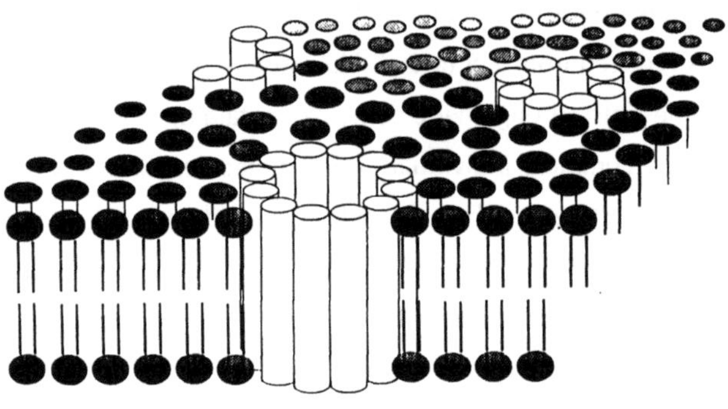

Figure 4: Schematic drawing of how GALA peptides would organize to form a transbilayer pore in a membrane at low pH. Individual GALA helices are depicted as cylinders.

The quenching of the tryptophan fluorescence when GALA incorporates into bilayers at low pH suggests that tryptophan residues from adjacent monomers can interact. This would lend support to a parallel orientation of the monomers when they assemble into a pore.

Complexes containing 8 or 12 GALA subunits were symmetrically arranged to form a circular pore by interactive computer graphics. The starting point for these studies was to place GALA in an ideal α-helix. Pairs of GALA units were positioned in a parallel orientation to minimize contacts between adjacent monomers. Pairs were then transposed in a symmetrical arrangement such that glutamic acid residues were oriented to interact with the hydrocarbon chains of the bilayer. As observed for the model of the alamethicin chain el (Fox & Richards, 1982) and the model channel-forming peptides of Lear et al. (1988), there is little steric overlap of the solvent-accessible surfaces (Connolly, 1983) between helices.

To estimate the pore diameters of GALA assemblies, we calculated the minimum distance between the van der Waals surfaces of apposing glutamate residues lining the pore. An eight-helix bundle of GALA molecules would have a radius of about 8 Å and would permit the diffusion of molecules with a cross-sectional area of 200 Å^2. If we assume that the interior of the channel is lined by a single-shell hydration layer, then the channel radius becomes 5 Å and the cross sectional area would be about 80 Å^2. A channel composed of 12-GALA helices arranged in a similar fashion and lined by a single hydration layer would have a radius of 10 Å and a cross-sectional area of 310 Å^2. A size dependence of leakage induced by GALA is consistent with a channel diameter between 5 and 10 Å since NADH (MW 660) has a Stokes radius of about 6 Å (Peters, 1986) and is permeant while fluorescein dextran (MW 4100) and inulin (MW 5200) have Stokes radii of 12 and 13 Å, respectively (Peters, 1986) and are not permeant. The molecular weight cutoff for release of solutes and the computed channel diameter are similar to those described for the staphyloccal α-toxin. The similarities of the channel properties of the membrane lytic protein staphyloccal α-toxin to those formed by GALA reinforce the concept that simple peptide sequences can mimic the actions of more complicated biological proteins (DeGrado et al., 1989).

Less detailed studies have recently provided evidence for pore formation by the peptide HIV_{arg} interacting with negatively charged liposomes composed of POPG

(Nieva et al., 1994) or neutral liposomes (J.L. Nieva, S. Nir and J. Wilschut, unpublished), and the all or none mechanism of leakage, which is indicative for pore formation was observed with several (but not all) analogues of pardaxin (D. Rapaport, S. Nir and Y. Shai, unpublished).

3. Pore-Forming Peptides Induce Rapid Phospholipid Flip-Flop in Membranes

This section summarizes the results in Fattal et al. (1994) where detailed references can be found. The formation and maintenance of phospholipid asymmetry in biological membranes is crucial for membrane function. Under most conditions the transbilayer flip-flop of phospholipids is exceptionally slow with half-times for flip-flop ranging from hours to weeks. Membrane proteins termed flippases have been identified that catalyze the transbilayer flip-flop of certain phospholipids yielding half times of less than 5 minutes. The low rates of transbilayer reorientation of phospholipids can be increased significantly by the incorporation in the membrane of channel forming antibiotics such as amphotericin B. We tested if peptides that interact with membranes and self-associate to form aqueous pores could mediate rapid phospholipid flip-flop. We used GALA, melittin, and three other peptides. The ability of these five different peptides to mediate transbilayer phospholipid flip-flop as a function of peptide/lipid ratios has been compared under conditions where vesicle integrity is retained.

A model that employs mass-action, was used to simulate and predict the extent and kinetics of flip-flop induced by pore-forming peptides, as well as peptides that merely perturb the membrane. When a peptide aggregate within a membrane has reached a critical size of M peptides, a channel or a pore is created within the membrane, which results in a fast flip-flop of probe molecules (as well as other molecules in the area of the pore and its vicinity). Thus, a relatively fast flip-flop occurs leading to equilibration of the lipid composition on the two monolayers. We denote this process as pore mediated flip-flop. In the case where the probe can transfer between vesicles, the presence of acceptor vesicles results in the removal of the probe molecules from the outer monolayer of prelabeled vesicles. For the purpose of generality, we consider that in addition to this major mode of flip-flop, a slower flip-flop mode can occur in vesicles including less than M peptides in an aggregate. This mode of flip-flop can reflect a perturbation in the membrane that proceeds for a limited duration or as long as the peptide resides in the membrane, and will be denoted perturbation-mediated flip-flop. Ordinarily in unperturbed phospholipid vesicles, including those employed in the current study, the flip-flop process is extremely slow in the absence of peptides or other perturbants. Times required for equilibration by such flip-flop are of the order of several hours to weeks. Hence on a time scale of minutes to hours, we ignore this type of "intrinsic" flip-flop. In the calculations, we have considered two types of flip-flop as described above, but in certain ideal cases only the "pore-mediated flip-flop" has to be considered on a time scale of up to several minutes.

In addition to demonstrating that pore forming peptides such as GALA or melittin could induce very rapid flip-flop with half times of the order of seconds, we demonstrate the existence of two other classes of peptides: those that mediate flip-flop by a perturbation mode and those that intercalate into the bilayer with little effect on phospholipid flip-flop.

4. Induction of Liposome Fusion by Peptides

We will focus here on recent studies employing a 23 amino acid peptide (HIV_{arg}) at the N-terminus of HIV-1 gp41.

One of the central issues addressed in this study is the relationship between pore formation by the peptides and induction of vesicle fusion. For HIV_{arg} and POPG liposomes, pore formation and fusion turned out to be conflicting processes requiring different structures of the peptide in the membrane (Nieva et al., 1994). For brevity, we will only present part of the Discussion.

4.1 Introduction

The amphipathic peptide mellintin which induces leakage and flip-flop (see previous section) can also induce liposome fusion.

Various peptides corresponding to the N-terminal "fusion domain" of viral envelope glycoproteins (White et al., 1983) can induce the fusion of certain types of liposomes. A 20-amino acid N-terminal peptide from the HA2 subunit of influenza hemagglutinin induces the fusion of SUV composed of POPC or egg PC (Murata et al., 1987; Rafalski et al., 1991) but no fusion of LUV. However, extensive pH-dependent fusion of large PE/PC/cholesterol (1:1:1) vesicles was observed, suggesting that the peptide may be inducing inverted micellar fusion intermediates, since certain species of PE can form non-bilayer structures (J.L. Nieva and J. Wilschut, unpublished data). On the other hand, a 17-amino acid HA2 peptide induces neither aggregation nor fusion of PC or PS LUV, although it causes the destabilization of PC LUV, suggesting that the fusion activity may be related to the length of the peptide (Duzgunes and Shavnin, 1992).

Peptides of varying length corresponding to the N-terminus of the SIV envelope glycoprotein gp32 were shown to induce fusion of SUV of various compositions, while fusion of LUV was observed only if the membranes contained PE (Martin et al., 1993).

For peptide-induced membrane fusion, the initial aggregation step must be considered as a critical factor. It has been pointed out that the insertion of viral fusogenic peptides into the target membrane does not (necessarily) cross-link two membranes (liposomes), whereas insertion of a segment of a viral envelope protein into the target membrane does cross-link the viral and target membranes (Epand et al., 1992). Clearly, vesicle aggregation at a primary minimum becomes more difficult with an increase in vesicle size (Nir et al., 1983). Thus, the peptide GALA can induce the fusion of PC SUV, but cannot aggregate PC LUV (Parente et al., 1988), although it can induce leakage from LUV, and it induces rapid flip-flop.

When PC or PG LUV are aggregated by a lectin or Ca^{2+}, respectively, the peptide GALA can induce liposome fusion (R. Parente and F. Szoka, unpublished). Similarly, the neurotoxic peptide pardaxin and several of its analogues could induce the fusion of SUV composed of PC and PS (Rapaport et al., 1993), but fusion of large PS liposomes was limited, unless liposome aggregation was promoted by Mg^{2+} (2.5 mM) (D. Rapaport, S. Nir, Y. Shai, unpublished), whereas Mg^{2+} could not induce liposome fusion or leakage without the peptide (Wilschut et al., 1981).

4.2. Discussion of Results

This study demonstrated peptide-induced fusion of large POPG vesicles by showing membrane mixing, and significant increase in vesicle sizes to 680 nm average diameter, and by showing aqueous content retention and mixing. All these assays gave consistent results for the increase in fusion activity with peptide to lipid ratios.

The values of the fusion rate constants increase dramatically with an increase in peptide to lipid ratios, particularly when going from 1:25 to 1:12.5. The values obtained, f=0.004 to 0.06 s^{-1} are in the range found in several studies of virus-liposome fusion (Nir et al., 1986b, 1990) and virus-cell fusion (Nir et al., 1986a, 1990). Of particular interest is the fact that the above f-values are comparable to those found for fusion of HIV-1 with liposomes (Larsen et al., 1993). In the latter study it was also reported that physiologically relevant concentrations of calcium stimulate HIV-1 fusion with erythrocyte ghost membranes and with liposomes of several compositions, whereas no such enhancement was observed with influenza virus (Stegmann et al., 1985) or Sendai virus (D. Hoekstra, personal communication). It has to be emphasized that the fusion of HIV with membranes does not require Ca^{2+}, but is promoted by it. We have pointed out that Ca^{2+} reduces, or at a low lipid:peptide ratio, it even inhibits the leakage induced by the peptide. Clearly this action of Ca^{2+} does not stem from significantly reducing the binding of the peptide to the vesicles, since the fusion is promoted under the same conditions.

Our results suggest that Ca^{2+} seals POPG membranes by causing structural changes which are not compatible with pore formation. We can conclude that part of the peptides remain in such structures even after the removal of Ca^{2+}, since the results indicate that upon the addition of EDTA the rate of leakage is not restored to its value prior to the addition of Ca^{2+}. This could mean that with Ca^{2+} the peptides associated with vesicular membranes adopt certain types of conformations or form aggregates that are irreversible on a time scale of several min. If EDTA is added at an earlier time to POPG + Ca^{2+} + peptide, the leakage is progressively restored to the level of POPG + peptide, indicating that the intermediates of the "Ca^{2+} structure" can revert to the pore forming state. It is interesting to note that addition of EDTA 15 sec after initiation of fusion does not arrest the process of membrane mixing during the course of about 2 min.

One of the most intriguing results is the effect of the order of addition of Ca^{2+} and peptide on the fusion of POPG vesicles. When Ca^{2+} is added 30 sec following the addition of the peptide, a very low rate and limited extent of membrane mixing are observed. On the other hand, a high degree of leakage of ANTS/DPX has occurred by that stage, indicating that a high degree of surface aggregation of the peptide which leads to pore formation has occurred.

Consequently, from the combination of the leakage and fusion results we conclude that the HIV_{arg} peptide associated with POPG vesicles forms two different types of structures. In the absence of Ca^{2+} a pore structure is predominant, which yields leakage, but is not susceptible to fusion. In the pore structure, in which the bound peptides span the membrane, they resemble in a sense transmembrane sequences of integral proteins which apriori are not likely to play an active role in membrane fusion. In the presence of Ca^{2+}, i.e., when fusion is promoted, another type of structure is formed. We do not know yet whether this structure is the fusion competent one, or perhaps its intermediates mediate the fusion, but this structure yields initially a smaller degree of leakage. Once these final structures, i.e., the pore structure or the "Ca structure", have been established they are relatively long living. Furthermore, it was found that once a pore structure has been established practically no dissociation of peptides from vesicles occurs.

The FTIR results demonstrate indeed the existence of two types of structures. In the absence of Ca^{2+} the spectrum is compatible with a majority of an α-helix conformation of the peptide. This conformation was proposed to be associated with pore formation by peptides in vesicle membranes (Lear et al., 1988; DeGrado et al., 1989; Parente et al., 1990a). In contrast, in the presence of Ca^{2+} the peptide mostly adopts an extended antiparallel β-structure. It must be cautioned though that part of the peptide still remains in an α-helix conformation even in the presence of Ca^{2+}. Hence, at higher peptide to lipid (e.g., 1:25) ratios there might be a sufficient number of peptides that can form pores in the presence of Ca^{2+}.

It might be tempting to propose that the extended antiparallel β-structure of the peptide is the one in which the vesicles are susceptible to fusion. In this arrangement the peptides may be visualized to reside on the surfaces of the vesicles. Recently, Epand et al. (1992) deduced that the measles fusion peptide shows mostly β-structure when immersed in a lipid environment. The majority of studies with fusion peptides have proposed that the α-helical conformation is the fusion competent one (Lear & DeGrado, 1987; Harter et al., 1989; Takahashi, 1990). Gallaher et al. (1992) raised questions about the helical structure as the functional conformation of fusion peptides.

The fusion activity of a 51-amino acid cationic peptide towards PS or PC LUV correlated with its hydrophobicity rather than the formation of an amphiphilic α-helix (Yoshimura et al., 1992). The rate of fusion of PS liposomes by means of pardaxin analogues does not correlate with the percent of helical structure (D. Rapaport, S. Nir and Y. Shai, unpublished).

Based on the sum of our findings, we cannot eliminate the possibility that the minority peptides in an α-helix conformation might be those responsible for fusion in a stage before the formation of a pore, or perhaps some intermediate stages prior to the formation of the "Ca^{2+} structure" might be the fusogenic ones, since the FTIR results are recorded under equilibrium conditions, when most of the fusion has already occurred. What seems clear is that there are indeed two different structures and corresponding modes of interaction of HIV_{arg} in POPG membranes, one in the presence of Ca^{2+} that promotes fusion, and one in the absence of Ca^{2+} that promotes pore formation and leakage. Once a given structure has been established, it is essentially irreversible.

5. Acknowledgements

Supported by NIH Grant GM30163 (F. Szoka) and NIH Grant AI25534 (N. Düzgünes and S. Nir), Elias Fattal was a fellow of Fondation pour la Recherche Therapeutique (France).

6. Abbreviations

ANTS, 8-aminonaphthalene-1,2,3-trisulfonic acid; Ap_5A, diadenosine pentaphosphate; DPX, p-xylylenebis[pyridinium bromide]; DOPC, Dioleoylphosphatidylcholine; DOPE, Dioleoylphosphatidylethanolamine; EDTA, ethylenediaminetetraacetic acid; FITC-dextran, fluorescein isothiocyanate dextran; FTIR, Fourier Transform Infrared Spectroscopy; GALA, peptide with the sequence, WEAALAEALAEALAEHLAEALAEALEALAA; HIV_{arg}, N-terminal sequence (23 aa) of HIV-1 (LAV_{1a}) gp41; LAGA, peptide with the sequence, WEAALAEAEALALAEHEALALAEAELALAA; LUV, large unilamellar vesicles; NADH, micotinamide adenine dinucleotide; NBD-PE, N-(7-nitro-benz-2-oxa-1,3-diazol-4-yl)phosphatidylethanolamine; PC, phosphatidylcholine; PE, phosphatidylethanolamine; POPG, 1-palmitoyl-2-oleoylphosphatidylglycerol; REV, reverse-phase evaporation vesicles; Rh-PE, N-(lissamine Rhodamine B sulfonyl) phosphatidylethanolamine; TES, 2-[[tris(hydroxymethyl)]amino]ethanesulfonic acid.

7. References

Bentz J, Nir S (1981) Kinetic and equilibrium aspects of reversible aggregation. J Chem Soc, Faraday Trans 1 77:1249-1275

Bentz J, Nir S, Covell D (1988) Mass action kinetics of virus-cell aggregation and fusion. Biophys J 54:449-462

Connoly ML (1983) Solvent-accessible surfaces of proteins and nucleic acids. Science 221:709-713

DeGrado WF, Wasserman ZR, Lear JD (1989) Protein design, a minimalist approach. Science 243:622-628

Düzgüneş N, Gambale F (1988) Membrane action of synthetic peptides from influenza virus hemagglutinin and its mutants. FEBS Lett 227:110-114

Düzgüneş N, Shavnin SA (1992) Membrane destabilization by N-terminal peptides of viral envelope proteins. J Membrane Biol 128:71-80

Düzgüneş N, Lima MCP, Stamatatos L, Flasher D, Alford D, Friend DS, Nir S (1992) Fusion of influenza virus with human promyelocytic leukemia and lymphoblastic leukemia cell, and murine lymphoma cells: Kinetics of low pH-induced fusion monitored by fluorescence dequenching. J Gen Virol 73:27-37

Ellens H, Bentz J, Szoka FC (1984) pH-induced destabilization of phosphatidylethanolamine-containing liposomes: role of inter-bilayer contact. Biochemistry 23:1532-1538

Epand RM, Cheetham J, Epand RF, Yeagle PL, Richardson CD, Rockwell A, DeGrado, WF (1992) Peptide models for the membrane destabilizing actions of viral fusion proteins. Biopolymers 32:309-314

Fattal E, Nir S, Parente RA, Szoka FC Jr (1994) Pore-forming peptides induce rapid phospholipid flip-flop in membranes. Biochemistry 33:6721-6731

Fox RO Jr, Richards FM (1982) A voltage-gated ion channel inferred from the crystal structure of alamethicin at 1.5Å resolution. Nature 300:325-330

Gallaher WR, Segrest JP, Hunter E (1992) Are fusion peptides really "sided" insertional helices? Cell 70:531-532

Gething MJ, White J, Waterfield M (1978) Purification of the fusion protein of Sendai virus: analysis of the NH_2-terminal sequence generated during precursor activation. Proc Natl Acad Sci (USA) 75:2737-2740

Goormaghtigh E, de Meutter J, Szoka F, Cabiaux V, Parente R, Ruysschaert J-M (1991) Secondary structure and orientation of the amphipathic peptide GALA in lipid structures. Eur J Biochem 195:421-429

Harter C, James P, Bächi T, Semenza G, Brunner J (1989) Hydrophobic binding of the ectodomain of influenza hemagglutinin to membranes occurs through the "fusion peptide". J Biol Chem 264:6459-6454

Hoekstra D, Kok JW (1989) Entry mechanisms of enveloped viruses: Implications for fusion of intracellular membranes. Biosci Rep 9:273-305

Huschilt JC, Millman BM, Davis JH (1989) Orientation of α-helical peptides in a lipid bilayer. Biochim Biophys Acta 979:139-141

Larsen C, Nir S, Alford D, Jennings M, Lee K, Düzgüneş N (1993) Human immunodeficiency virus type 1 (HIV-1) fusion with model membranes: Kinetic analysis and the role of lipid composition, pH and divalent cations. Biochim Biophys Acta 1147:223-236

Lear JD, DeGrado WF (1987) Membrane binding and conformational properties of peptides representing the NH_2 terminus of influenza HA-2. J Biol Chem 262:6500-6505

Lear JD, Wasserman ZR, DeGrado WF (1988) Synthetic amphiphilic peptide models for protein ion channels. Science 240:1177-1181

Martin I, Defrise-Quertain F, Decroly E, Vandenbranden M, Brasseur R, Ruysschaert J-M (1993) Orientation and structure of the NH_2-terminal HIV-1 gp41 peptide in fused and aggregated liposomes. Biochim Biophys Acta 1145:124-133

Murata M, Sugahara Y, Takahashi S, Ohnishi S-I (1987) pH-dependent membrane fusion activity of a synthetic twenty amino acid peptide with the same sequence as that of the hyhdrophobic segment of influenza virus hemagglutinin. J Biochem 102:957-962

Nieva JL, Nir S, Muga A, Goni FM, Wilschut J (1994) Interaction of the HIV-1 fusion peptide with phospholipid vesicles: Different structural requirements for fusion and leakage. Biochemistry 33:3201-3209

Nir S, Bentz J, Wilschut J, Düzgüneş N (1983) Aggregation and fusion of vesicles. Prog Surface Sci 13:1-124

Nir S, Klappe K, Hoekstra D (1986a) Kinetics of fusion between Sendai virus and erythrocyte ghosts: Application of mass action kinetic model. Biochemistry 25:2155-2161

Nir S, Klappe K, Hoekstra D (1986b) Mass action an analysis of kinetics and extent of fusion between Sendai virus and phospholipid vesicles. Biochemistry 25:8261-8266

Nir S, Düzgüneş N, Pedroso de Lima MC, Hoekstra D (1990) Fusion of enveloped viruses with cells and liposomes. Cell Biophys 17:181-201

Parente RA, Nir S, Szoka FC Jr (1988) pH-dependent fusion of phosphatidylserine small vesicles. J Biol Chem 263:4724-4730

Parente RA, Nir S, Szoka FC Jr (1990a) Mechanism of leakage of phospholipid contents induced by the peptide GALA. Biochemistry 29, 8720-8728.

Parente RA, Nadasdi L, Subbarao NK, Szoka FC Jr (1990b) Association of a pH-sensitive peptide with membrane vesicles: Role of amino acid sequence. Biochemistry 29:8713-8719

Peters R (1986) Fluorescence microphotolysis to measure nucleocytoplasmic transport and intracellular mobility. Biochim Biophys Acta 864:305-359

Pownall HJ, Gotto AM, Sparrow JT (1984) Thermodynamics of lipid-protein association and the activation of lecithin: cholesterol acyl transferase by synthetic model apolipopeptides. Biochim Biophys Acta 793:149-156

Rafalski M, Ortiz A, Rockwell A, Van Ginkel L, Lear J, DeGrado W, Wilschut J (1991) Membrane fusion activity of the influenza virus hemagglutinin: Interaction of HA2 N-terminal peptides with phospholipid vesicles. Biochemistry 30:10211-10220

Ramalho-Santos J, Nir S, Düzgüneş N, Carvalho AP, Lima MCP (1993) A common mechanism for virus fusion activity and inactivation. Biochemistry 32:2771-2779

Rapaport D, Hague GR, Pouny Y, Shai Y (1993) pH- and ionic strength-dependent fusion of phospholipid vesicles induced by pardaxin analogues or by mixtures of charge-reversed peptides. Biochemistry 32:3291-3297

Richardson CD, Choppin PW (1983) Oligopeptides that specifically inhibit membrane fusion by paramyxoviruses: Studies on the site of action. Virology 131:518-532.

Smoluchowski MV (1917) Investigation into a mathematical theory of the kinetics of coagulation of collolidal solutions. Z Physik Chem (Leipzig) 92:129-168

Subbarao NK, Parente RA, Szoka FC, Nadasdi L, Pongracz K (1987) pH-dependent bilayer destabilization by an amphipathic peptide. Biochemistry 26:2964-2972

Stegmann T, Hoekstra D, Scherphof G, Wilschut J (1985) Kinetics of pH-dependent fusion between influenza virus and liposomes. Biochemistry 24:3107-3113

Stegmann T, White J, Helenius A (1990) Intermediates in influenza induced membrane fusion. EMBO J 9:4231-4241

Takahashi S (1990) Conformation of membrane fusion-active 20- residual peptides with or without lipid bilayers. Implication of α-helix formation for membrane fusion. Biochemistry 29:6257-6264

Weinstein JN, Klausner RD, Innerarity T, Ralston E, Blumenthal R (1981) Phase transition release, a new approach to the interaction of proteins with lipid vesicles. Biochim Biophys Acta 647:270-274

White SH, King GI (1985) Molecular packing and area compressibility of lipid bilayers. Proc Natl Acad Sci (USA) 82:6532-6536

White J, Kielian M, Helenius A (1983) Membrane fusion proteins of enveloped animal viruses. Q Rev Biophys 16:151-195

White JM, Wilson IA (1987) Anti-peptide antibodies detect steps in a protein conformational change: low pH activation of the influenza virus hemagglutinin. J Cell Biol 105:2887-2896

Wilschut J, Düzgüneş N, Papahadjopoulos D (1981) Calcium/magnesium specificity in membrane fusion: Kinetics of aggregation and fusion of phosphatidylserine vesicles and the role of bilayer curvature. Biochemistry 20:3126-3133

Yoshimura T, Goto Y, Aimoto S (1992) Fusion of phospholipid vesicles induced by an amphiphilic model peptide: close correlation between fusogenicity and hydrophobicity of the peptide in an α-helix. Biochemistry 31:6119-6126

Molecular Mechanisms of Membrane Fusion

Nejat Düzgüneş§
Department of Microbiology
University of the Pacific
2155 Webster Street
San Francisco, CA 94115
USA

INTRODUCTION

A crucial step in the trafficking of intracellular membranes is membrane fusion. Endocytotic vesicles or phagosomes form via fusion of the apposed exoplasmic leaflets of the invaginated plasma membrane. The endocytotic vesicles are then thought to fuse with endosomes. Later in the endocytotic pathway, endosomes and phagosomes eventually fuse with lysosomes, in a process involving the interaction and fusion of the cytoplasmic leaflets of their membranes. Following endocytosis, receptor-containing vesicles undergo a "fission" or "budding" process to facilitate the transport of receptors back to the plasma membrane. This budding involves the interaction and fusion of the membrane leaflets facing the lumen of the flaccid vesicle. Likewise, in the Golgi apparatus, transport vesicles are thought to bud off from one compartment and fuse with the next one along the *cis, medial* and *trans* regions of the organelle.

How do membranes transiently lose their barrier properties at the precise sites of intermembrane attachment in an exquisitely controlled reaction, and then become stable again? How is membrane fusion between lipid bilayers controlled at the molecular level? In what ways are point defects, molecular shapes, intermembrane forces and hydrophobic interactions related to membrane fusion. How do proteins modulate membrane fusion? How is the fusion of intracellular membranes controlled by proteins? What are fusion pores? What can we learn from simpler fusion proteins of enveloped viruses, such as the hemagglutinin of the influenza virus envelope? These are some of the questions we shall attempt to answer in this chapter.

§ This chapter is dedicated, with gratitude, to my father, Prof. Orhan Düzgüneş.

NATO ASI Series, Vol. H 91
Trafficking of Intracellular Membranes
Edited by M.C. Pedroso de Lima N. Düzgüneş and D. Hoekstra

BIOPHYSICAL CONCEPTS IN MEMBRANE FUSION

The molecular mechanisms of membrane fusion have been studied in a number of experimental systems. These include planar bilayers, liposomes, viruses, intracellular membrane vesicles, and exocytosis (Düzgüneş, 1985, 1993a,b; Chernomordik et al., 1987; Prestegard & O'Brien, 1987; Plattner, 1989; Wilson et al.,1991; White, 1992; Zimmerberg et al., 1993). Although it is possible to draw general conclusions about the mechanisms of membrane fusion from these systems, each of these systems necesarily focuses on a particular aspect of the fusion process, which may not be directly applicable to the mechanisms of fusion in another system. For example, osmotic stress across the membrane may be important for the divalent cation-induced fusion of liposomes with each other (Ohki, 1984) or with lipid bilayers (Akabas et al., 1984) in certain cases. However, exocytosis of mast cell granules (Zimmerberg et al., 1987), virus-induced cell-cell fusion (White et al., 1981; Lifson et al., 1986), or the fusion of many types of liposomes (Nir et al., 1983; Düzgüneş, 1985; Wilschut, 1991; Düzgüneş & Nir, 1994) appear to proceed in the absence of an osmotic gradient. Nevertheless, certain molecular or physicochemical concepts can be drawn from a particular system, leading to the formulation of hypotheses regarding mechanisms in a different system.

Intermembrane forces. Biological membranes or liposomes composed of biologically relevant phospholipids do not approach each other to a close enough distance to undergo fusion in physiological concentrations of NaCl, because of electrostatic repulsion between the surfaces. Liposomes or biological membrane vesicles will form stable dimers only when there is a local minimum in the free energy of interaction between the particles, as defined by the sum of the attractive Van der Waals and repulsive electrostatic forces (Nir et al., 1983; Düzgüneş et al., 1985) (Figure 1). Dimerization, or aggregation, of liposomes or vesicles can occur in either the primary minimum, at intermembrane separations of < 10 Å, or in the secondary minimum, which may be located 20-100 Å from the membrane surface (Nir et al., 1981, 1983). Membranes must approach each other within the primary minimum for membrane fusion to occur. At these distances membranes appear to experience an additional "hydration," "structural," or "steric" repulsive force (McIntosh et al., 1987; Israelachvili & McGuiggan, 1988; Parsegian and Rand, 1991), which must be overcome or bypassed if membrane fusion is to occur.

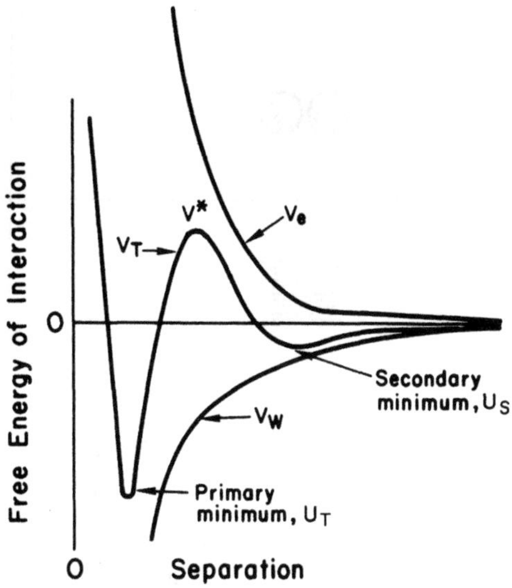

Figure 1. Free energy of interaction between two membranous vesicles as a function of the distance of separation between their surfaces. The total potential V_T is the sum of the electrostatic repulsive term V_e and the Van der Waals attractive term V_W. V^* is the height of the potential maximum, U_T is the depth of the primary minimum, and U_S is the depth of the secondary minimum. Reproduced, with permission, from Nir et al. (1983).

Point defects. The formation of point defects in membrane structure was postulated as a possible mechanism of fusion of phosphatidylserine liposomes in the presence of divalent cations (Papahadjopoulos et al., 1977). Electron microscopic and NMR evidence for point defects at sites of intermembrane attachment was found in certain liposomes after freezing and thawing (Hui et al., 1981). More recent studies on the mechanisms of fusion of influenza virus with target membranes have indicated that the N-terminal segment of the cleaved hemagglutinin (HA2) can insert into the membrane, and perturb the lipid bilayer structure (Harter et al., 1989; Düzgüneş & Gambale, 1988; Tsurudome et al., 1992; Weber et al., 1994; Rafalski et al., 1991). These local perturbations (perhaps "point defects") of bilayer structure are thought to lead to the initial step of lipid intermixing between apposed membranes (Düzgüneş & Shavnin, 1992). Molecular packing defects in membrane structure could also be induced by intermembrane contact and the localized phase separation of non-contacting lipids from those that form an interbilayer bridge (Düzgüneş & Papahadjopoulos 1983; Papahadjopoulos et al., 1990) (Figure 2). Following intermembrane contact

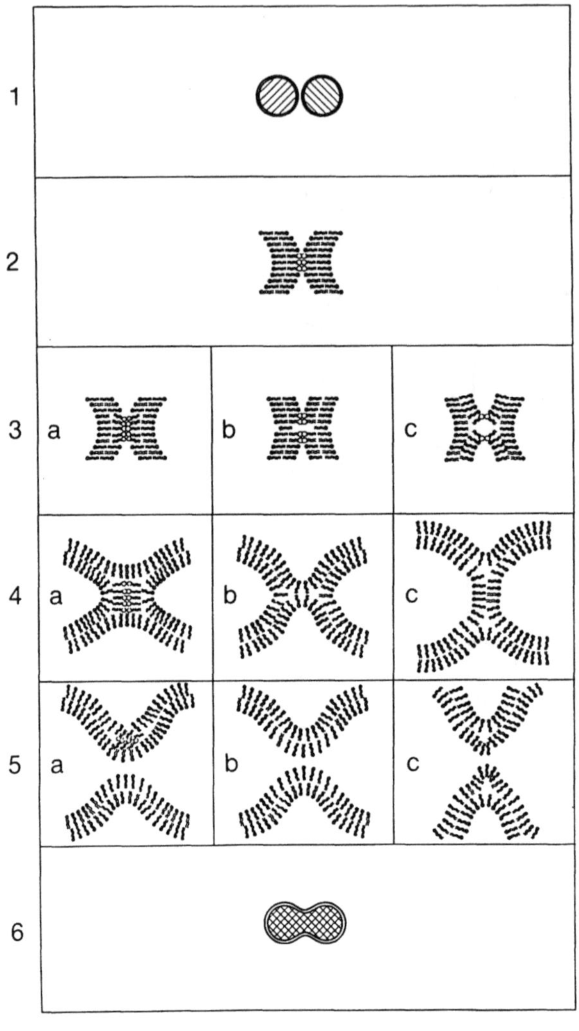

Figure 2. Proposed molecular rearrangements occurring during the fusion of two unilamellar liposomes. 1: Two liposomes before intermembrane contact. 2: Intermembrane contact. 3: Contact-induced defects (see text for details) 4: Subsequent intermediate phases in the membrane fusion reaction. 5: Formation of aqueous "pores" or a "curved bilayer annulus" between the internal contents of the liposomes. 6. Completion of the fusion reaction. Reproduced from Papahadjopoulos et al. (1990), with permission.

(Figure 2, panel 2), fusion can proceed in a number of different paths, depending on the phospholipid composition and the environmental factors. A dehydrated interbilayer complex can form (panel 3a), as in the case of phosphatidylserine liposomes in the presence of Ca^{2+}, with high-affinity Ca^{2+} binding (Portis et al., 1979; Ekerdt & Papahadjopoulos, 1982, Nir, 1984, Feigenson, 1986). This induces the formation of defects in molecular packing around the dehydrated complex. Local dehydration, or clustering due to ion complexation, may cause fluctuations in molecular packing density, with clusters of molecules more densely packed and areas with "point defects." (panel 3b). It is likely that Mg^{2+}-induced fusion of phosphatidylserine/phosphatidylethanolamine membranes proceeds by this mechanism (Düzgüneş et al., 1981, 1984a). Thermal fluctuations of lipids with "negative spontaneous curvature" (see Figure 3) in apposed regions of liposomes may overcome interbilayer repulsion and interact (panel 3c) (Chernomordik et al., 1985, 1987). An extension of the point defects concept is the proposal that

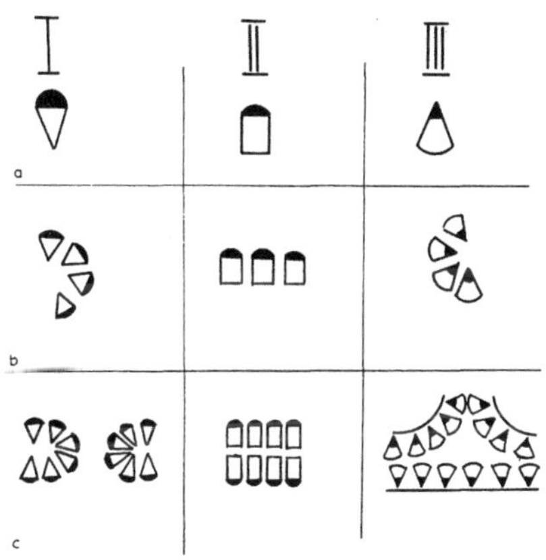

Figure 3. Schematic diagram of the shapes of different types of lipid molecules (a), energetically favorable structures formed by a monolayer of lipids of different shapes (b), and the hypothetical structure of defects in a bilayer membrane (c). Ia: A lipid with with the polar head-group area exceeding the cross-sectional area of the acyl chains (e.g. lysophosphatidylcholine). Ib: A lipid monolayer of "positive spontaneous curvature." Ic: A hydrophilic pore in a lipid bilayer. IIa: A cylindrically shaped lipid with equivalent head-group and hydrocarbon chain cross-sectional area (e.g. phosphatidylcholine). IIb: A planar monolayer. IIc: A well-packed bilayer without defects. IIIa: A lipid with the polar head-group area smaller than that of the cross-section of the hydrocarbon tails (e.g. phosphatidyl-ethanolamine). IIIb: A monolayer of "negative spontaneous curvature." IIIc: Local "bulging defect" in a bilayer. Reproduced from Chernomordik et al. (1985), with permission.

thermal fluctuations of phospholipids out of the plane of the lipid bilayer (to form non-bilayer transients) result in the formation of inter-membrane stalks as intermediates in membrane fusion (Markin et al., 1984; Chernomordik et al., 1985, 1987).

Molecular shapes. Another concept related to the formation of intermembrane stalks is that of molecular shapes. The formation of lamellar, hexagonal I (micellar) and hexagonal II (inverted tubular) phases by lipids in aqueous solutions is thought to be related to their molecular shapes (Cullis et al., 1991; Chernomordik et al., 1985). Thus, lipids, such as certain phosphatidylethanolamine species, with headgroups of smaller cross-sectional area than that of the acyl chains tend to form non-bilayer structures corresponding to the hexagonal II phase (Figure 3). In contrast, lipids, such as lysophosphatidylcholine, with a

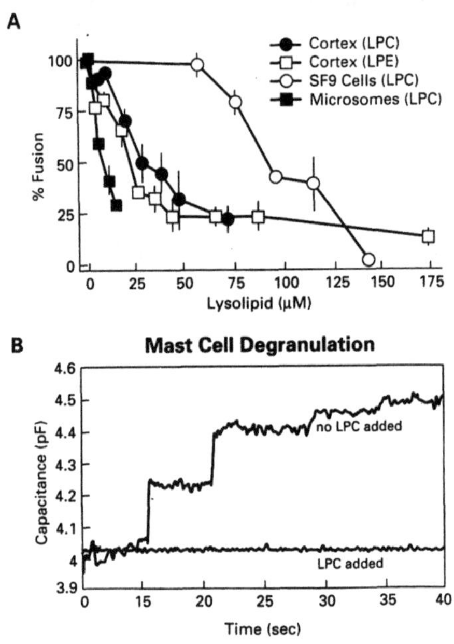

Figure 4. Inhibition of biological membrane fusion by lysolipids. A: Inhibition of Ca^{2+}-induced cortical exocytosis. Filled circles: Oleoyllysophosphatidylcholine. Open squares: Oleoyllysophosphatidylethanolamine. Open circles: Inhibition of low pH induced fusion of baculovirus-infected insect cells by oleoyllysophosphatidylcholine. Filled squares: Inhibition of GTP-induced fusion of microsomes by oleoyllysophosphatidylcholine. B: Inhibition of exocytosis in mast cells by 9.6 μM lysophosphatidylcholine, measured as the membrane capacitance in the whole-cell configuration. Reproduced, with permission, from Chernomordik et al. (1993).

larger head-group area than the acyl chains would form micellar structures. Lipids with equivalent head-group and acyl chain areas tend to have an overall cylindrical shape, and would form stable bilayers, an example being phosphatidylcholine. The spontaneous tendency of individual, or domains of, lipids to form non-bilayer phases most likely affects the outcome of interbilayer interactions. Thus, lipids that have the tendency to form the hexagonal II phase promote fusion (Cullis & Hope, 1978; Verkleij et al., 1980; Düzgüneş et al., 1981; Chernomordik et al., 1985), while lipids with the tendency to form the opposite curvature inhibit it (Chernomordik et al., 1985, 1993). This observation has led to the hypothesis that a membrane "stalk" is the intermediate in the fusion of apposed membranes (Markin et al., 1984; Chernomordik et al., 1985, 1993; Zimmerberg et al., 1993). Evidence to support this proposal has been obtained in several biological membrane fusion systems (Chernomordik et al., 1993); lysophosphatidylcholine was found to inhibit the fusion of intracellular organelles, virus-cell fusion and exocytosis (Figure 4). It is also possible, however, that lysolipids inhibit fusion by binding to hydrophobic fusion peptides of proteins involved in fusion (Zimmerberg et al., 1993).

Local dehydration and hydrophobic interactions. Early studies on divalent cation-induced fusion of liposomes lead to the concept of local dehydration of the membrane surface as a driving force for membrane fusion (Portis et al., 1979; McIver, 1979; Düzgünes et al., 1981; Hoekstra, 1982). Results obtained in studies of the fusion of liposomes with monolayers were interpreted in terms of the involvement of membrane surface hydrophobicity (Ohki & Düzgüneş, 1979), which was later analyzed quantitatively (Ohki & Ohshima, 1984; Ohki, 1988). Studies on forces between phospholipid bilayers supported on mica surfaces have also suggested that it is the hydrophobic attraction between exposed acyl chains in interacting membranes that mediates fusion (Helm et al., 1989). Local bilayer deformations arising from internal bilayer stresses may allow repulsive steric or hydration forces to be bypassed via a sudden instability or hydrophobic interactions (Helm et al., 1992; Helm & Israelachvili, 1993). Interaction of Ca^{2+} with negatively charged phospholipids in mixed membranes can lead to condensation of the lipid head-groups and cause the exposure of hydrophobic groups at domain boundaries (Düzgüneş et al., 1984a; Leckband et al., 1993).

The role of hydrophobic interactions in viral fusion also became apparent with the demonstration that the hemagglutinin of influenza virus exposes a hydrophobic segment at mildly acidic pH where the virions become fusion-active, and that the ectoplasmic domain of the protein binds to liposomes (Skehel et al., 1982; Doms et al., 1985). Mutations in this hydrophobic segment substituting glutamic acid residues for glycine, particularly at the N-terminus, impairs the fusion activity of the hemagglutinin expressed in cultured cells (Gething et al., 1986), and inhibits the interaction with phospholipid membranes of synthetic peptides derived from this region (Düzgüneş & Gambale, 1988; Rafalski et al., 1991).

FUSION OF PHOSPHOLIPID MEMBRANES

Role of anionic phospholipids and cation specificity. Liposomes constitute a convenient model system to study the role of individual membrane components in membrane fusion (Nir et al., 1983; Wilschut, 1991; Düzgüneş & Nir, 1994). Liposomes composed of anionic phospholipids can undergo fusion in the presence of Ca^{2+} or other divalent cations. This observation is particularly relevant, since membrane fusion appears to require Ca^{2+} in a number of biological systems, such as sperm-egg fusion (Yanagimachi, 1988), myoblast fusion (Wakelam, 1988), and exocytosis (Baker, 1988), albeit at varying concentrations. The susceptibility of acidic liposomes to Ca^{2+}-induced fusion decreases in the order phosphatidic acid > phosphatidylserine > phosphatidylglycerol >> phosphatidylinositol (Düzgüneş et al., 1985). The difference in fusion susceptibility between these phospholipids most likely arises from the binding affinity of Ca^{2+} for the phospholipid head-group, the proximity of the Ca^{2+}-binding site to the water-hydrocarbon interface (as with phosphatidic acid), and the presence of other hydrated bulky groups (such as in the case of phosphatidylinositol. These factors in turn contribute to the degree of dehydration of the membrane surface, and the close approach of the interacting membranes. The fusion of large unilamellar phosphatidylserine liposomes shows a striking specificity for Ca^{2+} over Mg^{2+}, in that Mg^{2+} is unable to induce fusion despite mediating aggregation at temperatures between 25-30°C (Wilschut et al., 1981; Shavnin et al., 1988). This specificity has been attributed to the ability of Ca^{2+} (and the inability of Mg^{2+}) to dehydrate the membrane surface. However, the presence of cholesterol in the membrane can render phosphatidylserine liposomes susceptible to fusion with Mg^{2+} at or above 30°C (Shavnin et al., 1988). Divalent cation induced fusion of liposomes can be further modulated by phosphate ions (Fraley et al., 1980) and by polyamines (Schuber et al., 1983; Meers et al., 1986).

Membranes containing anionic phospholipids can also undergo fusion with those containing cationic lipids (such as the synthetic lipid N-[1-(2,3-dioleyloxy)propyl]-N,N,N-trimethylammonium; "DOTMA"), in the absence of divalent cations (Stamatatos et al., 1988; Düzgüneş et al., 1989). Since cationic liposomes can fuse with each other in the presence of multivalent anions (Düzgünes et al., 1989), it is likely that the surface of a membrane containing negatively charged phospholipids presents itself as a multivalent anion to a positively charged membrane.

Phosphatidylethanolamine vs phosphatidylcholine. Although large unilamellar liposomes composed of pure phosphatidylserine do not fuse in the presence of Mg^{2+} at 25°C, the inclusion of phosphatidylethanolamine in these liposomes imparts susceptibility to fusion (Düzgüneş et al., 1981). Although pure phosphatidylinositol liposomes are resistant to fusion in the presence of Ca^{2+}, mixed phosphatidylinositol/phosphatidylethanolamine liposomes become susceptible to fusion (Sundler et al., 1981). In contrast, the presence of 50 mole % phosphatidylcholine in phosphatidylserine liposomes completely inhibits fusion by Ca^{2+} or Mg^{2+} (Düzgüneş et al., 1981). The difference between phosphatidylethanolamine and phosphatidylcholine with respect to the modulation of membrane fusion may be due to (i) the lower affinity of phosphatidylethanolamine for water (Jendrasiak & Hasty, 1974); (ii) the propensity of phosphatidylcholine to retain a lamellar structure and that of phosphatidylethanolamine to form non-bilayer phases (Reiss-Husson, 1967; Cullis et al., 1991), (iii) the ability of phosphatidylethanolamine bilayers to approach each other closer than phosphatidylcholine bilayers due to differences in repulsive hydration forces (Parsegian & Rand, 1991); and (iv) the electrostatic and hydrogen bonding interactions of the head-group of phosphatidylethanolamine and the absence of these interactions in phosphatidylcholine (Hauser et al., 1981).

Phosphatidic acid vs phosphatidylinositol. Phosphatidic acid liposomes can undergo fusion in the presence of low concentrations of Ca^{2+}, around 0.2 mM (Sundler & Papahadjopoulos, 1981), compared to a threshold concentration of 2 mM for large unilamellar phosphatidylserine liposomes (Wilschut et al., 1980). In contrast, phosphatidylinositol liposomes can aggregate but do not fuse in the presence of Ca^{2+}. As pointed out above, however, mixed phosphatidylinositol/

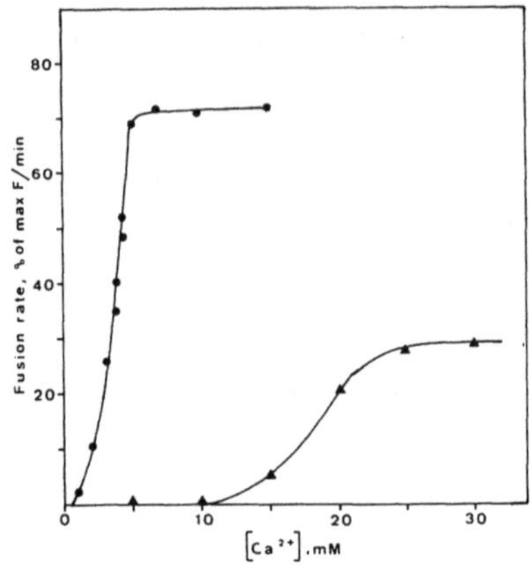

Figure 5. Phosphatidic acid/phosphatidylinositol specificity in membrane fusion. The effect of substituting phosphatidic acid (circles) for phosphatidylinositol (triangles) in liposomes also containing phosphatidylethanolamine, phosphatidylserine and phosphatidylcholine, on the initital rate of fusion, as a function of the Ca^{2+} concentration. Reproduced from Sundler & Papahadjopoulos (1981), with permission.

phosphatidylethanolamine liposomes are susceptible to fusion. A striking difference in the threshold Ca^{2+} concentration for fusion is observed between two types of liposomes differing only in their phosphatidic acid or phosphatidylinositol content, and otherwise composed of a mixture of phosphatidylcholine/phosphatidylethanolamine/phosphatidylserine (Sundler et al., 1981) (Figure 5). This observation has implications for the possible control of membrane fusion by the conversion of phosphatidylinositol to phosphatidic acid during the activation of various cells to exocytose (Cockcroft & Allan, 1985; Berridge, 1992). In this regard, it is interesting to note that phospholipase D can enhance the Ca^{2+}-induced fusion of liposomes (Park et al., 1992).

Role of membrane fluidity and lateral phase separation. The membrane phase state has a very significant effect on membrane fusion. If the membrane is below the gel-liquid crystalline phase transition temperature, the Ca^{2+}-induced fusion of large unilamellar phosphatidylserine liposomes is practically arrested (Wilschut et al., 1985). Phase separation of fusion-promoting lipids from fusion-inhibitory lipids can alter the susceptibiity of the membrane to fusion. Although liposomes composed of phosphatidylserine/dipalmitoylphosphatidylcholine are resistant to fusion in the presence of Ca^{2+} at temperatures where the membrane

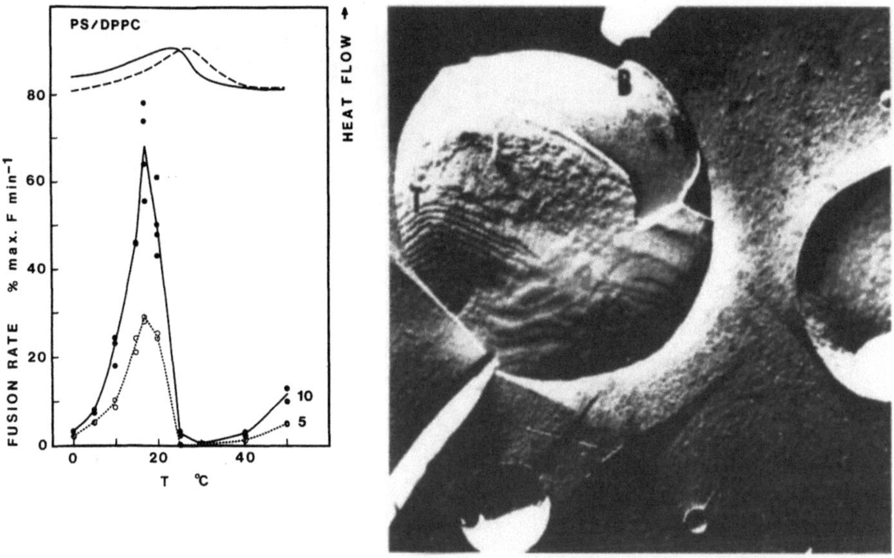

Figure 6. Left panel: Fusion of phosphatidylserine/dipalmitoylphosphatidylcholine (1:1) large unilamellar liposomes in the presence of 5 mM (open circles) or 10 mM Ca^{2+} (closed circles) at various temperatures, and the differential scanning calorimetry thermograms in the absence (solid line) and presence of 10 mM Ca^{2+} (dashed line). Reproduced from Düzgünes et al. (1984a), with permission. Right panel: Freeze fracture electron micrograph of phosphatidylserine/dipalmitoylphosphatidylcholine (1:1) liposomes quenched from 18°C, showing co-existing terraced, banded and smooth fracture faces, arising from lipid domains of varying composition. Reproduced from Stewart et al. (1979), with permission.

is in the liquid crystalline state, they are susceptible to fusion at temperatures within the relatively broad phase transition region (Düzgüneş et al., 1984a) (Figure 6). Molecular clusters rich in phosphatidylserine are thought to form in this temperature range (Stewart et al., 1979), and mediate fusion when they interact with similar clusters in apposed membranes.

Lipidic particles and the hexagonal II phase. Under fusion-inducing conditions, some lipids form inverted micelles called lipidic particles, which have been found at intermembrane attachment sites. Lipidic particles have been proposed as an intermediate structure in the fusion of certain types of liposomes (Verkleij et al., 1979, 1980; Hope et al., 1983; Verkleij et al., 1984). However, freeze-fracture

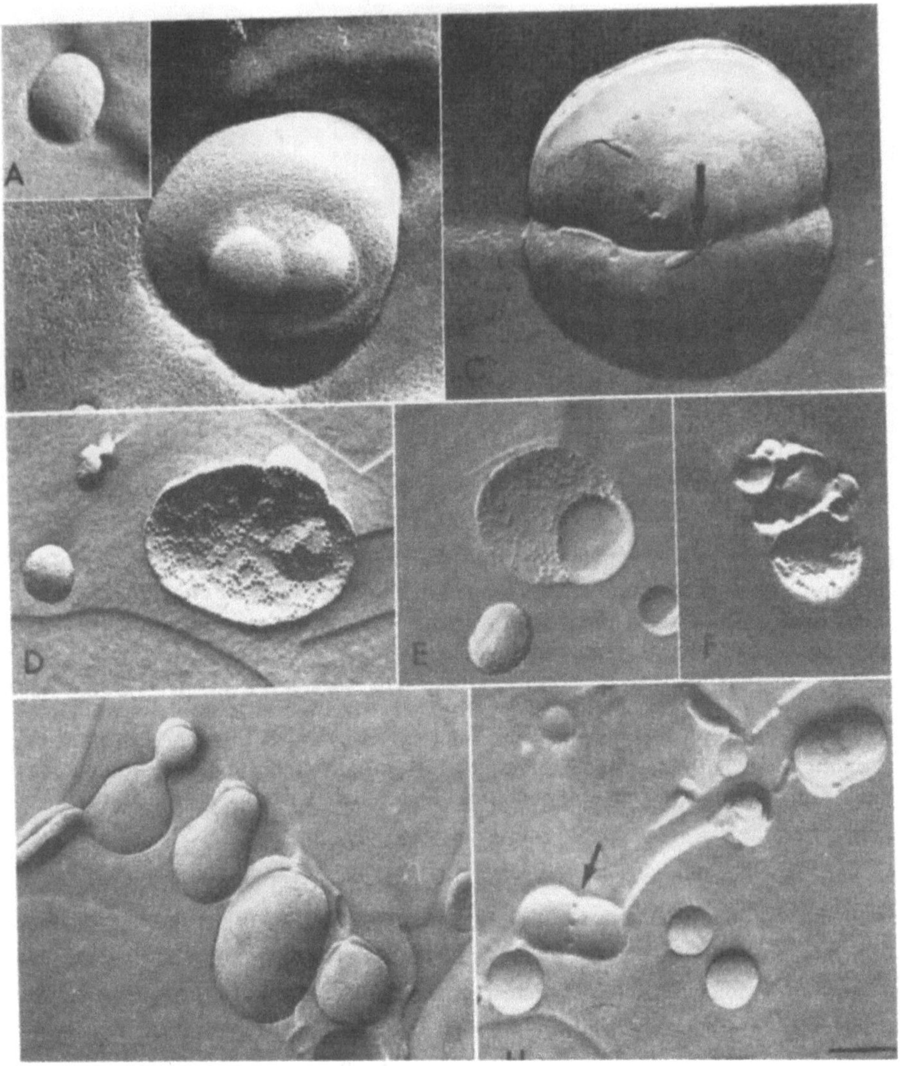

Figure 7. Freeze fracture electron micrographs of quick-frozen liposomes. A liposome before stimulation with Ca^{2+} (A). Cardiolipin/phosphatidylcholine (1:1) liposomes 1 s after the addition of 5 mM Ca^{2+} (B) Phosphatidylserine/ phosphatidylethanolamine (1:1) liposomes after a 1 s stimulation with Ca^{2+} (C). Cardiolipin/phosphatidylcholine liposomes incubated for 2 h in Ca^{2+} and glycerol (D & E), Ca^{2+} alone (F), or incubated for 2 h in glycerol, and then treated with Ca^{2+} for 1 s (G). Some lipidic particles are observed occasionally in glycerol-treated samples even in early times (H). Bar = 0.2 μm. Reproduced from Bearer et al. (1982) with permission.

electron microscopy of liposomes, composed of cardiolipin/phosphatidylcholine or phosphatidylserine/phosphatidylethanolamine (1:1), quick-frozen immediately following stimulation with Ca^{2+} reveals smooth elongated bridges or a tight lip structure between fusing liposomes (Figure 7) (Bearer et al., 1982). Only prolonged incubation of the vesicles with Ca^{2+}, or the presence of glycerol as a cryoprotectant leads to the formation of lipidic particles. In Figure 7G, the two vesicles at the lower right corner are aggregated, while at the top left fused liposomes remain separated by a curved bilayer annulus. The annulus appears to have widened in the two central fusion products. Similar observations have been made by other laboratories (Verkleij et al., 1984; Hui et al., 1988). Therefore, lipidic particles do not appear to be involved in the membrane fusion reaction *per se*, even in membrane systems in which they are observed after long periods of incubation. As pointed out earlier (Bearer et al., 1982; Düzgüneş et al., 1985), non-bilayer structures probably occur at the sites of fusion at rates too fast to be visualized by morphological studies, since the rate of the membrane fusion reaction in the Ca^{2+}/phosphatidylserine system has been estimated to be on the order of msec (Nir, et al., 1982, 1983; Miller & Dahl, 1982). These non-bilayer intermediates could transform in time to more stable structures, such as lipidic particles, the hexagonal II phase or the crystalline bilayer, depending on the phospholipid composition of the liposomes. The half-life of such non-bilayer intermediates have been predicted to be on the order of 1 ms (Siegel, 1986).

Although the bilayer to hexagonal II phase transition has been observed under conditions which also induce the fusion of certain membranes, and has been proposed as the driving force for membrane fusion (Cullis and Hope, 1978; Cullis and Verkleij, 1979), considerable evidence has accumulated against this hypothesis (Düzgüneş et al., 1981, 1984a, 1987a; Ellens et al., 1989). For example, Mg^{2+} can induce the fusion of phosphatidylserine/egg phosphatidylethanolamine liposomes without causing a transition into the hexagonal II phase (Düzgüneş et al., 1984a). Liposomes composed of N-methylated phosphatidylethanolamine can form a stable cubic phase at mildly acidic pH, at temperatures below the bilayer-hexagonal II phase transition, and undergo membrane fusion without the loss of aqueous contents. These observations have suggested that the formation of the cubic phase may be involved in the fusion reaction for this membrane composition (Ellens et al., 1989), and electron microscopic evidence for such structures between fusing membranes has been obtained (Siegel et al., 1989). However, liposomes composed of egg phosphatidylethanolamine fusing at low pH did not reveal such structures. Calculations of the free energies of proposed intermediate structures have indicated that a modified form of the intermembrane stalk is a more likely intermediate than the inverted micelle in a wide range of lipid compositions (Siegel, 1993).

PROTEIN-MEDIATED MEMBRANE FUSION

The different roles that proteins can play in membrane fusion are depicted schematically in Figure 8. Proteins can modulate or mediate membrane fusion

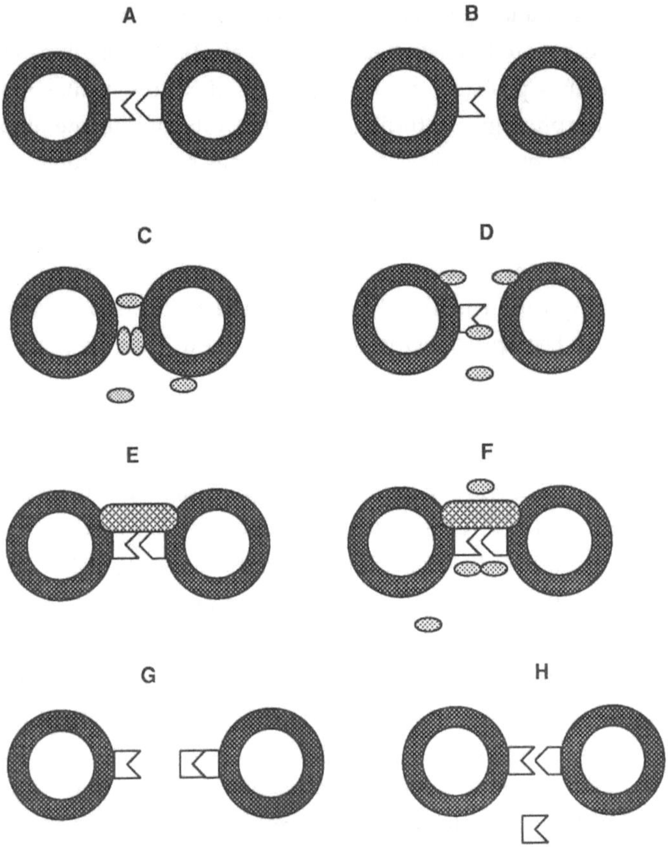

Figure 8. Role of proteins in membrane fusion. See text for a description of the individual cases.

in a number of ways. Proteins on both membranes may participate in both mutual recognition and membrane fusion (A). They may be actual membrane fusion proteins, such as the influenza hemagglutinin, which act on different types of membranes, including liposomes without specific receptors (B) (Pedroso de Lima et al., 1995; Düzgüneş & Nir, 1994). The ability of secretory granules to fuse with liposomes without specific receptors for granule proteins has prompted the hypothesis that fusion proteins are present only on the granule membrane and not on the plasma membrane with which they fuse (Zimmerberg et al., 1993). Proteins involved in fusion may also be cytosolic proteins that associate with membranes in the presence of inducers such as Ca^{2+}, and mediate the close approach of membranes, as in the case of annexins (C) (Creutz, 1992). Membrane-bound fusion molecules may cooperate with cytosolic factors in both close approach and fusion (D). Fusion inducing proteins may work in concert with recognition or docking proteins that ensure the interaction of specific intracellular membranes (E); cytosolic proteins may also participate in this process (F). Cytosolic factors may block a fusion protein on one of the fusion partners (G), as suggested by Baker (1988). If the blocking protein is modified and can no longer interact with one of the fusion proteins, membrane fusion may proceed (H). Proteins may also enzymatically generate fusion-inducing molecules, as in the case of phospholipase C (Nieva et al., 1989; Burger et al., 1991).

Lectins. The incorporation of glycolipids with head-groups containing three or more carbohydrate groups in phospholipid vesicles inhibits Ca^{2+}-induced fusion (Düzgüneş et al., 1984b; Hoekstra & Düzgüneş, 1986). Lectins specific for the glycolipids can mediate intermembrane attachment or lateral phase separation of the inhibitory glycolipids, and enhance the rate of fusion (Düzgüneş et al., 1984b; Hoekstra et al., 1985; Hoekstra & Düzgüneş, 1986, 1989). Depending on the glycolipid head-group, the threshold concentration of Ca^{2+} required to induce fusion can also be reduced drastically (Sundler & Wijkander, 19843 Hoekstra & Düzgünes, 1986). The interaction of Ca^{2+} with phospholipids may also induce the reorientation of the glycolipid head-groups perpendicular to the plane of the membrane, thereby enhancing lectin-mediated agglutination, but also inhibiting the fusion reaction (Hoekstra & Düzgüneş, 1989). Similarly, the incorporation of glycophorin in phosphatidylserine liposomes inhibits Ca^{2+}-induced fusion, but lectins can overcome part of this inhibitory effect (deKruijff et al., 1991). Membrane aggregating proteins may facilitate the fusion activity of Ca^{2+}, whose binding to membranes is enhanced dramatically upon intermembrane contact (Ekerdt & Papahadjopoulos, 1982; Nir, 1984; Feigenson, 1986).

Annexins. Annexins are Ca^{2+}-dependent, phospholipid-binding, cytosolic proteins which may be involved in intracellular membrane trafficking (Klee, 1988; Creutz, 1992). Synexin (annexin VII) mediates the aggregation of secretory vesicles in the presence of Ca^{2+} and facilitates their fusion if unsaturated free fatty acids such as arachidonic acid are present (Creutz, 1981). It also enhances the overall rate of fusion of liposomes (consisting of both the aggregation

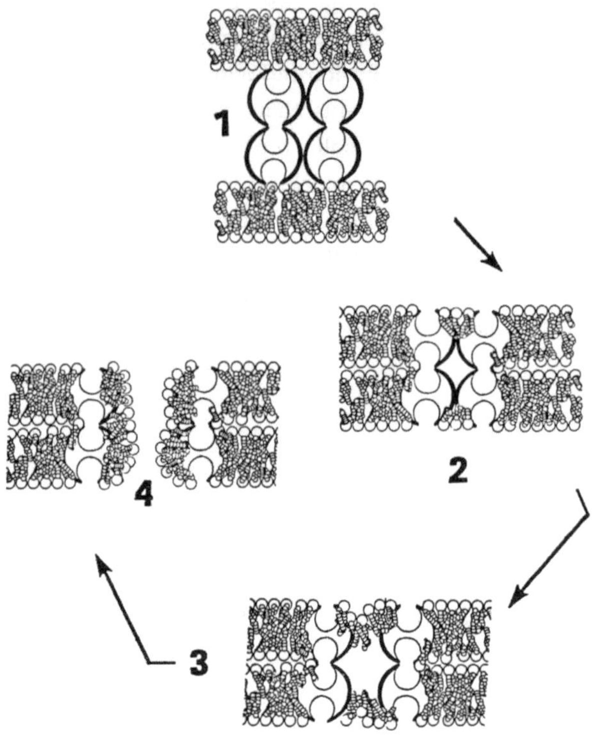

Figure 9. Hypothetical mechanism of action of synexin. In this hypothesis, synexin forms a dimeric hydrophobic bridge between two interacting membranes. 1: A tetramer of synexin binding to phospholipids of two adjacent membranes. 2: Migration of phospholipids over the hydrophobic bridge. 3: Dissociation of the synexin tetramer due to dissipation of hydrophobic forces driving polymerization. 4: Migration of phospholipids over the newly exposed hyrophobic face of synexin, and completion of the fusion reaction. Reproduced from Pollard et al. (1991), with permission.

and fusion steps) (Düzgüneş et al., 1980; Hong et al., 1981), and reduces the Ca^{2+} concentration necessary for fusion (Hong et al., 1982). It appears that synexin acts primarily by enhancing the initial aggregation step of the overall fusion reaction (Meers et al., 1988). An alternative hypothesis concerning the mechanism of synexin-mediated fusion is that it forms a hydrophobic bridge through which lipids migrate between two membranes (Figure 9) (Pollard et al., 1991). In the presence of calelectrin (annexin VI), oleic acid reduces the Ca^{2+} concentration threshold for fusion of secretory granules by about an order of magnitude (Zaks & Creutz, 1988). It has been proposed that the interaction of a GTP-binding protein with a lipase that produces fatty acids could activate annexins even at resting cytoplasmic Ca^{2+} concentrations (Creutz, 1992). Annexins also mediate the fusion of liposomes with neutrophil secretory granules or plasma membranes (Meers et al., 1987; Oshry et al., 1991). Neutrophil annexin I enhances the Ca^{2+}-dependent fusion of PS liposomes (Blackwood et al., 1990), and reduces the Ca^{2+} concentration threshold for fusion of liposomes composed of phosphatidic acid and phosphatidylethanolamine (Francis et al., 1992). Monomers of annexin I are thought to be able to contact two membranes simultaneously (Meers et al., 1992), and the protein has been suggested to be part of a multiprotein complex involved in intermembrane contact during the

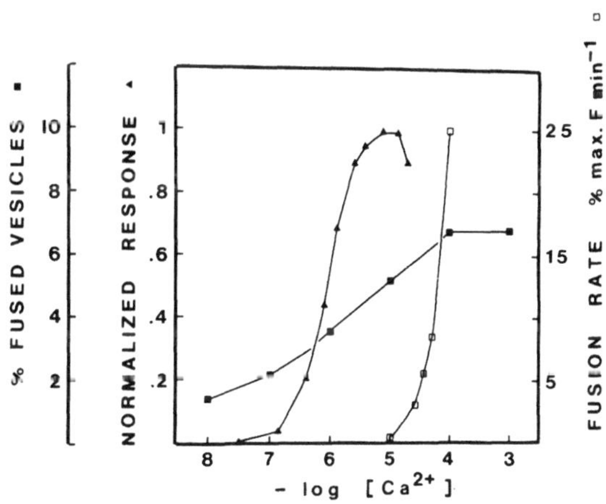

Figure 10. Ca^{2+} concentration dependence of exocytosis in leaky adrenal medullary cells (triangles; data from Baker et al., 1980), fusion of isolated secretory vesicles from adrenal medulla (solid squares; data from Dahl et al., 1979), and fusion of phosphatidic acid/phosphatidylethanolamine liposomes in the presence of synexin and Mg^{2+} (open squares; data from Hong et al., 1982). Reproduced with permission from Düzgüneş (1985).

exocytosis of specific granules in neutrophils (Meers et al., 1993). The phosphorylation state of annexin I may also affects its ability to aggregate secretory granules or liposomes (Wang & Creutz, 1992; Johnstone et al., 1993).

The Ca^{2+} concentration requirement for fusion of secretory vesicles or liposomes in the presence of annexins is higher than the overall intracellular concentration achieved during exocytosis (Figure 10). The affinity of annexins for Ca^{2+} may be different under intracellular conditions, due to interactions with other protein or lipid cofactors (Creutz, 1992). In addition, the Ca^{2+} concentration near the plasma membrane or intracellular stores of Ca^{2+} may be transiently higher than the average intracellular concentration during stimuli for secretion. Microdomains of Ca^{2+} adjacent to the cytoplasmic face of pre-synaptic membranes have been estimated to have concentrations as high as 200-300 μM (Llinas et al., 1992). Half-maximal rate constants of secretion have been shown to occur at about 200 μM Ca^{2+} (Heidelberger et al., 1994).

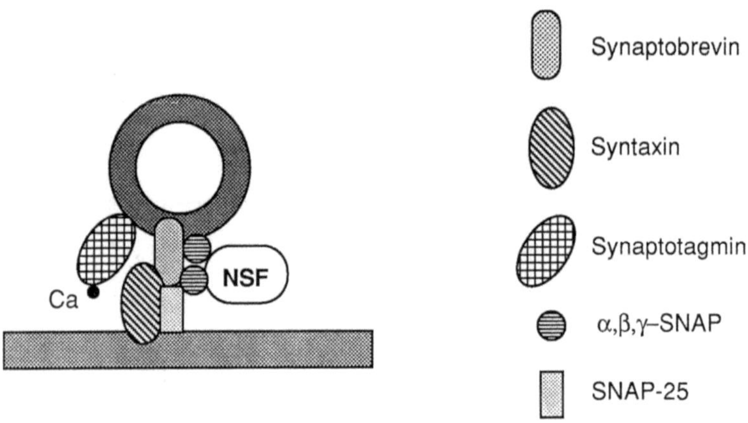

Figure 11. A model for the interaction of various proteins thought to be involved in synaptic exocytosis, based on models by Kelly (1993) and Südhof et al. (1993).

Synaptic vesicle proteins and the NSF-SNAP complex. Many soluble and membrane proteins involved in the fusion of intracellular vesicles have been identified (Rothman & Orci, 1992; Südhof et al., 1993; Kelly, 1993), and their properties are reviewed elsewhere in this volume (Höhne-Zell & Gratzl, 1995). A family of synaptic vesicle proteins, called synaptobrevins (Baumert et al., 1989) or vesicle-associated membrane proteins (Bennett & Scheller, 1993) are thought to be involved in exocytotic fusion, and are the targets of various neurotoxins which inhibit exocytosis (Schiavo et al., 1992; Höhne-Zell & Gratzl, 1995). Another synaptic vesicle protein, synaptotagmin, is considered to act as a Ca^{2+} sensor (Brose et al., 1992), and binds to syntaxin on the presynaptic membrane (Bennett et al., 1992). NSF, an "N-ethylmaleimide-sensitive factor" required for the fusion of certain intracellular membranes, and soluble NSF attachment proteins (SNAPs) (Rothman & Orci, 1992) were found to bind to receptors on membranes, which were identified as synaptobrevin, synaxins and a synapto-somal-associated protein, SNAP-25 (Söllner et al., 1993). These proteins may be involved both in the specific targeting of intracellular vesicles (Söllner et al., 1993), or in the fusion of these vesicles (Südhof et al., 1993). A model of how these proteins might interact, based on the schemes presented by Kelly (1993) and Südhof et al. (1993), is shown in Figure 11.

Fusion pores. Electron microscopic studies of exocytosis have revealed the opening of a pore between the exocytotic vesicle and the plasma membrane (Chandler, 1988). The conductance of the fusion pore can be monitored by patch-clamping in the whole cell configuration, and indicates that the initial pore is considerably smaller than that observed in electron microscopy (Breckenridge &

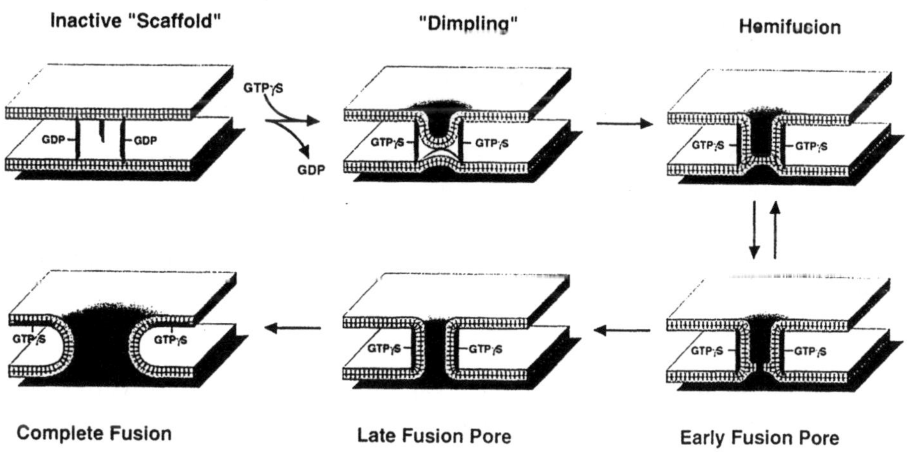

Figure 12. The lipidic pore model of exocytotic fusion. See text for a description of the different steps. Reproduced from Monck & Fernandez (1992), with permission.

Almers, 1987; Monck & Fernandez, 1992). Influenza virus hemagglutinin, perhaps the best studied fusion protein, also forms pores between erythrocytes and cells expressing the protein, following mild acidification (Spruce et al., 1989).

Several models have been proposed for the structure of fusion pores. Figure 12 shows the lipidic pore model proposed by Nanavati et al. (1992) and Monck & Fernandez (1992). Here, two membranes are initially separated by a protein scaffold. GTP- and Ca^{2+}-binding proteins may be part of the scaffold, since GTPγS and Ca^{2+} can trigger exocytotic fusion. In response to the fusion trigger (and the exchange of GDP for GTPγS), the scaffold mediates the formation of a dimple in the plasma membrane towards the secretory vesicle. The intermixing of outer monolayers of the two membranes is thought to occur at this stage due to the tension in the membrane, forming a hemifusion intermediate†. A small pore develops within the larger neck (or hour-glass structure) seen in electron microscopy. The pore then expands to allow the release of the contents of the secretory vesicle.

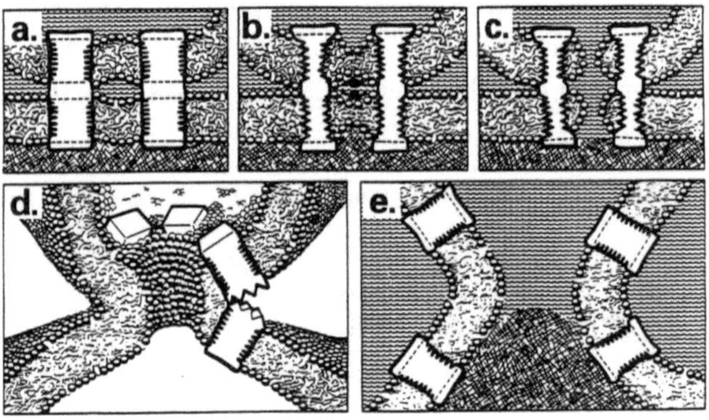

Figure 13. The lipid-protein model for a fusion pore. Reproduced from Zimmerberg et al. (1993), with permission.

In the lipid-protein pore model, exposure of hypothetical hydrophobic surfaces of proteins provides a conduit for phospholipids in two apposed mem-

† Hemifusion has also been termed "semi-fusion" (Düzgüneş, 1985) or monolayer fusion (Chernomordik et al., 1987). The intermixing of the outer (contacting) monolayers of two interacting membranes, has been observed previously in a number of phospholipid membrane fusion systems (Neher, 1974; Rosenberg et al., 1983; Markin et al., 1984; Bondeson & Sundler, 1985; Ellens et al., 1985; Düzgüneş et al., 1987b).

branes to intermix (Figure 13) (Zimmerberg et al., 1993). Here, insertion of hydrophobic regions of fusion proteins of one membrane into the other mediates the close approach of the membranes (Figure 13a). The proteins then undergo a conformational change which exposes hydrophobic regions over which the lipids equilibrate (b). The lipids then rearrange to maintain the hydration of their head-groups and to coat the hydrophobic segments of the pore proteins, forming a lipid-lined pore (c). The phospholipids then coat the entire surface of the proteins as a monolayer (d). The proteins subsequently dissociate to allow for complete membrane fusion (e). In contrast to the lipidic pore model presented in Figure 12, where the pore is lined by a bilayer of phospholipid, this model proposes a monolayer of lipid lining the pore.

Fusion of secretory vesicles with the plasma membrane can be transient, in that the fusion pore can first open and then close, in a process termed "flicker" (Breckenridge & Almers, 1987; Zimmerberg et al., 1987). Electrochemical measurements have indicated that small quantities of the contents of secretory vesicles can be released during this transient fusion process without the complete fusion of the vesicle and plasma membranes (Alvarez de Toledo et al., 1993).

Influenza virus hemagglutinin. The main features of the mechanism of fusion of influenza virus with cellular membranes is discussed by Pedroso de Lima et al. (1995) in this volume. Influenza HA-induced fusion has been proposed to proceed via an inverted micellar intermediate formed between three HA trimers, leading to a pore lined by a monolayer of phospholipids (Figure 14a) (Bentz et al., 1990). Contradicting this hypothesis is the observation that HA-mediated fusion can proceed even when the membrane containing the HA or the target membrane is composed of lipids that do not form non-bilayer phases under physiological conditions (Stegmann, 1993). Figure 14b depicts a possible proteinaceous fusion pore formed by the HA inserting into a target membrane, based on observations that a a pore is formed between HA-expressing cells and erythrocytes before the intermixing of lipids between the two membranes (Tse et al., 1993). HA anchored to glycosylphosphatidylinositol mediates the intermixing of lipids between cells expressing the modified HA and erythrocytes, but not the intermixing of aqueous contents, suggesting that semi-fusion has occurred (Figure 14c) (Kemble et al., 1994). The proposed "stalk" mechanism for HA-mediated fusion is presented in Figure 14d (Stegmann, 1994). It is assumed

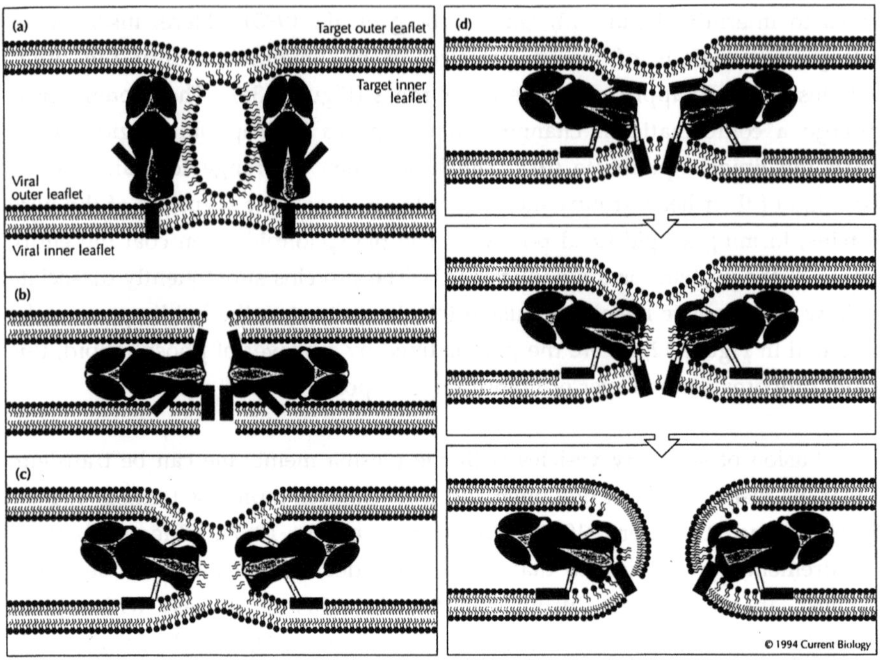

Figure 14. Proposed mechanims for membrane fusion induced by influenza virus hemagglutinin. The different models are discussed in detail in the text. Reproduced from Stegmann (1994), with permission.

in this model that the HA trimers must bend to be able to insert their "fusion peptides" into the target membrane, and produce a defect in lipid packing, as discussed above in the section on point defects. The defects are then thought to lead to the formation of an intermembrane stalk through which the apposed monolayers can intermix. This stalk will then have to rupture to mediate the opening of a lipid-lined fusion pore.

In an alternative hypothesis (Figure 15) (Zimmerberg et al., 1993), following the contact of the HA with the target membrane (a), triggering of membrane fusion by acidification causes the hydrophobic fusion peptide ($HA2_{1-21}$) and the adjoining "tethering sequence" ($HA2_{22-35}$) to extend outwards towards the target membrane, without requiring the initial bending of the HA trimers (c). The insertion of the fusion peptide into the target membrane and its transformation into an α-helical structure has been proposed to cause a contrac-

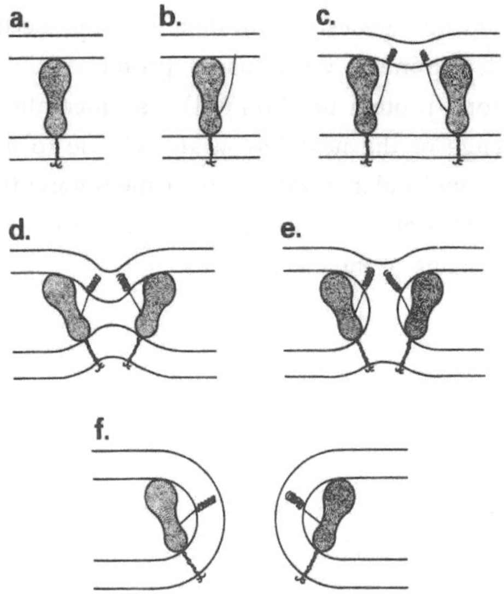

Figure 15. The "cast and retrieve" model of fusion induced by the influenza virus hemagglutinin. See text for the description of the sequence of events. Reproduced from Zimmerberg et al. (1993), with permission.

tion of the protruding HA2 segment, resulting in the bending of the target membrane and the viral membrane toward each other (d), until the two membranes form a stalk (e) and then fuse (f), rather than stay in an energetically unfavorable highly curved form.

PERSPECTIVE

The studies described in this chapter show the large variety of membrane systems that undergo fusion under different environmental conditions. We have indicated some of the emerging concepts in understanding membrane fusion, which are likely to apply to many different systems. However, the molecular details of, and the types of molecules involved in, each system differ, ranging from pure phospholipid vesicles to the rather "simple" (yet still not adequately understood) influenza HA, to the complex array of proteins regulating synaptic exocytosis. Understanding the mechanisms of these fusion events requires a detailed analysis of both the protein and lipid components involved. It will be of great interest to correlate alterations in these proteins (produced by site-directed mutagenesis) with details of the electrophysiology of fusion pores, and to determine if viral fusion proteins other than HA (for example the HIV envelope protein gp120/gp41) also mediate fusion via pore formation. The challenge of the next few years will be to test the different hypotheses regarding the molecular mechanisms of membrane fusion, as well as to develop specific inhibitors of fusion events involved in viral entry into cells, based on a detailed knowledge of the mechanisms.

ACKNOWLEDGEMENTS

This work was supported by NATO Collaborative Research Grant CRG900333, and NIH Grant AI-32399.

REFERENCES

Akabas MH, Cohen FS, Finkelstein A (1984) Separation of the osmotically driven fusion event from vesicle-planar membrane attachment in a model system for exocytosis. J Cell Biol 98: 1063-1071

Alvarez de Toledo G, Fernández-Chacón R, Fernández JM (1993) release of secretory products during transient vesicle fusion. Nature 363: 554-557

Baker PF (1988) Exocytosis in electropermeabilized cells: Clues to mechanism and physiological control. In: Membrane Fusion in Fertilization, Cellular Transport and Viral Infection (Düzgünes N, Bronner F, eds), pp 115-138, Academic Press San Diego

Baker PF, Knight DE, Whitaker MJ (1980) Calcium and the control of exocytosis. In: Calcium Binding Proteins: Structure and Function (Siegel FL, Carafoli E, Kretsinger RH, MacLennan DH, Wasserman RH, eds), pp 47-55, Elsevier/North Holland New York

Baumert M, Maycox PR, Navone F, De Camilli P, Jahn R (1989) Synaptobrevin: an integral membrane protein of 18 000 daltons present in small synaptic vesicle of rat brain. EMBO J 8: 379-384

Bearer EL, Düzgüneş N, Friend DS, Papahadjopoulos D (1982) Fusion of phospholipid vesicles arrested by quick freezing. The question of lipidic particles as intermediates in membrane fusion. Biochim Biophys Acta 693: 93-98

Bennett MK, Calakos N, Scheller RH (1992) Syntaxin: A synaptic protein implicated in docking of synaptic vesicels at presynaptic active zones. Science 257: 255-259

Bennett MK, Scheller RH (1993) The molecular machinery for secretion is conserved from yeast to neuron. Proc Natl Acad Sci USA 90: 2559-2563

Bentz J, Ellens H, Alford D (1990) An architecture for the fusion site of influenza hemagglutinin. FEBS Lett. 276: 1-5

Berridge MJ (1993) Inositol triphosphate and calcium signalling. Nature 361: 315-325

Blackwood RA, Ernst JD (1990) Characterization of Ca^{2+}-dependent phospholipid binding, vesicle aggregation and membrane fusion by annexins. Biochem J 266: 195-200

Bondeson J, Sundler R (1985) Lysine peptides induce lipid intermixing but not fusion of phosphatidic acid-containing vesicles. FEBS Lett 190: 283-287

Breckenridge LJ, Almers W (1987) Currents through the fusion pore that forms during exocytosis of a secretory vesicle. Nature 328: 814-817

Brose N, Petrenko AG, Südhof TC, Jahn R (1992) Synaptotagmin: A calcium sensor on the synaptic vesicle surface. Science 256: 1021-1025

Burger KNJ, Nieva JL, Alonso A, Verkleij AJ (1991) Phospholipase C activity-induced fusion of purel lipid model membranes. A freeze-fracture study. Biochim Biophys Acta 1068: 249-253

Chandler DE (1988) Exocytosis and endocytosis: Membrane fusion events captured in rapidly frozen cells. In: Membrane Fusion in Fertilization, Cellular Transport and Viral Infection (Düzgünes N, Bronner F, eds), pp 169-202, Academic Press San Diego

Chernomordik LV, Kozlov MM, Melikyan GB, Abidor IG, Markin VS, Chizmadzhev YA (1985) The shape of lipid molecules and monolayer membrane fusion. Biochim Biophys Acta 812: 643-655

Chernomordik LV, Melikyan GB, Chizmadzhev YA (1987) Biomembrane fusion: a new concept derived from model studies using two interacting planar lipid bilayers. Biochim Biophys Acta 906: 309-352

Chernomordik LV, Vogel SS, Sokoloff A, Onaran HO, Leikina EA, Zimmerberg J (1993) Lysolipids reversibly inhibit Ca^{2+}-, GTP- and pH-dependent fusion of biological membranes. FEBS Lett 318: 71-76

Cockcroft S, Allan D (1985) Loss of phosphatidylinositol and gain in phosphatidate in neutrophhils stimulated with fmet-leu-phe. In: Inositol and Phosphoinositides (Bleasdale JE, Eichberg J, Hauser G, eds), pp 161-177, Humana Press Clifton New Jersey

Creutz CE (1981) *Cis*-unsaturated fatty acids induce the fusion of chromaffin granules aggregated by synexin. J Cell Biol 91: 247-256

Creutz CE (1992) The annexins and exocytosis. Science 258: 924-931

Cullis PR, Hope MJ (1978) Effects of fusogenic agent on membrane structure of erythrocyte ghosts and the mechanism of membrane fusion. Nature 271: 672-674

Cullis PR, Verkleij AJ (1979) Modulation of membrane structure by Ca^{2+} and dibucaine as detected by ^{31}P NMR. Biochim Biophys Acta 552: 546-551

Cullis PR, Tilcock CP, Hope MJ (1991) Lipid polymorphism. In: Membrane Fusion (Wilschut J, Hoekstra D, eds), pp 35-64, Marcel Dekker New York

Dahl G, Ekerdt R, Gratzl M (1979) Models for exocytotic membrane fusion. Symp Soc Exp Biol 33: 349-368

deKruijff B, de Gier J, van Hoogevest P, van der Steen N, Taraschi TF, de Kroon T (1991) Effects of an integral membrane glycoprotein on phospholipid vesicle fusion. In: Membrane Fusion (Wilschut J, Hoekstra D, eds), pp 209-229, Marcel Dekker New York

Doms RW, Helenius AH, White JM (1985) Membrane fusion activity of the influenza virus hemagglutinin: the low pH-induced conformational change. J Biol Chem 260: 2973-2981

Düzgünes N (1985) Membrane fusion. In: Subcellular Biochemistry, Vol. 11, (Roodyn DB, ed), pp 195-286, Plenum New York

Düzgüneş N (ed) (1993a) Membrane Fusion Techniques, Methods in Enzymology, Vol. 220. Academic Press San Diego

Düzgüneş N (ed) (1993b) Membrane Fusion Techniques, Methods in Enzymology, Vol. 221. Academic Press San Diego

Düzgüneş N, Papahadjopoulos D (1983) Ionotropic effects on phospholipid membranes: Calcium-magnesium specificity in binding, fluidity, and fusion. In: Membrane Fluidity in Biology, Vol. 2, (Aloia RC, ed), pp 187-216, Academic Press New York

Düzgüneş N, Gambale F (1988) Membrane action of synthetic peptides from influenza virus hemagglutinin and its mutants. FEBS Lett 227: 110-114

Düzgüneş N, Shavnin SA (1992) Membrane destabilization by N-terminal peptides of viral envelope proteins. J Memb Biol 128: 71-80

Düzgüneş N, Nir S (1994) Liposomes as tools for elucidating the mechanisms of membrane fusion. In: Liposomes as Tools in Basic Research and Industry (Philippot JR, Schuber F, eds), CRC Press Boca Raton (in press)

Düzgüneş N, Hong K, Papahadjopoulos D (1980) Membrane fusion: The involvement of phospholipids, proteins and calcium binding. In: Calcium Binding Proteins: Structure and Function (Siegel FL, Carafoli E, Kretsinger RH, MacLennan DH, Wasserman RH, eds), pp 17-22, Elsevier/North Holland New York

Düzgüneş N, Wilschut J, Fraley R, Papahadjopoulos D (1981) Studies on the mechanism of membrane fusion: role of head-group composition in calcium- and magnesium-induced fusion of mixed phospholipid vesicles. Biochim Biophys Acta 642: 182-195

Düzgüneş N, Paiement J, Freeman KB, Lopez NG, Wilschut J, Papahadjopoulos D (1984a) Modulation of membrane fusion by ionotropic and thermotropic phase transitions. Biochemistry 23: 3486-3494

Düzgüneş N, Hoekstra D, Hong K, Papahadjopoulos D (1984b) Lectins facilitate calcium-induced fusion of phospholipid vesicles containing glycosphingolipids. FEBS Lett 173: 80-84

Düzgüneş N, Wilschut J, Papahadjopoulos D (1985) Control of membrane fusion by divalent cations, phospholipid head-groups and proteins. In: Physical Methods on Biological Membranes and their Model Systems (Conti F, Blumberg WE, DeGier J, Pocchiari F, eds), pp 193-218, Plenum Press New York

Düzgüneş N, Hong K, Baldwin PA, Bentz J, Nir S, Papahadjopoulos D (1987a) Fusion of phospholipid vesicles induced by divalent cations and protons. Modulation by phase transitions, free fatty acids, monovalent cations, and polyamines. In: Cell Fusion (Sowers AE, ed), pp 241-267, Plenum Press New York

Düzgüneş N, Allen TM, Fedor J, Papahadjopoulos, D (1987b) Lipid mixing during membrane aggregation and fusion. Why fusion assays disagree, Biochemistry 26: 8435-8442

Düzgüneş N, Goldstein JA, Friend DS, Felgner PL (1989) Fusion of liposomes containing a novel cationic lipid, *N*-[1-(2,3-(dioleyloxy)propyl]-*N*,*N*,*N*-trimethylammonium: Induction by multivalent anions and asymmetric fusion with acidic phospholipid vesicles. Biochemistry 28: 9179-9184

Ekerdt R, Papahadjopoulos D (1982) Intermembrane contact affects calcium binding to phospholipid vesicles, Proc Natl Acad Sci USA 79: 2273-2277

Ellens H, Bentz J, Szoka FC (1985) H^{+}- and Ca^{2+} -induced fusion and destabilization of liposomes. Biochemistry 24: 3099-3106

Ellens H, Siegel DP, Alford D, Yeagle PL, Boni L, Lis LJ, Quinn PJ, Bentz, J (1989) Membrane fusion and inverted phases. Biochemistry 28: 3692-3703

Feigenson GW (1986) On the nature of calcium ion binding between phosphatidylserine lamellae. Biochemistry 25: 5819-5825

Fraley R, Wilschut J, Düzgüneş N, Smith C, Papahadjopoulos D (1980) Studies on the mechanism of membrane fusion: the role of phosphate in promoting calcium-induced fusion of phospholipid vesicles. Biochemistry 19: 6021-6029

Francis JW, Balazovich KJ, Smolen JE, Margolis DI, Boxer LA (1992) Human neutrophil annexin I promotes granule aggregation and modulates Ca^{2+}-dependent membrane fusion. J Clin Invest 90: 537-544

Gething MJ, Doms RW, York D, White J (1986) Studies on the mechanism of membrane fusion: Site-specific mutagenesis of the hemagglutinin of influenza virus. J Cell Biol 102: 11-23

Harter C, James P, Bächi T, Semenza G, Brunner J (1989) Hydrophobic binding of the ectodomain of influenza hemagglutinin to membranes occurs through the "fusion peptide." J Biol Chem 264: 6459-6464

Hauser H, Pascher I, Pearson RH, Sundell S (1981) Preferred conformation and molecular packing of phosphatidylethanolamine and phosphatidylcholine. Biochim Biophys Acta 650: 21-51

Heidelberger R, Heinemann C, Neher E, Matthews G (1994) Calcium dependence of the rate of exocytosis in a synaptic terminal. Nature 371: 513-515

Helm CA, Israelachvili JN (1993) Forces between phospholipid bilayers and relationship to membrane fusion. In: Membrane Fusion Techniques, Methods in Enzymology, Vol 220, (Düzgünes N, ed), pp 130-143, Academic Press San Diego

Helm CA, Israelachvili JN, McGuiggan PM (1989) Molecular mechanisms and forces involved in the adhesion and fusion of amphiphilic bilayers. Science 246: 919-922

Helm CA, Israelachvili JN, McGuiggan PM (1992) Role of hydrophobic forces in bilayer adhesion and fusion. Biochemistry 31: 1794-1805

Hoekstra D (1982) Role of lipid phase separations and membrane dehydration in phospholipid vesicle fusion. Biochemistry 21: 2833-2840

Hoekstra D, Düzgüneş N (1986) *Ricinus communis* agglutinin-mediated agglutination and fusion of glycolipid-containing phospholipid vesicles. Effect of carbohydrate headgroup size, calcium ions and spermine. Biochemistry 25: 1321-1330

Hoekstra D, Düzgüneş N (1989) Lectin-carbohydrate interactions in model and biological membrane systems. In: Subcellular Biochemistry, Vol 14 (Harris JR, Etamadi AH, eds), pp 229-278, Plenum New York

Hoekstra D, Düzgüneş N, Wilschut J (1985) Agglutination and fusion of globoside GL-4 containing phospholipid vesicles mediated by lectins and Ca^{2+}. Biochemistry 24: 565-572

Höhne-Zell B, Gratzl M (1995) Molecular analysis of exocytosis in neurons and endocrine cells. In: Trafficking of Intracellular Membranes: From Molecular Sorting to Membrane Fusion (Pedroso de Lima MC, Düzgünes N, Hoekstra D, eds), Springer Verlag Berlin (in press)

Hong K, Düzgüneş N, Papahadjopoulos D (1981) Role of synexin in membrane fusion. Enhancement of calcium-dependent fusion of phospholipid vesicles. J Biol Chem 256: 3641-3644

Hong K, Düzgüneş N, Ekerdt R, Papahadjopoulos D (1982) Synexin facilitates fusion of specific phospholipid vesicles at divalent cation concentrations found intracellularly. Proc Natl Acad Sci USA 70: 4642-4644

Hope MJ, Walker DC, Cullis, PR (1983) Calcium and pH-induced fusion of small unilamellar vesicles consisting of phosphatidylethanolamine and negatively charged phospholipids: A freeze-fracture study. Biochem Biophys Res Commun 110: 15-22

Hui SW, Stewart TP, Boni LT, Yeagle PL (1981) Membrane fusion through point defects in bilayers. Science 212: 921-923

Hui S, Nir S, Stewart TP, Boni LT, Huang SK (1988) Kinetic measurements of fusion of phosphatidylserine-containing vesicles by electron microscopy and fluorometry. Biochim. Biophys. Acta 941: 130-140

Israelachvili JN, McGuiggan PM (1988) Forces between surfaces in liquids. Science 241: 795-800

Jendrasiak GL, Hasty JH (1974) The hydration of phospholipids. Biochim Biophys Acta 337: 79-91

Johnstone SA, Hubaishy I, Waisman DM (1993) Regulatiosn of annexin II-dependent aggregation of phospholipid vesicles. Biochem J 294: 801-807

Kelly RB (1993) Much ado about docking. Curr Biol 3: 474-476

Kemble GW, Danieli T, White JM (1994) Lipid-anchored influenza hemagglutinin promotes hemifusion, not complete fusion. Cell 76: 383-391

Klee CB (1988) Ca^{2+}-depenent phospholipid- (and membrane-) binding proteins. Biochemistry 27: 6645-6652

Leckband DE, Helm CA, Israelachvili JN (1993) Role of calcium in the adhesion and fusion of bilayers. Biochemistry 32: 1127-1140

Lifson JD, Feinberg MB, Reyes GR, Rabin L, Banapour B, Chakrabarti S, Moss B, Wong-Staal F, Steimer KS, Engleman EG (1986) Induction of CD4-dependent cell fusion by the HTLV-III/LAV envelope glycoprotein. Nature 323: 725-728

Llinas R, Sugimori M, Silver RB (1992) Microdomains of high calcium concentration in a presynaptic terminal. Science 256: 677-679

Markin VS, Kozlov MM, Borovjagin VL (1984) On the theory of membrane fusion. The stalk mechanism. Gen Physiol Biophys 5: 361-377

McIntosh TJ, Magid AD, Simon SA (1987) Steric repulsion between phosphatidylcholine bilayers. Biochemistry 26: 7325-7332

McIver DJL (1979) Control of membrane fusion by interfacial water: a model for the actions of divalent cations. Physiol Chem Phys 11: 289-302

Meers P, Hong K, Bentz J, Papahadjopoulos D (1986) Spermine as a modulator of membrane fusion: interaction with acidic phospholipids. Biochemistry 25: 3109-3118

Meers P, Ernst JD, Düzgüneş N, Hong K, Fedor J, Goldstein IM, Papahadjopoulos D (1987) Synexin-like proteins from human polymorphonuclear leukocytes. Identification and characterization of granule-aggregating and membrane-fusing activities. J Biol Chem 262: 7850-7858.

Meers P, Bentz J, Alford D, Nir S, Papahadjopoulos D, Hong K (1988) Synexin enhances the aggregation rate but not the fusion rate of liposomes. Biochemistry 27: 4430-4439

Meers P, Mealy T, Pavlotsky N, Tauber AI (1992) Annexin I-mediated vesicular aggregation: Mechanism and role in human neutrophils. Biochemistry 31: 6372-6382

Meers P, Mealy T, Tauber AI (1993) Annexin I interactions with human neutrophil specific granules: fusogenicity and coaggregation with plasma membrane vesicles. Biochim Biophys Acta 1147: 177-184

Miller DC, Dahl GP (1982) Early events in calcium-induced liposome fusion. Biochim Biophys Acta 689: 165-169

Monck JR, Fernandez JM (1992) The exocytotic fusion pore. J Cell Biol 119: 1395-1404

Nanavati C, Markin VS, Oberhauser A, Fernandez JM (1992) The exocytotic fusion pore as a protein-supported lipidic structure. Biophys J 63: 1118-1132

Neher E (1974) Asymetric membranes resulting from the fusion of two black lipid bilayers. Biochim Biophys Acta 373: 327-336

Nieva JL, Goni FM, Alonso A (1989) Liposome fusion catalytically induced by phospholipase C. Biochemistry 28: 7364-7367

Nir S (1984) A model for cation adsorption in closed systems: Application to calcium binding to phospholipid vesicles. J Coll Interface Sci 102: 313-321

Nir S, Bentz J, Düzgüneş N (1981) Two modes of reversible vesicle aggregation: particle size and the DLVO theory. J Coll. Interface Sci 84: 266-269

Nir S, Wilschut J, Bentz J (1982) The rate of fusion of phospholipid vesicles and the role of bilayer curvature. Biochim Biophys Acta 688: 275-278

Nir S, Bentz J, Wilschut J, Düzgüneş N (1983) Aggregation and fusion of phospholipid vesicles. Prog Surface Sci 13: 1-124

Ohki S (1984) Effects of divalent cations, temperature, osmotic pressure gradient and vesicle curvature on phosphatidylserine vesicle fusion. J Memb Biol 77: 265-275

Ohki, S. (1988) Surface tension, hydration eenergy and membrane fusion. In: Molecular Mechanisms of Membrane Fusion (Ohki S, Doyle D, Flanagan T, Hui SW, Mayhew E, eds), pp 123-138, Plenum Press New York

Ohki S, Düzgüneş N (1979) Divalent cation-induced interaction of phospholipid vesicle and monolayer membranes. Biochim Biophys Acta 552: 438-449

Ohki S, Ohshima H (1984) Divalent cation-induced surface tension increase in acidic phospholipid membranes. Ion binding and membrane fusion, Biochim Biophys Acta 776: 177-182

Oshry L, Meers P, Mealy T, Tauber AI (1991) Annexin-mediated membrane fusion of human neutrophil plasma membranes and phospholipid vesicles. Biochim Biophys Acta 1006: 239-244

Papahadjopoulos D, Vail WJ, Newton C, Nir S, Jacobson K, Poste G, Lazo R (1977) Studies on membrane fusion. III. The role of calcium-induced phase changes. Biochim Biophys Acta 465: 579-598

Papahadjopoulos D, Nir S, Düzgünes N (1990) Molecular mechanisms of calcium-induced membrane fusion. J Bioenerget Biomembr 22: 157-179

Park J-B, Lee TH, Kim H (1992) Fusion of phospholipid vesicles induced by phospholipase D in the presence of calcium ion. Biochem Int 27: 417-422

Parsegian VA, Rand RP (1991) Forces governing lipid interaction and rearrangement. In: Membrane Fusion (Wilschut J, Hoekstra D, eds), pp 65-85, Marcel Dekker New York

Pedroso de Lima MC, Ramalho-Santos J, Düzgünes N, Flasher D, Nir S (1995) Entry of enveloped viruses into host cells: Fusion activity of the influenza hemagglutinin. In: Trafficking of Intracellular Membranes: From Molecular Sorting to Membrane Fusion (Pedroso de Lima MC, Düzgünes N, Hoekstra D, eds), Springer Verlag Berlin (in press)

Plattner H (1989) Regulation of membrane fusion during exocytosis. Int Rev Cytol 119: 197-286

Pollard HB, Rojas E, Pastor RW, Rojas EM, Guy HR, Burns AL (1991) Synexin: Molecular mechanism of calcium-dependent membrane fusion and voltage dependent calcium channel activity. Ann NY Acad Sci. 635: 328-351

Portis A, Newton C, Pangborn W, Papahadjopoulos D (1979) Studies on the mechanism of membrane fusion: evidence for an intermembrane Ca^{2+}-phospholipid complex, synergism with Mg^{2+} and inhibition by spectrin. Biochemistry 18: 780-790

Prestegard JH, O'Brien MP (1987) Membrane and vesicle fusion. Ann Rev Phys Chem 38: 383-411

Rafalski M, Ortiz A, Rockwell A, Van Ginkel LC, Lear JD, DeGrado WF, Wilschut J (1991) Membrane fusion activity of the influenza virus hemagglutinin: Interaction of HA2 N-terminal peptides with phospho-lipid vesicles. Biochemistry 30: 10211-10220

Reiss-Husson F (1967) Structure des phases liquid-crystallines de différents phospholipides, monoglycerides, sphingolipides, anhydrides ou en présence d'eau. J Mol Biol 25: 363-382

Rosenberg J, Düzgünes N, Kayalar C (1983) Comparison of two liposome fusion assays monitoring the intermixing of aqueous contents and of membrane components. Biochim Biophys Acta 735: 173-180

Rothman JE, Orci L (1992) Molecular dissection of the secretory pathway. Nature 355, 409-415

Schiavo G, Benfenati F, Poulain B, Rossetto O, Polverino de Lareto P, DasGupta BR, Montecucco C (1992) Tetanus and botulinum-B neurotoxins block neurotransmitter release by proteolytic cleavage of synaptobrevin. Nature 359: 832-835

Schuber F, Hong K, Düzgüneş N, Papahadjopoulos D (1983) Polyamines as modulators of membrane fusion: Aggregation and fusion of liposomes. Biochemistry 22: 6134-6140

Shavnin SA, Pedroso de Lima MC, Fedor J, Wood P, Bentz J, Düzgünes N (1988) Cholesterol affects divalent cation-induced fusion and isothermal phase

transitions of phospholipid membranes. Biochim. Biophys. Acta 946: 405-416.
Siegel DP (1986) Inverted micellar intermediates and the transitions between lamellar, cubic and inverted hexagonal lipid phases. II. Implications for membrane-membrane interactions and membrane fusion. Biophys J 49: 1171-1183
Siegel DP (1993) Energetics of intermediates in membrane fusion: Comparison of stalk and inverted micellar intermediate mechanisms. Biophys J 65: 2124-2140
Siegel DP, Burns JL, Chestnut MH, Talmon Y. (1989) Intermediates in membrane fusion and bilayer/nonbilayer phase transitions imaged by time-resolved cryo-transmission electron microscopy. Biophys J 56: 161-169
Skehel JJ, Bayley PM, Brown EB, Martin SR, Waterfield MD, White JM, Wilson AI, Wiley DC (1982) Changes in the conformation of influenza virus hemagglutinin at the pH optimum of virus-mediated membrane fusion. Proc Natl Acad Sci USA 79: 968-972
Söllner T, Whiteheart SW, Brunner M, Erdjument-Bromage H, Geromanos S, Tempst P, Rothman JE (1993) SNAP receptors implicated in vesicle targeting and fusion. Nature 362: 318-324
Spruce AE, Iwata A, White JM, Almers W (1989) Patch clamp studies of single cell-fusion events mediated by a viral fusion protein. Nature 342: 555-558
Stamatatos L, Leventis R, Zuckermann MJ, Silvius JR (1988) Interaction of cationic lipid vesicles with negatively charged phospholipid vesicles and biological membranes. Biochemistry 27: 3917-3925
Stegmann T (1993) Influenza hemagglutinin-mediated membrane fusion does not involve inverted phase lipid intermediates. J Biol Chem 268: 1716-1722
Stegmann (1994) Anchors aweigh. Curr Biol 4: 551-554
Stewart TP, Hui SW, Portis AR, Papahadjopoulos D (1979) Complex phase mixing of phosphatidylcholine and phosphatidylserine in multilamellar membrane vesicles. Biochim Biophys Acta 556: 1-16
Südhof TC, De Camilli P, Niemann H, Jahn R (1993) Membrane fusion machinery: Insights from synaptic proteins. Cell 75: 1-4
Sundler R, Papahadjopoulos D (1981) Control of membrane fusion by phospholipid head groups. I. Phosphatidate/phosphatidylinositol specificity. Biochim Biophys Acta 649: 743-750
Sundler R, Wijkander J (1983) Protein-mediated intermembrane contact specifically enhances Ca^{2+}-induced fusion of phosphatidate-containing membranes Biochim. Biophys. Acta 730: 391-394
Sundler R, Düzgüneş N, Papahadjopoulos D (1981) Control of membrane fusion by phospholipid head groups. II. The role of phosphatidylethanolamine in mixtures with phosphatidate and phosphatidylinositol Biochim Biophys Acta 649: 751-758.
Tse FW, Iwata A, Almers W (1993) Membrane flux through the pore formed by a fusogenic viral envelope protein during cell fusion. J Cell Biol 121: 543-552
Tsurudome M, Glück R, Graf R, Falchetto R, Schaller U, Brunner J (1992) Lipid interactions of the hemagglutinin HA2 NH2-terminal segment during influenza virus-induced membrane fusion. J Biol Chem 267: 20225-20232
Verkleij AJ, Mombers C, Gerritsen WJ, Leunissen-Bijvelt L, Cullis PR (1979) Fusion of phospholipid vesicles in association with the appearance of

lipidic particles as visualized by freeze-fracturing. Biochim Biophys Acta 555: 358-361

Verkleij AJ, van Echteld CJA, Gerritsen WJ, Cullis PR, de Kruijff, B (1980) The lipidic particle as an intermediate structure in membrane fusion processes and bilayer to hexagonal H_{II} transitions. Biochim. Biophys. Acta 600: 620-624

Verkleij AJ, Leunissen-Bijvelt J, de Kruijff B, Hope M, Cullis PR (1984) Non-bilayer structures in membrane fusion. In: Cell Fusion, Ciba Foundation Symposium, Vol. 103, pp 45-59 Pitman Books London

Wakelam MJO (1988) Myoblast fusion-A mechanistic analysis. In: Membrane Fusion in Fertilization, Cellular Transport and Viral Infection (Düzgünes N, Bronner F, eds), pp 87-112, Academic Press San Diego

Wang W, Creutz CE (1992) Regulation of chromaffin granule aggregating activity of annexin I by phosphorylation. Biochemistry 31: 9934-9939

Weber T, Paesold G, Galli C, Mischler R, Semenza G, Brunner J (1994) Evidence for H^+-induced insertion of influenza hemagglutinin HA2 N-terminal segment into viral membrane. J Biol Chem 269: 18353-18358

White JM (1992) Membrane fusion. Science 258: 917-924

White J, Matlin K, Helenius A (1981) Cell fusion by Semliki Forest, influenza and vesicular stomatitis viruses. J Cell Biol 89: 674-679

Wilschut J (1991) Membrane fusion in lipid vesicle systems. An overview. In: (Wilschut J, Hoekstra, eds), pp 89-126, Marcel Dekker New York

Wilschut J, Düzgüneş N, Fraley, R, Papahadjopoulos D (1980) Studies on the mechanism of membrane fusion: Kinetics of Ca^{2+}-induced fusion of phosphatidylserine vesicles followed by a new assay for mixing of aqueous vesicle contents. Biochemistry 19: 6011-6021

Wilschut J, Düzgüneş N, Papahadjopoulos D (1981) Calcium/magnesium specificity in membrane fusion: Kinetics of aggregation and fusion of phosphatidylserine vesicles and the role of bilayer curvature. Biochemistry 20: 3126-3133

Wilschut J, Düzgüneş N, Hoekstra D, Papahadjopoulos D (1985) Modulation of membrane fusion by membrane fluidity: Temperature dependence of divalent cation-induced fusion of phosphatidylserine vesicles. Biochemistry 24: 8-12

Wilson, DW, Whiteheart SW, Orci L, Rothman JE (1991) Intracellular membrane fusion. Trends Biochem Sci 16: 334-337

Yanagimachi R (1988) Sperm-egg fusion. In: Membrane Fusion in Fertilization, Cellular Transport and Viral Infection (Düzgünes N, Bronner F, eds), pp 3-43, Academic Press San Diego

Zaks WJ, Creutz CE (1988) membrane fusion in model systems for exocytosis: Characterization of chromaffin granule fusion mediated by synexin and calelectrin. In: Molecular Mechanisms of Membrane Fusion (Ohki S, Doyle D, Flanagan T, Hui SW, Mayhew E, eds), pp 325-340, Plenum Press New York

Zimmerberg J, Curran M, Cohen FS, Brodwick M (1987) Simultaneous electrical and optical measurements show that membrane fusion precedes secretory granule swelling during exocytosis of beige mouse mast cells. Proc Natl Acad Sci USA 84: 1585-1589

Zimmerberg J, Vogel SS, Chernomordik LV (1993) Mechanisms of membrane fusion. Annu Rev Biophys Biomol Struct 22: 433-466

ENTRY OF ENVELOPED VIRUSES INTO HOST CELLS: FUSION ACTIVITY OF THE INFLUENZA VIRUS HEMAGGLUTININ

Maria C. Pedroso de Lima, João Ramalho-Santos[1], Nejat Düzgünes [2], Diana Flasher [2], and Shlomo Nir [3]

Department of Biochemistry
University of Coimbra
3049 Coimbra Codex
Portugal

INTRODUCTION

Considerable progress has been achieved in understanding the mechanism of entry of enveloped viruses into host cells during the last few years. Viruses have long been known to be capabable of transporting their genome and accessory proteins into the cytosol of the host cell, thus causing infections of variable severity. It is clear that the medical consequences of viral diseases remain an important reason to study the cellular entry of viruses, particularly since a knowledge of the molecular details underlying this event may help to develop new antiviral strategies. Apart from their importance as pathogens, viruses, particularly enveloped viruses, have proven to be an excellent tool for studying protein synthesis, processing and sorted transport, thus greatly contributing to our current knowledge of the complex pathways of intracellular traffic. To infect a host cell, enveloped viruses have evolved a mechanism which involves membrane fusion. Although the molecular details of this process remain unclear, it has been well established that specific virally encoded membrane proteins are responsible for triggering the fusion reaction. Because of the simplicity of the viral membrane protein composition of most enveloped viruses, viral systems provide a valuable tool for studying protein-mediated membrane fusion. Elucidation of the molecular mechanisms underlying virus cell fusion may lead to a better understanding of the ubiquitous intracellular fusion events where specific proteins are also involved.

---

[1] Department of Zoology, University of Coimbra, 3049 Coimbra Codex, Portugal.

[2] Department of Microbiology, University of the Pacific, School of Dentistry, San Francisco, CA 94115, U.S.A.

[3] The Seagram Center for Soil and Water Sciences, Faculty of Agriculture, Hebrew University of Jerusalem, Rehovot 76100, Israel.

NATO ASI Series, Vol. H 91
Trafficking of Intracellular Membranes
Edited by M.C. Pedroso de Lima N. Düzgüneş and D. Hoekstra

The two general entry pathways of enveloped viruses into cells are illustrated in Figure 1. Infectious cellular entry is initiated by the binding of the virus to a cell surface receptor, followed by the fusion of the viral envelope with a host cell membrane (Hoekstra and Kok, 1989; Marsh and Helenius, 1989; Hoekstra and Pedroso de Lima, 1992). Depending on the virus family, fusion can take place either directly with the cell plasma membrane at neutral pH (e.g., Sendai virus and human immunodeficiency virus) or with the endosome membrane following internalization of intact virions through receptor mediated endocytosis (e.g. influenza virus, Semliki Forest virus and vesicular stomatitis virus). In the latter endocytic route of entry, fusion is triggered by the low pH generated in the endosome lumen by a membrane-associated ATP-driven proton pump. Fusion of low pH-dependent viruses can take place either in the early endosomes or in the late endosomes, depending on the pH at which the virus acquires its fusogenic competence. The pH-dependence of fusion is a property of the viral fusion protein, being related to the pH-dependence of changes in its conformation (White, 1990). Acid-triggered viruses can be induced to fuse with the cell plasma membrane by lowering the pH of the medium following binding. While this process does not represent the physiological pathway of entry, it may lead to cell infection.

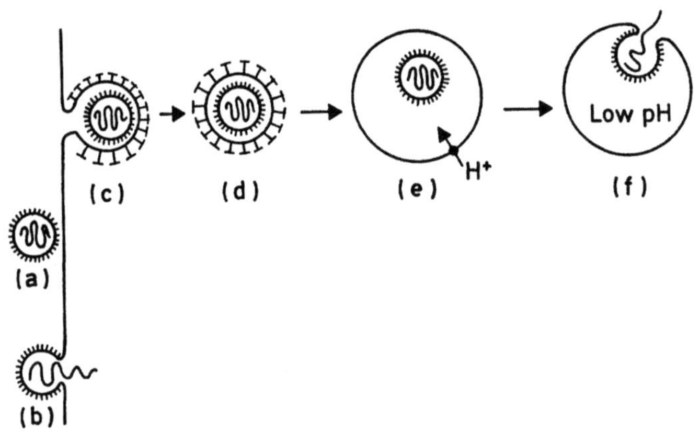

FIGURE 1. Pathways of entry of enveloped viruses. After attachment to the cell surface (**a**), the viral membrane fuses with the plasma membrane at neutral pH, and the nucleocapsid is released into the cytoplasm (**b**). Alternatively, virus particles are endocytosed through coated pits (**c**) and coated vesicles (**d**) into endosomes (**e**). After acidification of the latter compartment fusion occurs (**f**) with the limiting membrane from within, the nucleocapsid being released into the cytoplasm. (Reproduced with permission from Hoekstra and Pedroso de Lima, 1992).

INFLUENZA VIRUS

It has been demonstrated conclusively that influenza virus expresses its fusion activity only at acidic pH, thus utilizing the endocytic pathway to infect cells (White, 1990; Stegmann et al, 1989; Doms et al, 1990). Influenza virus is a negative-strand RNA virus belonging to the orthomyxovirus family. Figure 2 represents a schematic diagram of the structure of an influenza A virus particle shown in cross section. The virions are roughly spherical, ranging in diameter from 80 to 120 nm. The viral envelope consists of three integral membrane proteins: Hemagglutinin (HA), neuraminidase (NA) and M_2. HA is a homotrimeric molecule and the major glycoprotein of the envelope. HA recognizes cell surface receptors and contains the fusion potential that allows the virus to penetrate the cytoplasm of the host cell following endocytosis (see below). NA is a homotetrameric protein that has the capacity to partially remove the terminal sialic acid (N-acetyl neuraminic acid) residues present on membrane glycoproteins and/or glycolipids. Both HA and NA are glycoproteins that protrude from the viral membrane like spikes. M_2 is a homotetrameric protein which is present in relatively low amounts in the virus particle, but is extremely abundant in infected cells. It has been recently shown that M_2 forms a transmembrane channel capable of translocating protons and other monovalent cations across the membrane (Sugrue and Hay, 1991).

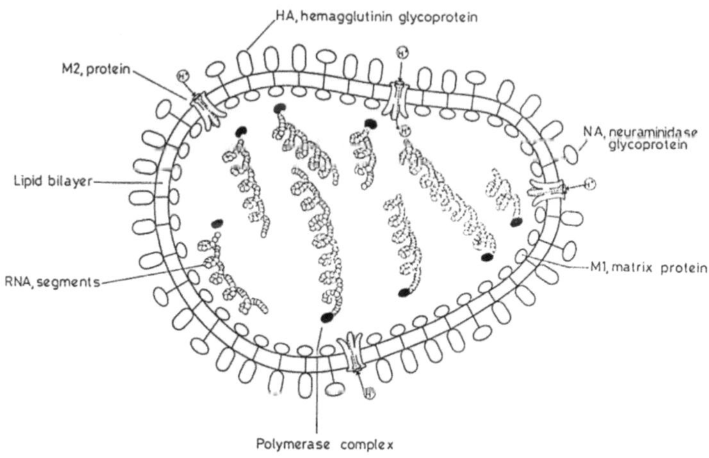

FIGURE 2. Schematic representation of the structure of an influenza A virus particle. (Adapted from Watson et al., 1987).

The virus genome is divided into eight segments of distinct sizes termed 1 to 8. Segments 1 through 6 each encode one gene product, and segments 7 and 8 comprise two genes encoding separate proteins. The eight segments are individually packaged into helical ribonucleoproteins

(RNPs). Each segment is associated with its own transcriptase enzyme which contains copies of three polymerase subunits (P_1, P_2 and PA). Surrounding the ribonucleoprotein core is the matrix protein, M_1, which is the most abundant viral protein. M_1 and RNPs assemble together to form the capsid structure (Lamb and Choppin, 1983; Lamb, 1989). M_1 also interacts with the envelope proteins, and therefore it is believed to serve as an adaptor between the RNPs and the viral envelope. Both replication of influenza virus and assembly of the RNPs take place in the nucleus. Following fusion, the viral RNPs and M_1 proteins dissociate from each other and the RNPs are then transported into the nucleus through the nuclear pores by an active mechanism (Martin and Helenius, 1991a; 1991b). Transcription and viral RNA synthesis occur in the nucleus (Lamb and Choppin, 1983). The ribonucleoprotein core "steals" short pieces of cellular mRNA (10-15 nucleotides) which are then used as primers for the synthesis of virus mRNAs. Transcription of RNA is accomplished by the action of the polymerase complex associated to the viral genome. The synthesized viral mRNA molecules move to the cytoplasmic ribosomes where they are translated. The viral envelope proteins, HA, NA and M_2 pass to the endoplasmic reticulum, Golgi apparatus and finally they are transported to the plasma membrane where they are inserted. The viral structural proteins, NP, M_1, P_1, P_2 and PA are all returned to the nucleus after their synthesis. The newly formed RNPs combine with the M_1 protein which thus covers the ribonucleoprotein core. The core migrates from the nucleus to the cytosol and then to the plasma membrane, where it combines with the viral envelope proteins which have already been inserted and the virion buds out of the cell. During the budding process, the host plasma membrane proteins are totally excluded from the viral membrane. Therefore, the viral envelope of the mature virions contains only the spike proteins and no host cell proteins. It is interesting to note that the lipid composition and transbilayer lipid distribution of the viral envelope, although similar, are not identical to those of the host cell plasma membrane (Lenard and Compans, 1974; Tsai and Lenard, 1975).

THE INFLUENZA VIRUS HEMAGGLUTININ

The infectious entry of influenza virus is mediated by HA. As mentioned before, HA contains both the capacity to bind to sialic acid-containing receptors on the cell surface and to catalyse the fusion reaction with the endosomal membrane (Wiley and Skehel 1987). It has been firmly demonstrated that the viral fusion activity resides exclusively on HA (White et al., 1982). HA has been the most extensively characterized viral fusion protein and it offers the best model for studying the protein-mediated membrane fusion. Bromelain treatment of viral HA yields a water soluble trimeric ectodomain termed BHA. This cleaved BHA fragment, which lacks the hydrophobic C-terminal membrane anchors, has been crystallized and its three dimensional structure at neutral pH has been determined from x-ray diffraction studies to a resolution of 3 Å (Wilson et al., 1981). Later it was found that BHA undergoes conformational changes at the pH optimum for virus fusion, which are biochemically and immunologically equivalent to those of the native HA (Skehel et al., 1982; Doms et al., 1985). Each monomer of the HA trimer is initially synthesized as a fusion-inactive precursor, denoted HA_0, consisting of a single polypeptide chain. A proteolytic processing by a host cell protease is required for viral infectivity and membrane fusion activity (Klenk et al., 1975; Lazarowitz and Choppin, 1975). This activation mechanism appears to occur later in the biosynthetic pathway prior to insertion of the protein into the host plasma membrane.

As a mature protein, HA consists of two polypeptides, HA_1 and HA_2, linked by a disulfide bridge (Figure 3). On the viral envelope the two chain monomers are associated noncovalently to form trimers. The trimeric form consists of a fibrous stem region protruding from the viral membrane and holding a globular head region. In the globular domain (head region) of the trimer, each HA_1 subunit, being entirely outside of the viral membrane, contains the sialic acid binding site and the major surface antigen against which neutralizing antibodies are produced. HA_1 is therefore involved in the binding of the virus to the host cell surface, while HA_2 constitutes the transmembrane subunit and is responsible for the actual fusion process (Wiley and Skehel, 1987). Each HA_2 subunit contains the so-called fusion peptide, an N-terminal conserved sequence of about 20 highly hydrophobic amino acids, which appears crucial for the expression of fusion activity of HA (Gething et al., 1986; Lear and DeGrado, 1987). In the native conformation (neutral pH) of HA the fusion peptides are tightly tucked by a network of hydrogen bonds at the trimer interface about 3.5 nm from the bottom and 10 nm from the tip of the trimer.

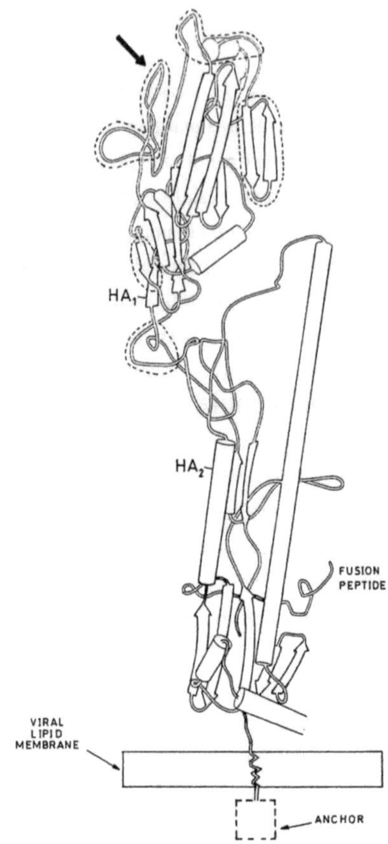

FIGURE 3. Schematic representation of the structure of the HA monomer at neutral pH. The HA_1 subunit contains the sialic acid binding site for HA indicated by the arrow as well as the major antigenic epitopes demarked by the dashed lines. The HA_2 subunit contains the fusion peptide which is exposed at acidic pH. (Adapted from Wilson et al., 1981).

ACID INDUCED CONFORMATIONAL CHANGES IN HEMAGGLUTININ

It is well demonstrated that at acidic pH, such as in an endosome, HA undergoes a major conformational change, leading to exposure of regions of the stalk of the trimer, including the previously buried N-terminus of the HA_2 subunit (Skehel et al., 1982; Doms et al., 1985; Wiley and Skehel, 1987). The pH at which the conformational change in HA occurs coincides with the pH of influenza virus fusion, thus providing evidence that the low pH-triggered conformational change resulting in exposure of the fusion peptides is required for membrane fusion (Skehel et al., 1982; Daniels et al., 1983; Wiley and Skehel, 1987; White, 1992). Extensive studies have been performed on the conformational changes of HA at low pH by using a variety of techniques, including enzyme susceptibility, circular dichroism, electron microscopy and antibody reactivity (Skehel et al., 1982; Ruigrok et al., 1986; Wharton et al., 1986; White and Wilson, 1987). A cartoon for the low pH-triggered conformational change of HA that has been inspired by these studies is shown in Figure 4. Although the general features of these conformational changes are relatively well understood, their interpretation has been based primarily on the behavior of BHA. While BHA has proved to be a useful model for studying the fusion-inducing conformational change of HA, fusion is not induced by BHA itself (Doms and Helenius, 1988).

The studies of White and Wilson (1987) involving the use of a panel of anti-peptide antibodies suggested that the low pH-triggered conformational change of isolated BHA occurs in two major stages: (i) the fusion peptides are released from within the trimer stem; (ii) this is followed by the dissociation of the globular head domains. While the crucial role of the fusion peptides and hence their release in the fusion reaction has been unequivocally demonstrated (Gething et al., 1986; Harter et al., 1989), it has remained unclear whether the unfolding of the HA trimer is required for fusion to occur, or, conversely, whether it leads to irreversible inactivation of the fusion activity of HA.

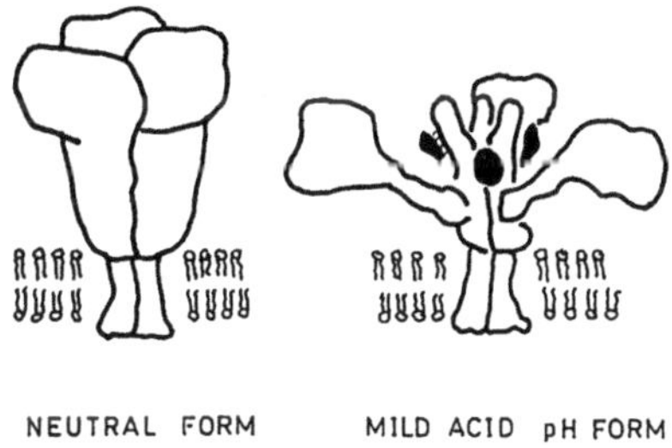

FIGURE 4. Schematic representation of the conformational change in HA. At mild-acidic pH, the fusion peptides, colored black, become exposed. (Adapted from Doms and Helenius, 1988).

It was first proposed by White et al. (1986) that exposure of the fusion peptides, although necessary, was not sufficient for fusion, and that the unfolding of the globular head regions of HA was required to enable the interaction of the fusion peptides with both the viral and the target membrane, thus bringing the two interacting membranes into intimate contact and facilitating fusion. However, later studies (Stegmann et al., 1990) have shown that HA can mediate fusion at low temperature without unfolding of the trimers. These authors have proposed that such unfolding is not only unnecessary for fusion, but it may render HA incapable of mediating fusion. It is well known that influenza virus loses most of its fusion capacity when exposed to low pH and 37°C in the absence of the target membrane (Sato et al., 1983; Stegmann et al., 1986). This loss of activity is thought to be due to a clustering and/or deformation of the HA spikes as a result of hydrophobic interactions between the fusion peptides, within and between the HA trimers, that became exposed upon acidification (Junankar and Cherry, 1986). The proposal of Stegmann et al. (1990) was based on their observation that slow, but extensive and sustained fusion of influenza virus with ganglioside-containing liposomes and erythrocyte ghosts was observed at 4°C. Yet no detectable changes were observed at the top of the trimer under these low temperature conditions, as evidenced by using specific monoclonal antibodies (Stegmann et al., 1990). In contrast, at 37°C influenza virus was rapidly inactivated during the low pH treatment, under which conditions dissociation of the top domains of HA was demonstrated to occur. The proposed model for HA-mediated fusion (Stegmann et al., 1990) involves the tilting of the entire HA trimer structure with respect to the plane of the viral membrane to allow the fusion peptides to penetrate into the target membrane without the opening of the top of the trimer. Other alternative models have been proposed for the mechanism of HA-mediated membrane fusion, which, although sharing some of the features of the above model, do not involve the energetically unfavourable bending of the intact HA trimer (White, 1990; Ellens et al., 1990; Bentz et al., 1990). However, which of these models, if any, may represent the actual mechanism for HA-mediated membrane fusion remains to be elucidated.

FUSION ACTIVITY OF INFLUENZA VIRUS

Experimental procedure:

In order to gain insight into the mechanism of fusion of influenza virus with cellular membranes, we have investigated the kinetics of low pH-induced fusion of influenza virus with the plasma membrane of several types of living cells as a model to study the fusion of the virus with the endosome membrane. Fusion was monitored with a fluorescence membrane lipid mixing assay based on the self-quenching properties of octadecylrhodamine B (R18) (Hoekstra et al., 1984). Influenza virus (A/PR/8/34) was labeled with R18 at a self-quenching concentration and fusion was monitored as the increase in R18 fluorescence due to dequenching of the probe as it dilutes into the target membrane.

Figure 5 shows an example of the time course of R18 fluorescence dequenching observed during incubation of influenza virus with human lymphocytic leukemia (CEM) cells (Düzgünes et al., 1992). Influenza virus was incubated with the cells for 10 minutes at neutral pH to allow virus binding to the cell surface, and fusion with the cellular membrane was induced by lowering the pH of the medium to 5. At neutral pH, there is no detectable increase in R18 fluorescence. However at pH 5, a rapid and extensive fusion is observed revealed as the increase in R18 fluorescence. Control experiments have shown that R18 dequenching is specific for the fusion reaction, thus ruling out nonspecific transfer of the probe between virus particles and the target membrane (Düzgünes et al., 1992).

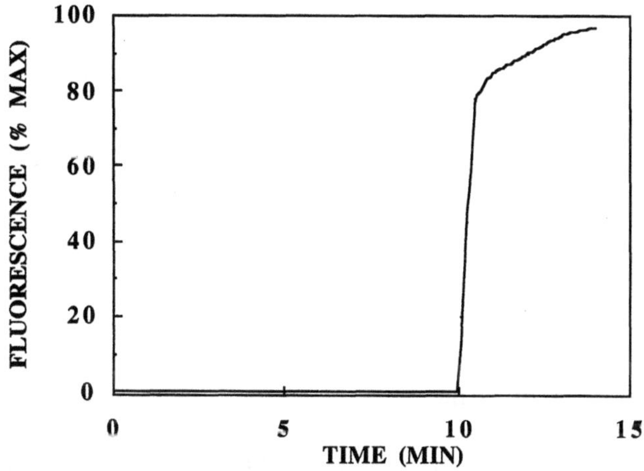

FIGURE 5. Fusion of influenza virus with CEM cells as monitored by fluorescence dequenching of R18. The time course of R18 fluorescence during incubation of 5 μg viral protein/ml influenza virus (A/PR/8/34 strain) with 2 x 10^7/ml CEM cells. The virus was incubated with the cells at pH 7.6 and 37°C for 10 min, and the pH was then lowered to 5. (Reproduced with permission from Düzgünes et al., 1992).

Role of cell surface sialic acid

As mentioned before, the initial event in the entry of a virus into host cells is attachment to specific receptors. Sialic acid residues present on membrane glycoproteins or glycolipids at the cell surface have been considered to be the primary receptors for influenza virus (Bergelson et al., 1982; Paulson et al., 1986; Suzuki et al., 1985). It may be speculated that sialic acid residues not only act as points of attachment for the virus, but also modulate the membrane fusion reaction.

In order to address the role of sialic acid residues in the virus-cell membrane fusion step, the effect of the removal of cell membrane sialic acid on the fusion activity of influenza virus (A/PR/8/34 strain) toward CEM cells was investigated (Pedroso de Lima et al., unpublished results). Figure 6 shows that viral fusion activity toward cells that have been pretreated with neuraminidase is significantly reduced as compared to untreated (control) CEM cells. As it would be expected, when the virions are prebound to control cells (Figure 6A) the kinetics and extent of fusion are higher compared to the case of no prebinding (Figure 6B), since the bound virus can start to fuse with the cells immediately following acidification. In contrast, for neuraminidase-treated cells, a slight decrease in both the rate and extent of fusion is observed with prebinding, suggesting that the presence of sialic acid residues is essential to confer biologically relevant viral fusion activity. In addition, virus prebound to neuraminidase-treated cells may inactivate faster upon acidification than virus bound to control cells (see below).

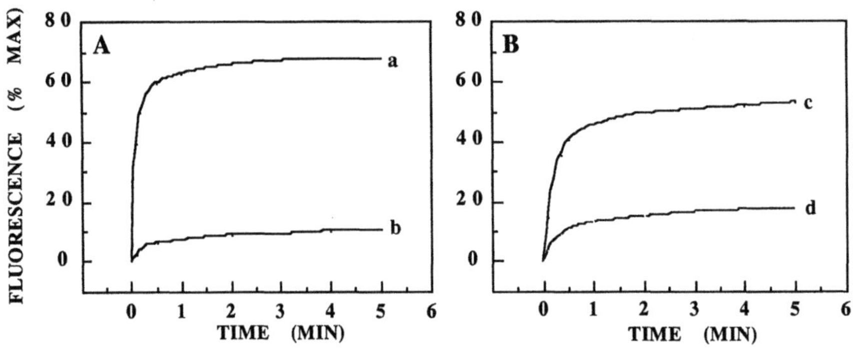

FIGURE 6. Effect of neuraminidase pretreatment of CEM cells on the fusion activity of influenza virus, with or without viral prebinding to the cells. Influenza virus (2 μg/ml) was added to 4×10^7 control (**a**, **c**) or neuraminidase-treated (**b**, **d**) CEM cells in a final volume of 2 ml and the extent of R18 fluorescence dequenching was measured after 5 min incubation at pH 5.0, with (**A**) or without (**B**) viral prebinding to the cells. In the case of viral prebinding (**A**) the virus was incubated with cells 5 min at neutral pH before acidification to pH 5.0. In the absence of viral prebinding (**B**) the pH of the cell suspension was adjusted to 5.0 just prior to the addition of virus.

The marked inhibition observed for fusion of influenza virus with neuraminidase-treated cells can be attributed to a decrease in virus binding due to the unavailability of viral receptors. Indeed, neuraminidase-pretreatment of CEM cells results in a considerable inhibition of virus-cell binding at neutral pH (Table 1), stressing the importance of sialic acid residues as primary virus receptors. Although virus binding to control cells does not vary significantly with the preincubation conditions (Table 1), it is enhanced when the virus is incubated with neuraminidase-treated cells at low temperatures, compared to that at 37°C. These results indicate that the degree of reversibility of virus binding at 37°C is more pronounced in neuraminidase-treated cells than in control cells, since the rate constant of dissociation is generally very small at low temperatures (Nir et al., 1983).

TABLE 1 Effect of neuraminidase pretreatment of CEM cells on influenza virus-cell binding at neutral pH [a]

Conditions	Cell associated virus (%)	
	Control	Neuraminidase
37°C	61.9 ± 5.3	18.4 ± 2.4
0°C	69.2 ± 2.5	52.3 ± 2.1

[a] Influenza virus (2 μg/ml) was added to $4x10^7$ control or neuraminidase-treated CEM cells in a final volume of 2 ml at pH 7.4 and incubated at 37°C (5 min) or at 0°C (30 min). The cells were sedimented to separate bound and unbound virus and binding was quantified from fluorescence values in the supernatant and pellet, following the addition of detergent. Values represent means of at least 5 experiments ± standard deviation.

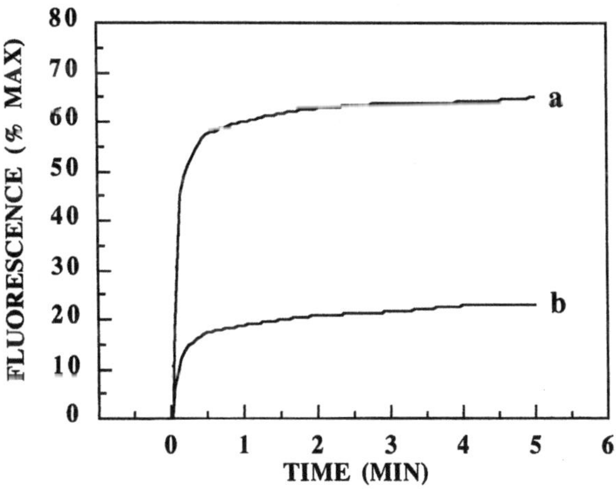

FIGURE 7. Effect of neuraminidase pretreatment of CEM cells on prebound influenza virus fusion activity. Influenza virus (2 μg/ml) was added to $4x10^7$ control (**a**) or neuraminidase-treated (**b**) CEM cells in a final volume of 2 ml. Following an initial preincubation of virus and target cells at pH 7.4 for 30 min at 0°C, unbound virus was removed by centrifugation and R18 dequenching was monitored at 37°C and pH 5.0.

We have also studied the fusion activity of influenza virus that has been bound to the plasma membrane of neuraminidase-treated and untreated cells (Figure 7). Although the receptor binding step is bypassed, the fusion of virions prebound to neuraminidase-treated cells is inhibited compared to fusion with control cells. The analysis of these data with the mass action kinetic model (see next chapter by Nir et al.) shows that removal of sialic acid residues from the cell surface leads to a decrease in the fusion rate constant and an increase in the inactivation rate constant. Therefore, influenza virus is only able to fuse efficiently without becoming significantly inactivated, when it is bound to sialic acid-containing receptors. Although not being able to fuse efficiently with neuraminidase-treated cells, it is interesting to note that influenza virus can associate extensively with neuraminidase-treated cells at low pH (Table 2). However, this enhanced association is most likely nonspecific, since it does not lead to efficient fusion. Furthermore, in contrast to control cells, where the time dependence of cell association closely parallels that of virus fusion, in the case of neuraminidase-treated cells a significant increase with time in virus-cell association at low pH is observed. This increase is not paralleled by an increase in viral fusion activity. These results suggest that influenza virus binds non-specifically to neuraminidase-treated cells at low pH, probably due to an increase in virus surface hydrophobicity (Ramalho-Santos et al., 1994).

TABLE 2 Effect of neuraminidase pretreatment of CEM cells on influenza virus-cell association at low pH [a]

Conditions	Cell associated virus (%)	
	Control	Neuraminidase
1 min	70.9 ± 2.2	41.2 ± 1.0
5 min	81.8 ± 2.1	61.7 ± 4.6

[a] Influenza virus (2 μg/ml) was added to $4x10^7$ control or neuraminidase-treated CEM cells in a final volume of 2 ml at pH 7.4 and at 37°C. After 5 min of preincubation the pH was lowered to 5.0 and the suspension was further incubated for 1 or 5 min. The cells were sedimented to separate virus not associated with the target membrane, and cell association was quantified from fluorescence values in the supernatant and pellet, following the addition of detergent. Values represent means of at least 5 experiments ± standard deviation.

In support of this interpretation, Figure 8 shows that the fluorescence of the virus associated with neuraminidase-treated cells at low pH is greatly quenched, whereas that of the virus associated with control cells under the same low pH conditions is essentially dequenched. This observation indicates that the virus is merely associated with cellular membranes from which sialic acid has been removed, but does no fuse with them.

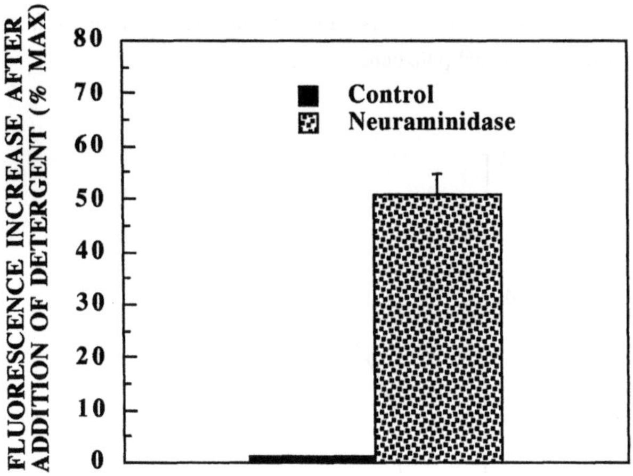

FIGURE 8. Effect of neuraminidase pretreatment of CEM cells on the fluorescence quenching of virus particles associated with target cells at low pH. Influenza virus (2 µg/ml) was added to $4x10^7$ control (dark bar) or neuraminidase-treated (checkered bar) CEM cells in a final volume of 2 ml at pH 7.4 and at 37°C. After 5 min of preincubation the pH was lowered to 5.0 and the suspension further incubated for 5 min. The cells were sedimented to separate virus not associated with the target membrane and R18 fluorescence in the pellet was registered before and after the addition of detergent. Values represent means of at least 5 experiments ± standard deviation.

In summary, these results demonstrate that in addition to mediating virus-cell binding, sialic acid-containing receptors are required for efficient fusion. By inhibiting low pH-induced viral inactivation, sialic acid residues on the cell surface may be involved in the formation of an active fusion complex.

The pH-dependence of fusion

Figure 9 illustrates the effect of pH on the fusion of influenza virus with HL-60 human promyelocytic leukemia cells. Like CEM cells, HL-60 cells do not internalize the virus to a significant extent during the relatively short times of pre-incubation before the reduction of the pH of the medium (Düzgünes et al., 1992). Following an initial virus-cell binding at 37°C, fusion with the plasma membrane was induced by lowering the pH to different values. The initial rate of fusion displays a maximum in the pH range 4.7-5.0, whereas the extent of fusion after a 5 minute incubation at pH 5 is maximal in the range 4.9-5.2. An interesting observation is that while the extent of fusion is slightly larger at pH 4.74 than at pH 5.01 up to 1 minute after lowering the pH, it is much lower at pH 4.74 than at pH 5.01 after a 5 minute incubation. This difference is explained by the mass action kinetic model which invoked a small increase in the fusion rate constant and in the inactivation rate constant at pH 4.74, thus showing that the virus fuses faster, but also inactivates faster upon lowering the pH (Düzgünes et al., 1992).

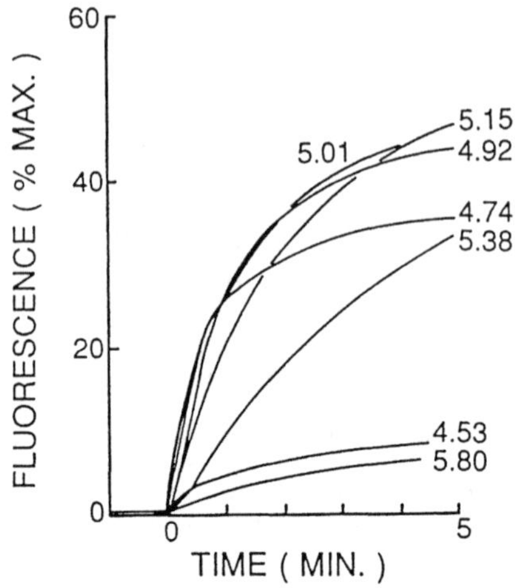

FIGURE 9. The effect of pH on the kinetics of fusion of influenza virus with HL-60 cells. 2 μg of influenza virus (A/PR/8/34 strain) were added to 2 x 10^7/ml HL-60 cells. After a preincubation of the virus with the cells at 37°C for 10 min, the pH was lowered to the indicated values. (Reproduced with permission from Düzgünes et al., 1992).

However, the pattern described above was more extreme in the case of influenza virus fusing with the plasma membrane of PC-12 cells (Ramalho-Santos et al., 1993). Figure 10 shows the effect of pH on influenza virus fusion activity at 37°C toward PC-12 cells, either with (A) or without (B) virus-cell prebinding at neutral pH. As it would be expected, the initial kinetics and extent of fusion are larger with virus-cell prebinding, since the virus can start fusing immediately upon lowering the pH. The short lag phase observed in the absence of virus-cell prebinding reflects the time required for the virus to adhere to the cell surface before starting to fuse. But, most interesting is the observation that whereas influenza virus exhibits its maximal fusion acitivity at pH 5.2 without virus-cell prebinding, in the presence of virus-cell prebinding, acidification of the medium below pH 5.2 does not lead to any decrease in fusion activity (Ramalho-Santos et al., 1993). Indeed, in this latter case, the extent of fusion following a 1 minute incubation at pH 4.5 is higher than at pH 5.2, while similar values for the extent of fusion are observed after 5 minutes at both pH values.

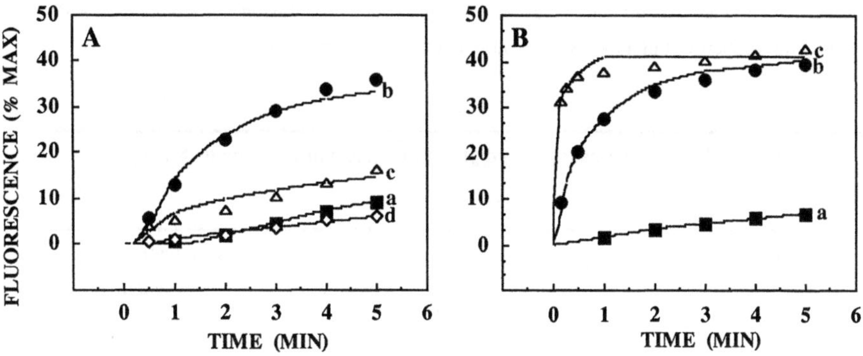

FIGURE 10- pH effect on influenza virus fusion activity towards PC-12 cells in the absence and presence of viral prebinding. **A-** Influenza virus (1 μg/ml viral protein) was added to 5 x 10^6 PC-12 cells in a final volume of 2 ml and R18 dequenching was monitored for 5 min at 37°C. The pH of the cell suspension was adjusted previously to pH 5.8 (**a**), 5.2 (**b**), 4.5 (**c**) and 4.0 (**d**). The values calculated with the mass action kinetic model are also presented for pH 5.8 (■), 5.2 (●), 4.5 (△) and 4.0 (◇).
B- Influenza virus (1 μg/ml viral protein) was added to 5 x 10^6 PC-12 cells in a final volume of 2 ml at 37°C and pH 7.4. After 5 min the pH was lowered to pH 5.8 (**a**), 5.2 (**b**) and 4.5 (**c**), and R18 dequenching was monitored for 5 min. When the pH was adjusted to 4.0 the dequenching was only slightly quicker and more extensive than the one obtained at pH 4.5 (not shown, see Table 3).The values calculated with the mass action kinetic model are also presented for pH 5.8 (■), 5.2 (●) and 4.5 (△). (Reproduced with permission from Ramalho-Santos et al., 1993).

These results are explained by the kinetic analysis by an increase in both the fusion and inactivation rate constants upon lowering the pH. Table 3 summarizes the results obtained from the analysis for the effect of temperature and pH on fusion and inactivation in the presence of virus-cell prebinding (Ramalho-Santos et al., 1993), and shows that both the fusion rate constant and inactivation rate constant increase significantly with decreasing pH and with increasing temperature.

When the virus is not prebound, a large fraction of virus becomes inactivated upon exposure to low pH by the time the virus is able to establish a direct contact with the cell surface sialic acids, resulting in an apparent optimal fusion activity at pH 5.2. With virus-cell prebinding, inactivation of prebound virus also occurs upon acidification, although at a slower rate than that of virus alone (Düzgünes et al., 1992). Since the virus fuses faster at lower pH values, larger values for the extent of fusion are observed at the early stages of fusion at pH 4.5 than at pH 5.2. However, since the virus also inactivates faster at pH 4.5 than at pH 5.2, similar values for the extent of fusion are registered after a 5 minute incubation. These results thus indicate that the dependence of the overall fusion rate on the pH may be obscured by the low pH viral inactivation. The comparison of these results with those obtained with the pH-dependence of fusion activity of influenza virus toward HL-60 cells (Figure 9) suggests that viral fusion activity does not depend solely on viral properties but appears to be modulated by the target membrane.

TABLE 3 Effect of temperature and pH on fusion and inactivation with virus-cell prebinding[a]

Conditions	pH	Fusion rate constant f (s^{-1})	Inactivation rate constant γ (s^{-1})
Effect of temperature			
37°C	5.2	0.03	0.016
20°C	5.2	0.0015	0
4°C	5.2	2.7×10^{-5}	0
Effect of pH at 37°C	4	0.37	0.14
	4.5	0.3	0.1
	5.2	0.03	0.016
	5.8	8×10^{-4}	0.002

[a] The estimate uncertainties in the parameters are: f (25%), γ (20%), unless a range is indicated. (Adapted from Ramalho-Santos et al., 1993).

Low pH viral inactivation

Figure 11 demonstrates that incubation of the virus alone at low pH (i.e. in the absence of cells) at 37°C results in a very rapid and extensive inactivation of influenza virus, irrespective of whether the low pH-treated virus had (B) or had not been (A) prebound to the PC-12 cells at neutral pH (Ramalho-Santos et al., 1993). Indeed, a 1 minute exposure of influenza virus to low pH and 37°C in the absence of PC-12 cells is sufficient to lead to an almost complete inhibition of the viral fusion activity assayed at pH 5.2 and 37°C (Figure 11A and B, curve b). Inactivation of influenza virus is temperature-dependent, being greatly inhibited as the temperature of low pH preincubation is decreased. Indeed, in contrast to the observations at 37°C, when the virus was incubated for 1 minute at 20°C, no significant viral inactivation was detected. To observe an inactivation at 20°C similar to that observed at 37°C, the preincubation period at pH 5.0 had to be extended for 5 minutes (Figure 11A, curve c). At 0°C, influenza virus had to be preincubated for at least 1 hour at pH 5 to allow a detectable virus inactivation (see below and Table 4). In marked contrast to the extensive and rapid viral inactivation at 37°C toward PC-12 cells, influenza virus is only partially inactivated in its fusion activity toward HL-60 or CEM cells (Düzgünes et al., 1992). This suggests the existence of different mechanisms of inactivation depending on the cellular plasma membrane. It is possible that fusion activity of influenza virus may be restored upon interaction of the viral glycoproteins with certain components at the plasma membrane of HL-60 cells or CEM cells but not in PC-12 cells. These components may partially dissociate the preformed clusters of the envelope glycoproteins at acidic pH, and they could well be sialic acid residues which, as shown above, inhibit low pH viral inactivation.

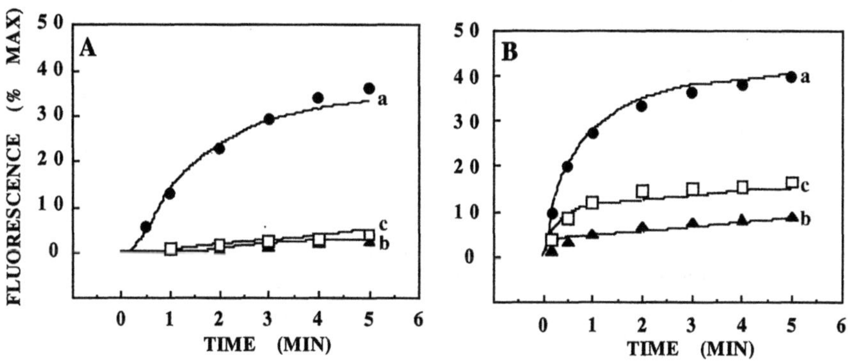

FIGURE 11- Inactivation of influenza virus assessed in the absence and presence of viral prebinding. The virus was incubated at pH 5.0 in the absence of the target membrane for various times and at different temperatures. Fusion activity of the preincubated virions was monitored either (**A**) for 5 min at pH 5.2 and 37°C, following addition of 1 μg/ml viral protein to 5 x 10^6 PC-12 cells (final volume 2 ml), or (**B**) by adding 1 μg/ml viral protein to 5 x 10^6 PC-12 cells (final volume 2 ml) at pH 7.4 and 37°C. After 5 min the pH was lowered to 5.2 and R18 dequenching followed for 5 min.
a- Control (no preincubation).
b- Virus preincubated for 1 min at 37°C.
c- Virus preincubated for 5 min at 20°C.
The values calculated with the mass action kinetic model are also presented for **a** (●), **b** (▲) and **c** (□). For both **A** and **B** viral preincubation for 1 min at 20°C or for 30 min on ice did not result in any change in fusion activity. For **B** viral preincubation for 30 min at 20°C resulted in similar activity as that registered for 1 min preincubation at 37°C (not shown). (Reproduced with permission from Ramalho-Santos et al., 1993).

As mentioned before, one of the focal points for the proposal of a revised model for the mechanism of HA-mediated membrane fusion (Stegmann et al., 1990) was the observation that low pH preincubation of influenza virus alone at 0°C, does not result in inactivation of the fusion capacity of the virus. Based on the results of our analysis on the effect of pH on the rate constants of fusion and inactivation, we designed critical experiments to demonstrate whether the lack of inactivation in the cold was indeed real or apparent (Ramalho-Santos et al., 1993).

The experiments consisted of extending the time of preincubation of the virus alone at 0°C to 1 hour at pH 4.0 instead of pH 5.0, since the kinetic analysis indicates that the value of the rate constant of inactivation is larger at 4.0 than at pH 5.0. Following binding of such pretreated virus to PC-12 cells at neutral pH for 10 minutes, the extent of fusion was measured at 4°C after 1 hour incubation at pH 4.0, where the fusion rate constant is also larger than at pH 5.0. Table 4 summarizes the results obtained in these experiments. Under these experimental conditions, a significant decrease in the extent of fusion as compared to the control is observed. Even at pH 5.2, viral inactivation is noticeable at 4°C. At 10°C and pH 5.2, a 30 minute incubation of the virus alone was sufficient to lead to significant viral inactivation. Table 4 also includes the values obtained for the percentage of virions bound to cells at neutral pH, and the total amount of virus associated with the cells after the initial binding at neutral pH followed by a subsequent reduction of the pH.

TABLE 4 Fusion and inactivation of influenza virus at low temperatures

pH	Experimental conditions[a]	Fusion[b] (% max)	Binding[c] (%)	Cell Association[d] (%)
5.2	Control (0°C)	5.2	53.9	68.0
	Inactivated (0°C)[d]	3.3	51.6	50.9
4.0	Control (0°C)	20.3	53.9	77.5
	Inactivated (0°C)[e]	13.5	51.6	77.7
5.2	Control (0°C)	7.9	49.7	50.9
	Inactivated (0°C)[f]	1.1	48.2	59.4

[a] The virus was bound to PC-12 cells for 10 min at pH 7.4, and the pH was lowered to the indicated value. [b] The extent of fusion 60 min after the reduction of pH. [c] Percent of added virus bound to cells after 10 min at pH 7.4. [d] Percent of added virus bound and fused after 10 min at pH 7.4 and 60 min at the indicated pH. [e] The virus was "inactivated" for 1 h at 0°C at the indicated pH. [f] The virus was "inactivated" for 30 min at pH 5.2 and 37°C.
(Reproduced with permission from Ramalho-Santos et al., 1993).

In contrast to the values for the extents of fusion, virus binding and cell association are relatively insensitive to pH. Table 5 lists the values of f and γ obtained from the kinetic analysis for the cases presented in Table 4. The values of f and γ at 37°C and 20°C are also included. As shown in Table 5, both f and γ values decrease with decreasing temperature, but it is clear that a low pH preincubation of influenza virus alone at 0°C does result in inactivation of the fusion capacity of the virus.

TABLE 5 Fusion and inactivation rate constants for influenza virus preincubated without cells at low pH [a]

Temperature (°C)	pH	f (s^{-1})	γ (s^{-1})
37	5.0	0.03	0.046
20	5.0	1.5×10^{-3}	6×10^{-3}
4/0[a]	5.0	2.4×10^{-5}	10^{-4}
4/0[a]	4.0	9.7×10^{-5}	1.5×10^{-4}

[a] Experimental conditions as in Table 4. The parameters determined were based on experimental values given in Table 4.
(Reproduced with permission from Ramalho-Santos et al., 1993).

Role of globular head domain dissociation of hemagglutinin

While the importance of fusion peptide exposure in the HA-mediated membrane fusion has been clearly demonstrated (Gething et al., 1986; Harter et al., 1989), no direct evidence exists for the involvement of the globular head regions of HA in the fusion process. The results of Stegmann et al. (1990) indicate that fusion of influenza virus can take place in the cold without unfolding of the HA trimers. While it is not known why these authors have not detected changes at the HA_1 top domains by using monoclonal antibodies, it is possible that the behavior of HA in response to low pH in the cold is different from that at physiological temperatures. As mentioned earlier, the proposal that the tops of the HA trimer do not dissociate during the fusion process and that their dissociation would result in inactivation of fusion capacity was based on the apparent lack of viral inactivation at 0°C and low pH. However, as described in the previous section, our results demonstrate the existence of a slow inactivation in the cold, thus implying that such unfolding occurs at 0°C.

Recently, Kemble al. (1992) have analysed the membrane fusion activity and conformational changes of a mutant HA (Cys-HA) in which the globular head region has been locked by disulfide bonds. These authors have shown that the fusion activity and low pH-induced conformational changes of Cys-HA-expressing cells are impaired. In addition, purified Cys-HA trimers are impaired in their capacity to undergo the low pH-induced conformational changes that lead to the release of the fusion peptides and changes in the globular head domain interface.

It is clear that in addition to the exposure of the fusion peptide, molecular rearrangements are required for fusion to occur. Our view is that fusion is not mediated by the final equilibrium low pH-induced conformation of HA, but rather by the process of conformational change of the protein at low pH. The strong correlation between the fusion rate constants and inactivation rate constants suggests that virus-cell fusion and viral inactivation are mediated by a common molecular mechanism. In agreement with previous suggestions (Doms and Helenius, 1986; Sarkar et al., 1989; Ellens et al., 1990; Stegmann et al., 1990) that more than one trimer is involved in the active fusion complex, our view is that the rate limiting step in the fusion of prebound influenza virus is the rearrangement of the HA trimers interacting with the target membrane.

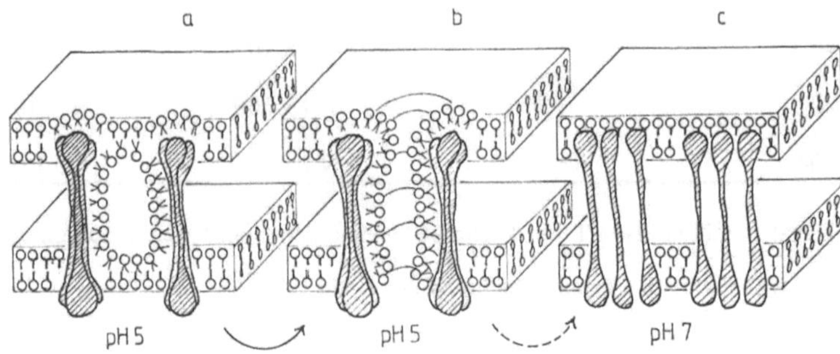

FIGURE 12 Shematic representation of the postulated stages during influenza virus HA-mediated fusion. (**a**): At pH 5, several unfolded HA trimers associate and interfacial water is expelled, thus leading to the formation of a fusogenic intermediate. (**b**): Following the mixing of the outer leaflets of the fusing bilayers, the inner leaflets will mix with a susequent formation of a small pose causing membrane fusion. (**c**): Upon neutralization, the dissociation of the oligomeric fusion complex would arrest the bilayer fusion. As referred to in the text, this latter stage will be rapid at 20°C or 37°C but it will be very slow at 0°C or 4°C. (Reproduced with permission from Pedroso de Lima and Hoekstra, 1994).

Figure 12 shows a highly schematic representation of the postulated stages during influenza virus HA-mediated fusion (a and b), incorporating the idea that more than one trimer may be required. This representation is based on a model proposed by Bentz et al. (1990) in which the HA mediated fusion starts with the formation of a small pore, as in the exocytotic fusion reaction (Breckenridge and Almers, 1987). In this model the HA trimers remain in an upright position throughout the fusion process. The fusion peptides, once exposed at low pH, do not insert into either the target or the viral membrane, but rather form a hydrophobic coat along the HA allowing for the flow of lipids between the two interacting membranes. We have found that fusion of influenza virus with PC-12 cells is arrested immediately upon neutralization both at 37°C and 20°C (Ramalho-Santos et al., 1993), which points to the requirement of continuous acidification during the fusion process, as previously observed (Stegmann et al., 1986). We suggest that under these neutral pH conditions, the oligomeric fusion complex dissociates rapidly both at 37°C and 20°C, probably due to a repulsion of adjacent charged amino acids at neutral pH. This view is schematically illustrated in Figure 12c. In contrast to this observation, Stegmann et al. (1990) have found that at 0°C virus-liposome complexes actively fused when neutralized beyond the lag phase at pH 5 and they suggested that the membrane merging is pH-independent. Our proposal is that fusion is pH-dependent, but it can proceed at 0°C upon neutralization since the dissociation of the fusion complex at this low temperature would be expected to be slow (Ramalho-Santos et al., 1993; Nir et al., 1983).

CONCLUDING REMARKS

The entry of enveloped viruses into host cells represents a typical example of protein-mediated membrane fusion. Influenza virus hemagglutinin is so far the best characterized viral fusion protein, both structurally and mechanistically, thus offering the best model for studying protein-mediated fusion in biological membranes. Although the molecular details of the acid-induced conformational change of HA are relatively well understood, it is not known how HA induces the perturbation of the lipid bilayer structure required to cause membrane fusion. As described, several models have been proposed for the mechanism of HA-mediated membrane fusion. These models which share several common features, involve different views for the various steps of the fusion process. Clearly, a detailed understanding of how HA mediates membrane fusion, will help to clarify how other proteins might mediate intracellular membrane fusion reactions.

ACKNOWLEDGEMENTS

This work was supported by JNICT (Portugal) and by a NATO Collaborative Research Grant, CRG 900333.

REFERENCES

Bentz J, Ellens H, Alford D (1990) An architecture for the fusion site of influenza hemagglutinin. FEBS Lett. 276: 1-5

Bergelson LD, Bukrinskaya AG, Provkazova NV, Shaposhnikjova GI, Kocharov SL, Shevchenko VP, Kornilaeva V, Fomina-Ageeva EV (1982) Role of gangliosides in reception of influenza virus. Eur. J. Biochem. 128: 467-474

Breckenridge LJ, Almers W (1987) Final steps in exocytosis observed in cell with giant secretory granules. Proc. Natl. Acad. Sci. 84: 1945-1949

Daniels RS, Douglas AR, Skehel JJ, Wiley DC (1983) Analysis of the antigenicity of influenza hemagglutinin at the pH optimum for virus-mediated membrane fusion. J. Gen. Virol. 64: 1657-1662

Doms R, Helenius A (1988) Properties of a viral fusion protein. In Molecular Mechanisms of Membrane Fusion (Ohki S, Doyle D, Flanagan TD, Hui SW, Mayhew E, eds), pp. 385-398, Plenum Press, New York

Doms RW, Helenius AH, White J (1985) Membrane fusion activity of the influenza virus hemagglutinin. The low-pH induced conformational change. J. Biol. Chem. 260: 2973-2981

Doms RW, White J, Boulay F, Helenius A (1990) Influenza virus hemagglutinin and membrane fusion. In Membrane Fusion (Wilschut J, Hoekstra D, eds), pp. 313-335, Marcel Dekker New York

Düzgünes N, Pedroso de Lima MC, Stamatatos L, Flasher D, Alford D, Friend DS, Nir S (1992) Fusion activity and inactivation of influenza virus: Kinetics of low-pH induced fusion with cultured cells. J. Gen. Virol. 73: 27-37

Ellens H, Bentz J, Zheng F, Mason D, White J (1990) The fusion site of influenza HA expressing fibroblasts requires more than one HA trimer. Biochemistry 59: 9697-9707

Gething MJ, Doms RW, York D, White J (1986) Studies on the mechanism of membrane fusion. Site-specific mutagenesis of the hemagglutinin of influenza virus. J. Cell Biol. 102: 11-23

Harter C, James P, Bächi T, Semenza G, Brunner J (1989) Hydrophobic binding of the ectodomain of influenza hemagglutinin to membranes occurs through the "fusion peptide". J. Biol. Chem. 264: 6459-6464

Hoekstra D, de Boer T, Klappe K, Wilshut J (1984) Fluorescence method for measuring the kinetics of fusion between biological membranes. Biochemistry 23, 5675-5681

Hoekstra D, Kok JW (1989) Entry mechanisms of enveloped viruses. Implications for fusion of intracellular membranes. Biosc. Rep. 9: 273-305

Hoekstra D, Pedroso de Lima MC (1992) Molecular mechanisms of enveloped viruses entry into host cells: Protein dynamics and membrane fusion. In: Advances in Membrane Fluidity vol. 6: Membrane Interactions of HIV: Implications for Pathogenesis and Therapy in AIDS (Aloia, RC, Curtain, CC, eds), pp. 71-97, Wiley-Liss, New York

Junankar PR, Cherry RJ (1986) Temperature and pH dependence of the hemolytic activity of influenza virus and of the rotational mobility of the spike glycoproteins. Biochim. Biophys. Acta 854: 198-206

Kemble GW, Bodian DL, Rosé J, Wilson IA, White JM (1992) Intermonomer disulfide bonds impair the fusion activity of the influenza hemagglutinin. J. Virol. 66: 4940-4950

Klenk HD, Rott R, Orlich M, Blodorn J (1975) Activation of influenza A virus by trypsin treatment. Virology 68: 426-439

Lamb RA, Choppin PW (1983) The gene structure and replication of influenza virus. Annu. Rev. Biochem. 52: 467-506

Lamb RA (1989) Genes and proteins of the influenza virus. In The Influenza Viruses (Krug RM, ed.), pp. 1-87, Plenum Press, New York

Lazarowitz S, Choppin PW (1975) Enhancement of the infectivity of influenza A and B viruses by proteolytic cleavage of the hemagglutinin polypeptide. Virology 68: 440-454

Lear JD, DeGrado WF (1987) Membrane binding and conformational properties of peptides representing the NH_2 terminus of influenza HA_2. J. Biol. Chem. 262: 6500-6505

Lenard J, Compans RW (1974) The membrane structure of lipid containing viruses. Biochim. Biophys. Acta 344: 51-94

Marsh M, Helenius A (1989) Virus entry into animal cells. Adv. Virus Research 36: 107-150

Martin K, Helenius A (1991a) Nuclear transport of influenza virus ribonucleoproteins: The viral matrix protein (M_1) promotes export and inhibits fusion. Cell 67: 117-130

Martin K, Helenius A (1991b) Transport of incoming influenza virus nucleocapsids into the nucleus. J. Virol. 65: 232-244

Nir S, Bentz J, Wilschut J and Düzgünes N (1983) Aggregation and fusion of vesicles. Prog. Surface Sci. 13: 1-124

Paulson GNR, Murayama, JI, Sze G, Martin E (1986) Biological implications of influenza virus receptor specificity. In Virus Attachment and Entry into Cells (Crowell RL, Lonberg-Holm K, eds), pp. 144-151, American Society for Microbiology, Washington

Pedroso de Lima MC, Hoekstra D (1994) Liposomes, viruses and membrane fusion. In Liposomes as Tools in Basic Research and Industry (Philippot JR, Schuber F, eds), pp. 137-156, CRC Press, New York

Pedroso de Lima MC, Ramalho-Santos J, Flasher D, Nir S, Düzgünes N (1994) Target membrane sialic acid modulates both binding and fusion activity of influenza virus (submitted for publication)

Ramalho-Santos J, Nir S, Düzgünes N, Carvalho AP, Pedroso de Lima MC (1993) A common mechanism for influenza virus fusion activity and inactivation. Biochemistry 32: 2771-2779

Ramalho-Santos J, Negrão, R, Pedroso de Lima MC (1994) Role of hydrophobic interactions in the fusion activity of influenza and Sendai viruses towards model membranes. Biosc. Rep. 14: 15-24.

Ruigrok RWH, Wrigley NG, Calder LJ, Cusack S., Wharton SA, Brown EB, Skehel JJ (1986) Electron microscopy of the low pH structure of the influenza virus hemagglutinin. EMBO J. 5: 41-49

Sarkar DP, Morris SJ, Eidelman O, Zimmerberg J, Blumenthal R (1989) Initial stages of influenza hemagglutinin-induced cell fusion monitored simultaneously by two fluorescent events: Cytoplasmic continuity and lipid mixing. J. Cell Biol. 109: 113-122

Sato SB, Kawasaki K, Ohnishi SI (1983) Hemolytic activity of influenza virus hemagglutinin glycoproteins activated in mildly acidic environments. Proc. Natl. Acad. Sci. 80: 3153-3157

Skehel JJ, Bayley PM, Brown EB, Martin SR, Waterfield MD, White J, Wilson IA, Wiley DC (1982) Changes in the conformation of influenza hemagglutinin at the pH optimum of virus-mediated membrane fusion. Proc. Natl. Acad. Sci. 79: 968-972

Stegmann T, Doms RW, Helenius A (1989) Protein-mediated membrane fusion. Annu. Rev. Biophys. Biophys. Chem. 18: 187-211

Stegmann T, Hoekstra D, Scherphof G, Wilschut J (1986) Fusion activity of influenza virus: A comparison between biological and artificial target membrane vesicles. J. Biol. Chem. 261: 10966-10969

Stegmann T, White JM, Helenius A (1990) Intermediates in influenza induced membrane fusion. EMBO J. 9:4231-4241

Sugrue RJ, Hay AJ (1991) Structural characteristics of the M_2 protein of influenza A viruses: Evidence that it forms a tetrameric channel. Virology 180: 617-624

Susuki Y, Matsunaga M, Matsumoto M (1985) N-acetylneuraminyllactosylceramide, $G_{M3\text{-}NeuAc}$, a new influenza A virus receptor which mediates the adsorption-fusion process of viral infection. J. Biol. Chem. 260: 1362-1365

Tsai KH, Lenard J (1975) Asymmetry of influenza virus membrane bilayer demonstrated with phospholipase C. Nature 253: 554-555

Watson JD, Hopkins NH, Roberts JW, Steitz JA, Weiner AM (eds) (1987) Molecular Biology of the Gene. The Benjamin Cummings Publishing Company Inc., Menlo Park

Wharton SA, Skehel JJ, Wiley DC (1986) Studies of influenza haemagglutinin-mediated membrane fusion. Virology 149: 27-35

White J (1990) Viral and cellular membrane fusion proteins. Annu. Rev. Physiol. 52: 675-697

White J (1992) Membrane fusion. Science 258: 917-924

White J, Doms R, Gething MJ, Kielian M and Helenius A. (1986) Viral membrane fusion proteins. In Virus Attachment and Entry into Cells (Crowell RL, Lonberg-Holm K, eds), pp. 54-59, American Society for Microbiology, Washington

White J, Helenius A, Gething MJ (1982) Haemagglutinin of influenza virus expressed from a cloned gene promotes membrane fusion. Nature 300: 658-659

White JM, Wilson IA (1987) Antipeptide antibodies detect steps in a protein conformational change: Low pH activation of the influenza virus hemagglutinin. J. Cell Biol. 105: 2887-2896
Wiley DC, Skehel JJ (1987) The structure and function of the hemagglutinin membrane glycoprotein of influenza virus. Annu. Rev. Biochem. 56: 365-394
Wilson IA, Skehel JJ, Wiley DC (1981) Structure of the haemagglutinin membrane glycoprotein of influenza virus at 3 Å resolution. Nature 289: 366-373

Mass action model of virus fusion

Shlomo Nir, Nejat Düzgüneş[1], Dick Hoekstra[2], João Ramalho-Santos[3], and Maria C. Pedroso de Lima[4]

The Seagram Center for Soil and Water Sciences
Faculty of Agriculture
Hebrew University of Jerusalem
Rehovot 76100
Israel

1. Introduction

The purpose of this presentation is to describe procedures of analysis of final extents and kinetics of virus fusion with target membranes. The presentation of results will focus on deductions from studies of final extents of fusion.

The mass action model for membrane fusion views the overall fusion reaction as a sequence of a second-order process of virus-liposome (Nir et al., 1986a), or virus-cell (Nir et al., 1986b; Bentz et al., 1988; Nir et al., 1990) adhesion or aggregation, followed

[1]Department of Microbiology, University of the Pacific, School of Dentistry, San Francisco, CA 94115, and Department of Pharmacy, University of California, San Francisco, Ca 94143, U.S.A.
[2]Department of Physiological Chemistry, University of Groningen, Bloemsingel 10,9712 KZ Groningen, The Netherlands.
[3]Department of Zoology and Center for Neurosciences, University of Coimbra, 3049 Coimbra, Portugal.
[4]Department of Biochemistry and Center for Neurosciences, University of Coimbra, 3049 Coimbra, Portugal.

NATO ASI Series, Vol. H 91
Trafficking of Intracellular Membranes
Edited by M.C. Pedroso de Lima N. Düzgüneş and D. Hoekstra

by the first-order fusion reaction itself. In the application of fusion assays based onfluorescence, model calculations enable transformation of fluorescence increase values to the percentage of fused particles. The analysis of final extents of fluorescence intensity can yield the percentage of virions capable of fusing with certain target membranes at given pH and temperatures (Nir et al., 1986a,b,c; Nir et al., 1990; Larsen et al., 1993). It can also yield the number of virions that can fuse per single cell (Nir et al., 1986b).

The application of the analysis of kinetics of fusion allows the separation of the overall fusion reaction into the adhesion stage and the actual fusion reaction. Thus, the analysis enables elucidation of the extent to which the action of viral glycoproteins goes beyond the promotion of contact between apposed membranes, as well as details of virus inactivation (Nir et al., 1990; Düzgüneş et al., 1992; de Lima et al., 1992; Ramalho-Santos et al., 1993), or the effect of various additives, such as cholesterol sulfate (Cheetham et al., 1994) on the modulation of viral fusion activity. The analysis provides the following parameters: (i) $\beta (\beta = 1 - \alpha)$, the percentage of virions capable of fusing with the target membrane at the given conditions, e.g., pH and temperature; (ii) f(sec^{-1}), the fusion rate constant; (iii) C($M^{-1}.sec^{-1}$) the forward rate constant of adhesion; (iv) D(sec^{-1}), the rate constant of dissociation. In addition, it provides, when applicable, the rate constants describing the process of low pH inactivation. In the case of influenza virus a great deal of simplification in the analysis can be accomplished by prebinding the virus to cells at neutral pH, where fusion does not occur; then lowering the pH. This procedure can yield f-values directly without having to determine the additional parameters C and D. We can project that the same procedure will also be applicable for the case of Semliki Forest virus. However, pursuing the same line of thought in the case of Sendai virus by preincubating it with cells at pH 9 and then reducing the pH to neutral, was not applicable, since a certain degree of fusion did occur at pH 9. The other procedure of prebinding the virus to cells in the cold (Hoekstra et al. 1989), is not always optimal due to lag times involved in attaining full fusion activity following the elevation of temperature.

2. Final extents of virus-cell fusion

The final extents of fusion of virions with cells can be deduced from the final extents of fluorescence intensity increase, employing an assay based on mixing of membranes, such as the R_{18} dequenching assay (Hoekstra et al., 1984). Here, it is assumed that the virus is initially labeled.

Let I be the fractional increase in the final fluorescence intensity of R_{18} molecules due to their dilution as a result of fusion. I is given by

$$I = 1 - X \tag{1}$$

in which X is the relative surface concentration of R_{18} molecules, provided that their initial concentration is sufficiently low. Let us denote the surface areas of a single virus particle and a cell by S_v and S_e, respectively, and let M be the average number of virus particles per cell. If all virus particles have fused, then (Nir et al., 1986b)

$$I = 1 - MS_v/(MS_v + S_e) = S_e/(S_e + MS_v) = 1/(1 + MS_v/S_e) \tag{2}$$

If the fraction of virus particles capable of fusing is q then

$$I = q/(1 + q\,MS_v/S_e) \tag{3}$$

If S_v/S_e is known, then Eq (2) does not involve any parameter and can be used to generate predicted I values. In the case of Sendai virus fusing with erythrocyte ghosts at 37°C and pH 7.4, the use of Eq (3) indicated that $q>0.86$, i.e., practically all virus particles were fusion active. With a different strain of Sendai virus (Cantell) and a somewhat different procedure of preparation of erythrocyte ghosts (Cheetham et al., 1994) the value of q was also close to unity, but it was reduced to 0.73 when the ghosts included 20 mol% of cholesterol sulfate.

Eq (2) could be applied to a certain range of virus/cell ratios up to 100:1 for the case of Sendai virus fusing with erythrocyte ghosts at 37°C and pH 7.4. The application of Eq (2) to the case of influenza virus fusing with erythrocyte ghosts at 37°C and pH 5 yielded good simulations and predictions for I values for virus/cell numbers of up to about 400 (Nir et al., 1990). The surface area of influenza virus is about twofold smaller than that of Sendai virus, the respective average radii being 50 and 75 nm. At larger number ratios of Sendai virus/erythrocyte ghost, the overestimates obtained by the calculated values necessitated introduction of an assumption that there is a limit on the number of virions that can fuse with a single cell.

If N is the number of virions that can fuse per single cell and p, the number of virus particles per cell divided by N is greater than 1, then Eq (2) is modified to

$$I = 1/[p(1 + NS_v/S_e)] \tag{4}$$

The employment of Eq (4) to the fusion of Sendai virus with erythrocyte ghosts yielded good simulations and predictions for many number ratios of virions/cell. The predictions were reasonably good for different batches of virus and ghosts, despite large differences in the rates of fusion for the different batches. Furthermore, in a solution of 4% polyethylene glycol, the rate of fusion was significantly enhanced without affecting the number of virions fusing per single cell (Hoekstra et al., 1989). The use of Eq (4) yielded N = 100-200, i.e., about 100 Sendai virus particles fusing per single cell (depending on the estimated number of particles (Nir et al., 1986b; Hoekstra and Klappe, 1986), in contrast to about 1500 that can bind per cell (Hoekstra and Klappe, 1986).

It is desirable to avoid the use of excessive ratios of virus particles per cell, since it can lead to an erroneous conclusion that only a fraction of the virus can fuse. For cells fusing with virions, it may be difficult to retain cell viability for a couple of hours, until final extents of fusion are reached. So far, a rigorous test of Eq (4) has been carried out only with erythrocyte ghosts. However, the fusion studies of influenza virus with several types of suspension cells resulted in the conclusion that at pH 5 and 37°C all virions are capable of fusing (Düzgüneş et al., 1992). A similar conclusion could be reached regarding the fusion of Sendai virus with a variety of suspension cells at neutral pH, and also for its fusion with the adherent cell line, PC-12, which was established from rat adrenal pheochromocytoma (de Lima et al., 1992). In the latter case, it was estimated that about 1500 Sendai virus particles could fuse with a single cell. The surface area of a PC-12 cell is about 4-fold larger than that of an erythrocyte ghost. This means that the number of virions that can fuse per unit area of a PC-12 cell is about twice the value found for erythrocyte ghosts. We will present evidence that the number of virions that can fuse with a biological membrane is limited by the amount of viral glycoproteins that can be accommodated in the cellular membrane by fusion, rather than by a limit on the number of receptor sites where the bound virus can fuse. We will extend our arguments by referring to a host of results on virus-liposome fusion.

Influenza virus (Nir et al., 1986a, Stegmann et al., 1989), Sendai virus (Nir et al., 1986c; Klappe et al., 1986; Amselem et al., 1986; Cheetham et al., 1994), SIV (simian immunodeficiency virus) (Larsen et al., 1990) and HIV (human immunodeficiency virus) (Larsen et al., 1993), were shown to fuse with phospholipid vesicles of several compositions, including pure neutral, pure acidic or mixed. Hence, this may indicate that specific receptors are not a prerequisite for viral fusion, although there are differences in the tendency of the viruses to fuse with membranes of different compositions.

For all viruses tested, the fusion products consist of a single virus and several liposomes. The reason why virions cannot fuse with the fusion products may be due to a limit on the incorporation of viral glycoproteins in the target membrane. This could be the main reason for the limit set on the number of virions fusing with a single ghost, whose surface area is three orders of magnitude larger than that of a Sendai virus. We noted before that at least about 400 influenza virus particles, whose surface area is half that of Sendai virus, could fuse with a single ghost. We found that about 400 reconstituted Sendai virosomes could fuse with a single erythrocyte ghost (D. Hoekstra, K. Klappe and S. Nir, unpublished data), vs. 100 particles of the intact virus. This ratio corresponds to the inverse ratio of the surface areas, since the average radius of an intact virus is two-fold larger than that of a virosome. Thus, our results favor the conclusion that the limit on the number of virus particles that can fuse with a single ghost is due to a saturation limit rather than due to a limit on discrete fusion sites.

3. Final extents of virus-liposome fusion

Analysis of the results of final extents of virus-liposome fusion led to the conclusion that the fusion products consist of a single virus and several liposomes. This conclusion was found to hold for influenza virus (Nir et al., 1986a; Stegmann et al., 1989); Sendai virus (Klappe et al., 1986; Nir et al., 1986c; Amselem et al., 1986; Cheetham et al., 1994) and HIV-1 (Larsen et al., 1993). This mode of interaction yields an equation for the final extent of fusion of fully active virions (Nir et al., 1986a),

$$V_f / V_o = 1 - \exp\left(\frac{-L_o}{V_o}\right), \qquad (5)$$

in which L_0 and V_0 are initial molar concentrations of liposomes and virions, and V_f is the concentration of fused virions. This equation could yield predictions for experimental results obtained with liposome/virus ratios varying by two orders of magnitude (Nir et al., 1986a). In the latter case, the liposomes were labeled according to the RET assay (Struck et al., 1981).

Another demonstration for this mode of interaction where one virion fuses with several liposomes, was achieved by adding liposomes or virions to a system of virions and liposomes after long incubation times. The addition of unlabeled liposomes resulted in an increase in fluorescence intensity which was close to the final level predicted for the final virus/liposome ratio. This process was occasionally iterated for several rounds, indicating that the added liposomes fused with the fusion products (Klappe et al., 1986; Larsen et al., 1993).

A direct consequence of the fact that virus-liposome fusion products consist of a single virus and several liposomes is that a certain fraction of the virus population will not fuse, unless the liposome/virus ratio is large. Thus, for a 1/1 population, the fraction of fused virions is (1-exp(-1)), whereas for a number ratio $L_0/V_0 = 4$, most of the virions fuse.

In the general case, a certain fraction, α, of the virus would remain unfused even in the presence of a large excess of liposomes. Two models have been considered. If the binding of liposomes to inactive virions is reversible, then all the liposomes will eventually fuse, whereas V_f is given by:

$$V_f = V_o(1 - \alpha)\{1 - \exp - L_o/(V_o(1 - \alpha))\} \tag{6}$$

If the binding of liposomes to inactive virions is irreversible, then the concentration of fused liposomes, L_f, will be

$$L_f = L_o(1 - \alpha) \tag{7}$$

and

$$V_f = V_o\,(1 - \alpha)\,(1 - \exp - L_o/V_o) \tag{8}$$

In the case of Sendai virus fusing with liposomes, the assumption of an essentially irreversible binding of the fusion-inactive virus to liposomes, Eq (8), gave a better simulation and prediction of the experimental results. Eqs (6) or (8) reduce to Eq (5) in the case $\alpha = 0$. An illustration is given in Table 1.

In studies with HIV-1 (Larsen et al. 1993) detailed calculations ruled out the possibility that the final extents of fluorescence increase might be due to transfer of the probe rather than fusion. In the latter case, an average between the predictions of Eqs (6) and (8), or merely Eq (8), gave the best fits of calculated values to experimental values of fluorescence intensity increase.

Table 1. Final Intensity (I) Levels of Fluorescence[a] for Sendai Virus Fusing with Liposomes. Experimental and Calculated Values

Liposome composition	Liposomal phospholipid (μM)	pH	I(%) Exp.	I(%) Calc.	% fusion activity
CL	2.5	5.0	50.6	44.0	100
	10	5.0	85.8	78.0	100
	50	5.0	99.0	95.0	100
	2.5	7.4	27.6	26.3	60
	10	7.4	44.0	46.9	60
	50	7.4	62.7	56.8	60
CL/	2.5	7.4	13.0	13.1	30
DOPC	10.0	7.4	21.0	23.5	30
(1:1)	50	7.4	29.6	28.4	30
PS	50	7.4	24.0	25.0	25

[a]Large unilamellar liposomes of about 100 nm in diameter were used. The studies with Sendai virus employed the R_{18} probe. The values given are taken from Nir et al. (1986c) where more cases are presented. The concentration of Sendai virus used was 20 μg of protein/2 mL, or 2.8 μM with respect to phospholipid. The calculations employed Eq (8).

3.1 A model for partial fusion activity

All three viruses studied, influenza virus, Sendai virus and HIV-1, exhibit a phenomenon of partial fusion activity toward liposomes, with values of α ranging from 0 to 0.75 at 37°C. All three viruses exhibit complete fusion activity towards liposomes composed of the acidic phospholipid CL at pH 5. On the other hand, influenza virus yields 100% fusion with PS liposomes, which are also negatively charged at neutral pH, but Sendai virus yields only 30% fusion. As discussed previously (Nir et al. (1988)), one possible explanation for these findings is that the virus preparation is heterogeneous.

Specifically, it is possible that a certain fraction of virions cannot form sufficiently close contacts with given liposomal membranes, and cannot induce target membrane destabilization, although binding can occur.

An alternative possibility is that the membrane of each virion might include a limited number of fusion-capable sites, consisting of well-defined combinations of glycoproteins at various spatial arrangements (e.g., the F and HN glycoproteins of Sendai virus). If virus binding to a liposome does not occur at an "active" (viral) site, fusion will not occur. If binding is essentially irreversible under the given conditions, as the analysis has indicated for the case of Sendai virus (Nir et al., 1986c) and HIV-1 (Larsen et al., 1993), all of the virions may be bound, but a certain fraction may remain unfused. Experiments were carried out to resolve this issue of the gross mechanism of partial fusion activity in the cases of Sendai virus (Nir et al., 1990) and, recently, influenza virus (J. Ramalho-Santos, M. de Lima and S. Nir, unpublished data).

In the first study, the experimental procedure was designed to separate the nonfused Sendai virus from the liposomes and fusion products after a long period of incubation at pH 7.4 and 37°C. At that stage, the final extent of fusion, or R_{18} fluorescence increase, had been attained. Although the binding of the fusion-inactive virus to the liposomes was known to be essentially irreversible, the nonfused virus could be separated from the liposomes by employing sucrose gradients and prolonged centrifugation. The rationale for this experiment was that, if the collected unfused virions were inherently fusion-inactive, then their subsequent incubation with liposomes should not result in their fusion. In order to optimize the chances for viral fusion, liposomes were added to R_{18}-labeled virions in large excess. Liposomes consisting of PS were initially chosen since only 25% of the virions were fusion-active at neutral pH. No significant differences were observed between the final extents of fusion following 20 h or 3-4 h of incubation. Thus, 4 h of incubation could yield the final extent of virus fusion.

The collected "fusion-inactive" virions fused with added PS liposomes, the percentage of virions fusing in this second round being close to that in the first round. In the first round, 22% fusion was observed as the average of three batches of virus, whereas in the second round, the average was 19.5%. Preliminary results with liposomes composed of PS/DOPE/Cholesterol 40/20/40 (mole percent) exhibited the same phenomenon, that the "fusion-inactive" virions are capable of fusing, although the percentage of virions that fused in the second round was about ⅔ of the value found in the first round. A similar conclusion was recently reached for the case of influenza virus interacting with PC/PE (2/1), PC/PE/Chol (2/1/1), and PC/PE/PS/Chol (1/1/1/1) LUV.

These results demonstrate that, in the case of Sendai virus and influenza virus fusing with liposomes of a variety of compositions, the existence of a majority of unfused virions does not imply that most of the individual virions are incapable of fusing. Apparently, in most of the encounters between the virus and liposomes, attachment of the virus to the liposome occurs at an "inactive" site on the viral membrane, such that the virus does not fuse, although it binds to the liposome. This binding is essentially irreversible under normal conditions. If such an attached virus is released from the liposome (e.g., by means of a sucrose gradient and centrifugation), it has the same chance as any other virus in the population to fuse with the liposomes by forming an attachment via an active site consisting of a proper combination and arrangement of glycoproteins.

We found that the final extents of fusion of Sendai virus (Fonteijn et al., 1992) with didodecyl phosphate (DDP) vesicles and those of influenza virus interacting with liposomes (J. Ramalho-Santos, M. de Lima, and S. Nir, unpublished data) increase with temperature. The analysis of kinetics of the overall fusion process (Fonteijn et al., 1992) indicated a dramatic increase in the rate constant of dissociation in this temperature range. It may be that, at higher temperatures, some fraction of the improperly bound virions could be released and thus have an increased chance of fusing in later encounters with the vesicles.

In most cases, influenza and Sendai virus have partial fusion activity toward liposomes but not cells. When the virus binds to the cellular surface via an inactive site on the virus, or if an inactive intermediate state is created in the vicinity of the attachment site, the ability of the cell membrane to wrap around the virus may facilitate the formation of a contact with the cellular surface via another active site.

It remains to be seen to what extent the general concept of irreversible binding of the virus at an inactive site on the virus, or at the target membrane can be operative in partial fusion activity in virus-cell systems.

4. Kinetics of virus-cell and virus-liposome fusion

We will present here the relatively simpler equations for the case of virus prebound to cells. The derivation of expressions for the kinetics of virus-liposome fusion for the case of a fully active virus is given in Nir et al. (1986a). Based on these equations an extension of the program to the case of partially active virus has also been described (Nir et al., 1986c). A further extension which accounts for low pH inactivation has been described in Nir et al. (1988) and Stegmann et al. (1989).

The equations for the kinetics of virus-cell fusion were first derived in Nir et al. (1986b), and analytically approximated by Bentz et al. (1988). An extension of the programs to account for low pH inactivation could not utilize the latter equations, but had to be based on the original equations in Nir et al. (1986b), as described in Nir (1993), de Lima et al. (1992) and Ramalho-Santos et al. (1993).

4.1. Kinetics of fusion of prebound virus: Analysis of inactivation.

Unlike the case of Sendai virus whose fusion activity toward erythrocyte ghosts (and cells) is optimal at 37°C and pH 7.5, no fusion of influenza virus occurred under these conditions with erythrocyte ghosts or suspension cells (Nir et al., 1990; Düzgüneş et al., 1992; Stegmann et al., 1990; Ramalho-Santos et al., 1993). However, a fraction, B, of the virus adhered to the cells at pH 7.4, typically 40-100%, following 5 to 20 min of preincubation with $\geq 2\times10^7$ cells/mL, at virus/cell number ratios of the order of 100. After a preincubation period of several minutes, the reduction of pH results in a rapid increase in fluorescence intensity. Since the bound virions occupy a small fraction of the cellular surface area, the increase in fluorescence during the first few seconds (e.g., 5 to 10 s) after reducing the pH may be considered to reflect essentially infinite probe dilution. If no viral inactivation occurs, then the fusion rate coefficient, f, is a constant, but in general, we set it as f(t) to allow for its time dependence.

Let F(t) be the fraction of the prebound virus that has fused following the reduction of the pH. F(t) satisfies the equation

$$dF/dt = (1 - F)\, f(t) \tag{9}$$

Since F(0) = 0, the general solution of this linear differential equation in F is

$$F(t) = 1 - \exp - \int_0^t f(u)du \tag{10}$$

In the special case where f is independent of time this equation reduces to

$$F(t) = 1 - \exp(-ft) \tag{11}$$

or

$$I(t) = [1 - \exp-(ft)]B \tag{12}$$

Ordinarily, virus inactivation due to exposure of the virus to low pH can be neglected during the earliest period of fluorescence intensity increase, and the fusion rate constant can be determined from Eq (12). During the first stage, i.e., $ft << 1$, I(t) should increase linearly with the time, t.

Virus inactivation can exhibit itself by some form of a decay of f(t) with time. The exact functional form of f(t) has not been determined. Nir et al. (1990) pointed out that with the amount of data available, it was not possible to discern between the models of inactivation given in Nir et al. (1988). In the general case, a residual fusion activity, i.e., non-vanishing value of f(t) is retained even after long times of exposure of the virus to low pH. This can be expressed by a partially reversible process of inactivation. However, depending on the amount of available data, the initial kinetics of fusion (within several minutes) can be expressed by ignoring the residual fusion activity after long times of exposure, i.e., utilizing just two parameters, $f \equiv f(0)$ and $\gamma_1 \equiv \gamma$, the rate constant of inactivation.

One functional form of decay of f is given by (Nir et al., 1988)

$$f(t) = f(0)\,[\exp - \gamma t + \gamma_2(1 - \exp -\gamma t)/\gamma] \tag{12}$$

in which $\gamma = \gamma_1 + \gamma_2$.

At t = 0, f(t) = f(0), whereas at larger t

$$f(t) = f(0)\gamma_2/\gamma \tag{13}$$

Thus, the ratio γ_2/γ gives the residual fusion activity. When $\gamma_2 = 0$, $\gamma = \gamma_1$, and Eq (13) reduces to

$$f(t) = f(0)\exp(-\gamma t) \tag{14}$$

A substitution of Eq (13) in Eq (10) gives (Nir et al. 1990)

$$I(t) = (1 - \exp[f(0)[(\gamma_1/\gamma^2)\exp(-\gamma t) - (\gamma_2/\gamma)t - \gamma_1/\gamma^2]])B \tag{15}$$

If the decay is assumed to be irreversible, i.e., $\gamma_2 = 0$, then $\gamma_1 = \gamma$ and Eq (16) reduces to

$$I(t) = [1 - \exp[f(\exp(-\gamma t) - 1)/\gamma]]B \tag{16}$$

Eq (13) can be viewed as a phenomenological expression that explains the kinetic data (Nir et al., 1988; 1990; Stegmann et al., 1989). This equation follows from a model that assumes that after a very short time the viral glycoproteins (or their aggregates) exist in an activated state which is compatible with the fusion state, whereas later they can transform reversibly to an inactivated state.

Another expression has been explicitly considered,

$$f(t) = f(0) / (1 + \gamma t)^2 \tag{17}$$

in which γ, the decay constant analogous to that in Eq (13), is implicitly assumed to depend on the surface concentration of HA molecules, surface viscosity, and potential barriers for close approach of macromolecules.

As in Eqs (13) and (14), a residual fusion activity can be retained if the equilibrium distribution of (HA) particles is taken into account (Nir et al., 1988, 1990).

A substitution of Eq (18) into Eq (10) yields in this case

$$I(t) = [1 - \exp[f(0)(1/(1 + \gamma t) - 1)/\gamma]]B \tag{18}$$

Our opinion is that it is still worth trying to resolve which functional form can best explain the experimental data, since such a determination can be helpful in suggesting mechanisms. In this context, Bentz (1992) derived expressions which considered intermediates in the kinetics of virus-membrane fusion. Unlike our approach which distinguishes between partial fusion activity and viral inactivation, such a distinction does not appear in the latter study. We have pointed out that unfused influenza virus particles that were separated from liposomes after long periods of inactivation could still fuse with added liposomes. More discussions of applications can be found in Nir et al. (1990), Düzgüneş et al. (1992), Ramalho-Santos et al. (1993) as well as elsewhere in this book in the chapters of M. de Lima and N. Düzgüneş.

5. Abbreviations

Chol, cholesterol; CL, cardiolipin; DOPC, dioleoylphosphatidylcholine; DOPE, dioleoylphosphatidylethanolamine; LUV, large unilamellar vesicles; PC, phosphatidylcholine; PE, phosphatidylethanolamine; PS, phosphatidylserine; R_{18}, octadecylrhodamine B chloride.

6. References

Amselem S, Barenholz Y, Loyter A, Nir S, Lichtenberg D (1986) Fusion of Sendai virus with negatively charged liposomes as studied by pyrene-labelled phospholipid liposomes. Biochim Biophys Acta 860:301-313

Bentz J, Nir S, Covell D (1988) Mass action kinetics of virus-cell aggregation and fusion. Biophys J 54:449-462

Bentz J (1992) Intermediates and kinetics of membrane fusion. Biophys J 63:448-459

Cheetham JJ, Nir S, Johnson E, Flanagan TD, Epand RM (1994) The effects of membrane physical properties on the fusion of Sendai virus with human erythrocyte ghosts and liposomes: Analysis of kinetics and extent of fusion. J Biol Chem 269:5467-5472

Düzgüneş N, Lima MCP, Stamatatos L, Flasher D, Alford D, Friend DS, Nir S (1992) Fusion of influenza virus with human promyelocytic leukemia and lymphoblastic leukemia cell, and murine lymphoma cells: Kinetics of low pH-induced fusion monitored by fluorescence dequenching. J Gen Virol 73:27-37

Fonteijn TAA, Engberts JBFN, Nir S, Hoekstra D (1992) Asymmetric fusion betwen synthetic DI-N-Dodecyl phosphate vesicles and virus membranes. Biochim Biophys Acta 1110:185-192

Hoekstra D, DeBoer T, Klappe K, Wilschut J (1984) Fluorescence method for measuring the kinetics of fusion between biological membranes. Biochemistry 23:5675-5681

Hoekstra D, Klappe K (1986) Sendai virus-erythrocyte membrane interaction: quantitative and kinetic analysis of viral binding, dissociation, and fusion. J Virol 58:87-95

Hoekstra D, Klappe K, Hoff H, Nir S (1989) Mechanism of fusion of Sendai virus: Role of hydrophobic interactions and mobility constraints of viral membrane proteins. Effects of polyethylene glycol. J Biol Chem 264:6786-6792

Klappe K, Wilschut J, Nir S, Hoekstra D (1986) Parameters affecting fusion between Sendai virus and liposomes. Role of viral proteins, liposome composition and pH. Biochemistry 25:8252-8260

Larsen CE, Alford DR, Young LJT, McGraw TP, Düzgüneş N (1990) Fusion of simian immunodeficiency virus with liposomes and erythrocyte ghost membranes: Effects of lipid composition, pH and calcium. J Gen Virol 71:1947-1955

Larsen CE, Nir S, Alford DR, Jennings M, Lee K, Düzgüneş N (1993) Human immunodeficiency virus type 1 (HIV-1) fusion with model membranes: Kinetic analysis and the role of lipid composition, pH and divalent cations. Biochim Biophys Acta 1147:223-236

Lima MCP, Ramalho-Santos J, Martins MF, Carvalho AP, Bairos VA, Nir S (1992) Kinetic modeling of Sendai virus fusion with PC-12 cells: Effect of pH and temperatue on fusion and viral inactivation. Eur J Biochem 205:181-186

Nir S, Stegmann T, Wilschut J (1986a) Fusion of influenza virus and cardiolipin liposomes at low pH: Mass action analysis of membrane lipid mixing and aqueous contents mixing. Biochemistry 25:257-266

Nir S, Klappe K, Hoekstra D (1986b) Kinetics of fusion between Sendai virus and erythrocyte ghosts: Application of mass action kinetic model. Biochemistry 25:2155-2161

Nir S, Klappe K, Hoekstra D (1986c) Mass action analysis of kinetics and extent of fusion between Sendai virus and phospholipid vesicles. Biochemistry 25:8261-8266

Nir S, Stegmann T, Hoekstra D, Wilschut J (1988) Kinetics and extents of fusion of influenza virus and Sendai virus with liposomes. in Molecular Mechanisms of Membrane Fusion (Ohki S, Doyle D, Flanagan TD, Hui SW, & Mayhew E, eds) Plenum Press New York

Nir S, Düzgüneş N, Lima MCP, Hoekstra D (1990) Fusion of enveloped viruses with cells and liposomes. Cell Biophys 17:181-201

Nir S (1993) Analysis of kinetics and extent of fusion of viruses with target membranes. in Methods in Enzymology 220, (Düzgüneş N, ed) Academic Press New York

Ramalho-Santos J, Nir S, Düzgüneş N, Carvalho AP, Lima MCP (1993) A common mechanism for virus fusion activity and inactivation. Biochemistry 32:2771-2779

Stegmann T, Nir S, Wilschut J (1989) Membrane fusion activity of influenza virus. Effects of gangliosides and negatively charged phospholipids in target liposomes. Biochemistry 28:1698-1704

Stegmann T, White J, Helenius A (1990) Intermediates in influenza induced membrane fusion. EMBO J 9:4231-4241

Struck DK, Hoekstra D, Pagano RE (1981) Use of resonance energy transfer to monitor memebrane fusion. Biochemistry 20:4093-4099

Viral Membrane Proteins as Tools to Study Protein Folding, Assembly, and Transport

Robert W. Doms, Stephen T. Abedon, and Thomas M. Richardson, Jr.
Department of Pathology and Laboratory Medicine
University of Pennsylvania
Philadelphia PA 19104

Abstract

The endoplasmic reticulum (ER) maintains a highly specialized environment that supports rapid and efficient protein folding and assembly. Viral membrane proteins have been used as tools to understand these processes. In this paper, we will review commonly used techniques that may be employed to monitor protein folding and assembly. Such approaches can be used to characterize folding factors in the ER as well as to develop a more complete understanding of viral membrane protein biosynthesis. Such information is required to fully evaluate the effects of mutations on viral membrane protein structure and function.

Introduction

Viral membrane proteins have long been used as tools to study basic questions in cell biology, including how proteins fold and assemble, are targeted in the cell, and are processed for antigen presentation. The reasons for this are in part historical: viral proteins can be expressed in any cell line that is permissive for infection by the appropriate virus. Thus, prior to the advent of modern molecular techniques and the development of expression systems, viral membrane protein processing could be studied in many cell types simply by infection. Furthermore, viral membrane proteins are often well characterized genetically, antigenically, and structurally. It is possible to select for various temperature sensitive mutants, and the proteins are expressed at high levels. In addition, viruses rely on the host cell for the machinery required for protein synthesis, processing, and transport. Thus, there is nothing inherently unique about viral membrane proteins--they are processed in the same way as cellular proteins, and so can be used as tools to dissect basic cellular functions.

Viral membrane proteins are also studied because they play several important roles during the virus life cycle, being responsible for attachment to the cell surface and for inducing

NATO ASI Series, Vol. H 91
Trafficking of Intracellular Membranes
Edited by M.C. Pedroso de Lima N. Düzgüneş and D. Hoekstra

a fusion reaction between the viral envelope and a cellular membrane. Thus, some viral membrane proteins serve as excellent models for receptor binding and membrane fusion activities. Other viral membrane proteins exhibit enzymatic activity and at least one, the influenza virus M2 protein, forms an ion channel (Pinto *et al.*, 1992). In this chapter, we will review approaches and techniques used to study viral membrane protein folding and assembly. The strengths and weaknesses of each approach will be emphasized rather than the technical details of each, which can be found in the indicated references.

Folding of viral membrane proteins

General life cycle. Most viral membrane proteins are translocated into the endoplasmic reticulum as either type 1 or type 2 integral membrane proteins. Translocation occurs cotranslationally and is initiated by either a cleaved (type 1) or non-cleaved (type 2) signal sequence. There are also several examples of type 3 viral membrane proteins, which have the orientation of type 1 membrane proteins (with the N-terminus of the protein on the exoplasmic side of the membrane) but lack a cleaved N-terminal signal sequence. At least some of these proteins are translocated in the ER posttranslationally (Wilson-Rawls *et al.*, 1994). Most viral membrane proteins span the bilayer one time, with the consequence that they are divided into three topologically distinct domains: an ectodomain that contains most of the protein's mass, a transmembrane region, and a cytoplasmic domain. In this chapter, we will concentrate on type I and II membrane proteins, specifically their ectodomains.

Viral membrane proteins are typically large, structurally complex, and form oligomeric complexes. The step-wise process by which these proteins acquire their final three dimensional structures from their extended, nascent conformations has been the subject of intense scrutiny. While the information required for a protein to attain its final structure lies in its primary sequence (Anfinsen, 1986), it has become clear that folding is not a spontaneous process. Rather, folding is aided by the unique environment within the lumen of the ER as well as by a battery of folding enzymes and molecular chaperones. Modifications such as glycosylation and disulfide bond formation also play an important role. Folding begins cotranslationally. Thus, by the time synthesis is complete, the N-terminal portion of the protein may already be extensively folded. Such vectorial folding is one way the number of possible folding reactions can be limited, perhaps increasing the efficiency of the process.

Certain protein modifications can also occur cotranslationally. Disulfide bonds can clearly form cotranslationally, aided both by a resident ER protein, protein disulfide isomerase, as well as by a redox potential sufficiently oxidizing to allow disulfide bonds to form (Bergman and Kuehl, 1979; Peters and Davidson, 1982; Braakman *et al.*, 1991; Hwang *et al.*, 1992; Segal *et al.*, 1992). The importance of disulfide bond formation for correct folding is indicated by the fact that cysteine residues are generally highly conserved. Also, addition of a reducing agent such as dithiothreitol prevents disulfide formation and, as a consequence, newly synthesized glycoproteins are retained in the ER in an unfolded state (Braakman *et al.*, 1992a; Braakman *et al.*, 1992b; deSilva *et al.*, 1993).

The second protein modification that can greatly influence folding is glycosylation. N-linked carbohydrate chains are added cotranslationally, and so are added before or during protein folding. Inhibition of N-linked glycosylation by addition of tunicamycin, for example, often results in misfolding and retention in the ER, though there are some exceptions (Gibson *et al.*, 1979; Machamer *et al.*, 1985; Doms, 1993). This indicates that the addition of large, hydrophilic carbohydrate chains is often required for a protein to fold correctly, perhaps by rendering folding intermediates more soluble and less prone to aggregation. That carbohydrate addition may play a general role in protein folding rather than a specific, local function, is indicated by a number of findings. For example, the effects of N-linked site elimination may be additive in some cases. For the influenza HA and VSV G glycoproteins, elimination of any single N-linked site is well tolerated, though the loss of both sites in VSV G or three or more sites in influenza prevents normal folding (Machamer *et al.*, 1985; Machamer and Rose, 1988b; Ng *et al.*, 1990; Gallagher *et al.*, 1992). In some instances, addition of a new N-linked consensus site can compensate for the loss of another, providing further evidence that there is some flexibility in the positions of carbohydrate chains in a protein (Guan *et al.*, 1985; Gallagher *et al.*, 1988; Machamer and Rose, 1988a).

Protein folding continues posttranslationally, and is aided by a number of folding enzymes such as protein disulfide isomerase and proline hydroxylase, and molecular chaperones such as GRP78/BiP and calnexin. GRP78/BiP is a lumenal protein that is retained in the ER by a C-terminal KDEL sequence and exhibits a relatively broad substrate specificity. It has been shown that GRP78-BiP recognizes a heptameric motif with alternating hydrophobic residues (Blond-Elguindi *et al.*, 1993). By binding to unfolded regions of the native protein that contain hydrophobic residues, GRP78/BiP may prevent protein aggregation and assist in folding. Calnexin, a more recently discovered chaperone, is an integral membrane protein that appears to bind only to glycoproteins (Hammond and Helenius, 1993; Bergeron *et al.*, 1994). Calnexin may aid in the retention of proteins in the ER until folding and assembly are complete. Both of these chaperones remain stably associated with proteins that fail to fold correctly. Thus, long term association of a molecular chaperone with a viral membrane protein may well indicate that it is misfolded.

One of the last folding events is oligomerization, probably because proteins must fold into relatively mature conformations before they can recognize other subunits as appropriate assembly partners. In the case of the trimeric influenza hemagglutinin (HA), subunits are recruited randomly from a mixed pool of monomers in the ER - there is no preferential association between subunits arising from the same polysome (Boulay *et al.*, 1988). Generally, proteins are rapidly cleared from the ER following assembly. Mutations that prevent normal folding and assembly almost always cause proteins to be retained in the ER where they are ultimately degraded. While many proteins fold and assemble with a $t_{1/2}$ of ≤ 10 min., some fold more slowly (Doms *et al.*, 1993). The Uukuniemi virus G2 protein, for example, folds with a $t_{1/2}$ of approximately 40 min. (Persson and Pettersson, 1991). Despite differences in folding kinetics, most viral membrane proteins fold, assemble, and are transported from the ER efficiently. Proteins that exit the ER have generally attained mature conformations. Carbohydrate processing in the Golgi may influence protein structure, and in some instances proteins may assemble into higher order structures in the Golgi or even undergo disulfide bond rearrangement.

<u>Reasons for studying.</u> There are at least two general reasons for studying viral membrane protein folding. First, as discussed above, viral membrane proteins are useful tools with which to study normal cellular processes. Studies using viral membrane proteins have contributed greatly to our understanding of protein folding, the contribution of N-linked carbohydrates to the folding process, the role of molecular chaperones such as calnexin, and the quality control mechanisms that regulate transport from the ER. Second, mutagenesis is commonly employed to study viral membrane protein structure and function. While mutations can affect a protein function such as receptor binding or membrane fusion directly, they can also influence function indirectly through alterations in protein structure. Thus, to accurately assess the effects of any given mutation on protein structure and function, it is important to know the kinetics and efficiency with which a protein is processed and transported.

Folding assays

A variety of approaches can be taken to follow the time course of protein folding, all of which rely on pulse-chase protocols. A detailed description of this technique is provided by Bonifacino et al. (Bonifacino, 1992). Briefly, cells are placed in methionine free medium for approximately an hour, after which [^{35}S]methionine is added to begin the labeling period. The labeling period is terminated by removing the media and replacing it with normal growth medium containing excess unlabeled methionine. There are two critical parameters: the amount of label used, and the length of the labeling period. The amount of label used will depend on the number of methionines in the protein of interest as well as the protein's expression level. We generally use 100 μCi/ml for labeling, but this can be increased if expression levels are low. The length of the labeling period depends on the process being studied. For example, disulfide bond formation, which begins cotranslationally and is generally complete by 2 to 3 min. after synthesis, requires pulse-labeling periods of < 5 min.

Disulfide bond formation. Most viral and cellular membrane proteins contain intrachain disulfide bonds. Disulfide bond formation is an early folding event, often beginning cotranslationally. Disulfide bond formation typically causes proteins to migrate more quickly in SDS-PAGE under non-reducing conditions. Thus, the mature, fully oxidized molecule will migrate more quickly than the fully reduced precursor. Species containing an incomplete complement of disulfide bonds may also be resolved (Machamer *et al.*, 1990; Braakman *et al.*, 1992a). To detect disulfide bond formation, the standard pulse chase protocol is employed with as short a pulse time as possible. Since a typical viral membrane protein is synthesized in only 1 to 2 minutes and disulfide formation begins cotranslationally, pulse times of 5 min or less are usually necessary. An oxidizing agent such as iodoacetamide is included in the lysis buffer to prevent disulfide formation from occurring after lysis of the cells. After immunoprecipitation the sample is subjected to SDS-PAGE under non-reducing conditions. One sample should be reduced prior to SDS-PAGE to confirm the position of the fully reduced molecule.

The major advantages of monitoring disulfide bond formation is that it is applicable for most proteins and does not require special reagents, such as conformation dependent antibodies. It is important to retain the stacking gels, since misfolded proteins may form high molecular weight, disulfide-linked complexes that may not enter the separating gel. Incompletely disulfide bonded molecules are often associated with molecular chaperones. In the case of VSV G protein, GRP78/BiP is associated with a large fraction of incompletely disulfide bonded molecules but only a small fraction of the fully oxidized protein .

Association with molecular chaperones. It appears that most viral membrane proteins associate with molecular chaperones such as GRP78/BiP and calnexin during folding. Binding to molecular chaperones is transient and often labile: special lysis conditions must sometimes be employed. To detect GRP78/BiP binding, for example, it is important to deplete the lysate of ATP and to keep it at 4°C . To detect chaperone binding, a pulse chase experiment is performed, with equal portions of each lysate being immunoprecipitated with antibodies to the viral protein and to the molecular chaperone. By comparing the amount of env protein co-immunoprecipitated by the chaperone antibody to the total amount of env available, the kinetics and efficiency of chaperone binding and release can be determined. Long term, stable chaperone binding can be taken as presumptive evidence that a protein is misfolded.

Antigenic changes. One of the most useful ways to monitor protein folding is to use antibodies to monitor the loss or gain of antigenic epitopes. Antibodies directed to conformational epitopes are especially useful for folding studies since these epitopes are composed of amino acids that are separated in the protein's primary sequence but are brought into close proximity during protein folding. Antibodies that may prove useful for folding studies typically immunoprecipitate native protein, but fail to react with the denatured molecule by western blotting (Doms *et al.*, 1988). Other antibodies may recognize epitopes that are present in the nascent protein, but are sequestered as the protein folds. Generally, these antibodies recognize linear epitopes and typically react with denatured protein by western blot, but fail to immunoprecipitate native protein. By using antibodies to different regions of the protein of interest, it may be possible to learn how quickly different domains fold relative to one another (Doms, 1990).

Acquisition of function. In some instances, acquisition of function (such as receptor binding or enzymatic activity) can be taken as a measure of protein folding. The HIV-1 env protein, for example, binds to CD4. Because the binding site is conformational in nature, nascent env protein will not bind to CD4. As the protein folds, however, it acquires the ability to bind CD4, which occurs approximately 15 min after synthesis (Earl *et al.*, 1991).

Protease resistance. Many viral membrane proteins are relatively resistant to proteases. Influenza HA, for example, is sensitive to trypsin immediately after synthesis, but becomes resistant to digestion with kinetics that parallel the rate of HA trimer formation (Copeland *et al.*, 1986; Gething *et al.*, 1986). There is no way to determine a priori which protease(s) might be useful. A panel of proteases with different specificities can be screened against the mature protein, such as that found in virus particles. Proteases to which the mature protein is resistant may then be used in pulse chase experiments to determine the kinetics of folding. This has been used, for example, to follow the folding of the influenza HA protein (Copeland *et al.*, 1986; Gething *et al.*, 1986).

Assembly assays

Most viral membrane proteins form homo- or hetero-oligomeric complexes that are covalently or noncovalently associated (Doms *et al.*, 1993). Oligomerization is typically a late folding event, since individual subunits must fold to the point where they can be recognized as appropriate assembly partners. Assembly is required for transport - mutations that prevent assembly cause the protein to be retained in the ER. A number of methods can be used to monitor assembly, as briefly discussed below.

Velocity gradient sedimentation. Newly synthesized viral membrane proteins are monomeric. If cells are pulse labeled for a brief period, lysed, and the lysate subjected to velocity gradient sedimentation, the protein will sediment in monomeric form. Centrifugation is generally performed using 5 to 20% sucrose gradients since over this concentration range the effects of viscosity are negligible. If a membrane protein is being studied, a nonionic detergent must be included at a concentration above the critical micelle concentration (we use 0.1% TX100 or 40mM octylglucoside). After centrifugation, fractions are collected and the distribution of the protein across the gradient determined by immunoprecipitation (it is important to use an antibody that recognizes all conformations of the protein for this step). Alternatively, if protein concentration is sufficient, fractions may be directly visualized by SDS-PAGE and western blotting. An example of velocity gradient sedimentation is shown in Figure 1. A secreted form of the human immunodeficiency virus type 1 envelope protein was placed atop a 5% to 20% sucrose gradient and centifuged for 20 hours at 4°C and 40,000 rpm in a SW40Ti rotor. Fractions were collected from the bottom of the gradient and the distribution of envelope protein determined by SDS-PAGE and western blotting. Because the envelope protein was reduced and denatured, it migrates as a single band at 140 Kd as seen in the figure. Since, in velocity gradient sedimentation, proteins with a higher molecular weight sediment more rapidly, envelope protein dimers and higher order oligomeric structures are recovered nearer to the bottom of the gradient, while monomeric envelope protein is found closer to the top. The positions of monomeric and dimeric protein are indicated.

Note that for protein characterization, the position of monomeric protein can be confirmed by denaturing it with SDS prior to centrifugation. Pulse-chase protocls can also be used. Immediately after pulse-labeling, most of the protein will be recovered in monomeric form. With increasing time of chase, the metabolically labeled protein will assemble and sediment further down the gradient (Copeland *et al.*, 1986; Gething *et al.*, 1986; Doms *et al.*, 1987; Einfeld and Hunter, 1988). To determine the position at which oligomeric protein sediments, purified virus particles can be solubilized and used as a source of mature oligomeric protein. After centrifugation, the sedimentation coefficients can be estimated by using standards in parallel gradients, and by the method of McEwen if the concentration of sucrose in the peak fraction is known (McEwen, 1967), and in turn related to the protien's molecular weight.

One of the drawbacks of velocity gradient centrifugation is the amount of time involved in preparing, running, and analyzing the gradients. A more significant problem, however, is that some noncovalently associated oligomers are not stable to detergent solubilization (e.g. the SFV spike protein) or centrifugation (the VSV G protein), and so are recovered as monomers (Doms *et al.*, 1987). If this occurs, several approaches may be tried. First, different

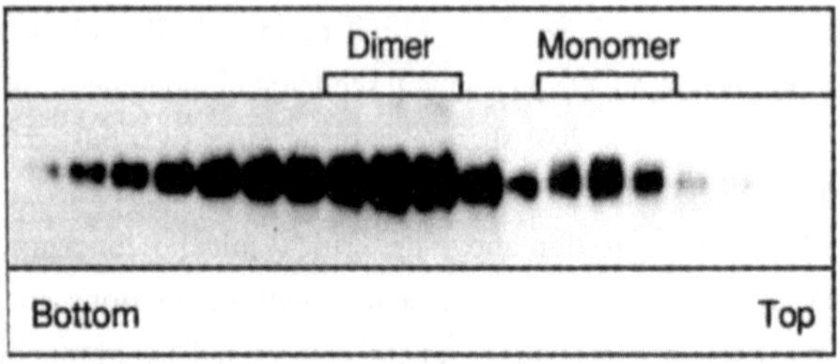

Figure 1. Velocity gradient sedimentation. A soluble form of the human immunodeficiency virus type 1 envelope protein was subjected to centrifugation on a 5%-20% sucrose velocity gradient. Fractions were collected from the bottom of the gradient and the distribution of envelope protein across the gradient determined by SDS-PAGE and western blot. The positions of dimeric and monomeric envelope protein are indicated.

detergents may be utilized. In our laboratory, we test Triton X-100, octyl glucoside, and CHAPS. The protein may also be chemically cross-linked prior to centrifugation (Earl *et al.*, 1990), though this technique has its own attendant problems as discussed below. Some proteins exist in distinct conformations, and the stability of these conformations may vary. For example, viral proteins that mediate membrane fusion at acid pH have two distinct conformations: the neutral pH conformation, and one that occurs at the pH threshold required for fusion, generally between pH 5 and 6. The VSV G protein is a trimer which undergoes such a pH-dependent conformational change. At neutral pH, G protein trimers are stable to detergent solubilization but dissociate during ultracentrifugation. However, the acid pH conformation of G protein is much more stable. Thus, G protein assembly can be monitored by velocity gradient centrifugation provided that the pH of the gradients is kept at or below pH 5.6 (Doms *et al.*, 1987). Proteins that have receptor binding properties may also be incubated with their ligand in an attempt to obtain a more stable complex. Finally, it is important to demonstrate that the membrane protein being studied sediments independently of other viral proteins. While not a problem when expression systems are used, during the course of a virus infection spike proteins may associate with matrix or capsid proteins.

While sucrose velocity gradient sedimentation is a useful technique for monitoring the kinetics of oligomerization, it is less useful for actually defining the oligomeric state of any given protein. In particular, there are two factors that make it difficult to define the oligomeric state of a protein using this technique alone. First, membrane proteins bind detergent causing them to sediment more slowly than one would predict based on size alone. The effects of detergent can be taken into account using equilibrium gradient sedimentation, though this is often not practical. Second, viral membrane proteins often have an elongated shape, which also causes them to sediment more slowly than expected. As a result, it is not advisable to predict the oligomeric state of a protein based on its sedimentation coefficient alone. Rather, gradient centrifugation should be used in conjunction with chemical cross-linking, as described below.

Chemical cross-linking. Chemical cross-linkers may be used both to define the oligomeric state of a protein as well as to monitor the rate at which oligomers assemble. Pierce Chemical Co. is a good source of cross-linkers and technical advice. Cross-linkers contain two reactive groups separated by a spacer arm of variable length. The reactive groups can be identical (homobifunctional) or different (heterobifunctional), and most react with primary amines. The solubility of cross-linkers varies, with some being membrane permeable. It is difficult to predict which cross-linker will work best for any given protein. One can try a variety of cross-linkers with different solubility properties and spacer lengths individually, or a number of cross-linkers can be pooled together. Cross-linking should be performed in buffers that lack primary amines, and most cross-linkers work best at mildly basic pH. We generally perform cross-linking reactions for 15 to 30 min, after which we add glycine to quench excess cross-linker. To avoid the problem of diffuse protein bands which often result from cross-linking, we prefer gradient over single concentration SDS-PAGE gels.

Results from cross-linking studies must be carefully interpreted: lack of cross-linking does not necessarily mean that a protein is monomeric, and a positive result does not necessarily mean that it is oligomeric. A negative result (lack of cross-linking) is more believable if a number of cross-linkers have been tried under a variety of conditions, and if other techniques (such as velocity gradient sedimentation) corroborate the result. If higher molecular weight complexes are observed after cross-linking, they may reflect the native oligomeric state of the protein, or they may be artefactual homo- or hetero-oligomeric protein complexes. The presence of a number of findings are indicative that a cross-linked complex reflects native oligomeric structure: if the complex is obtained at relatively low concentrations of cross-linker (1 mM or less); if the same complex is observed with several reagents; if cross-linking is quantitative; and if a ladder of progressively larger complexes is observed with increasing concentrations of cross-linker. A good example of a ladder can be seen in Figure 2. The soluble, dimeric form of the human immunodeficiency virus envelope protein shown in Figure 1 migrates as a 140 kD monomer under denaturing conditions in SDS-PAGE. Addition of the cross-linker ethylene glycolbis-[succinimidylsuccinate] (EGS) prior to SDS-PAGE cross-links the envelope protein dimers, preventing them from dissociating. As a result, the cross-linked dimers migrate as 280 kD complexes in SDS-PAGE. Cross-linking is first observed at very low concentrations of EGS, and is complete with 0.5 mM EGS. Three findings support the conclusion that prior to denaturation the envelope protein exists as a dimer: cross-linking of envelope protein dimers occured at low concentrations of EGS, cross-linking was quantitative at a relatively low concentration of EGS, and other cross-linked species were not observed. Similar results were obtained with other chemical cross-linkers, further supporting this conclusion.

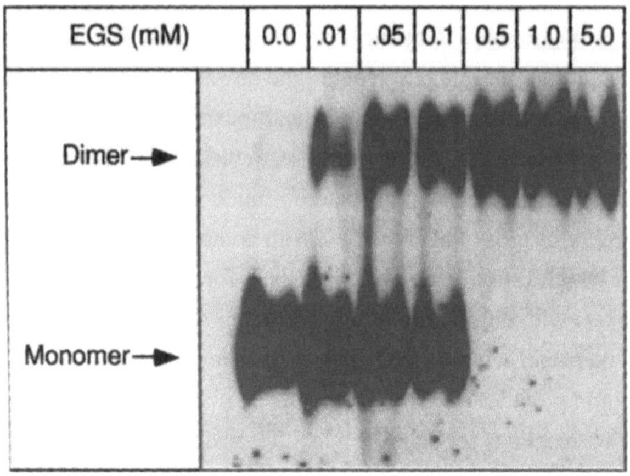

Figure 2. Chemical cross-linking. Dimeric HIV-1 envelope protein was incubated with the indicated concentrations of the chemical cross-linker EGS (Pierce). The cross-linked proteins were reduced, boiled, run on a 3%-10% SDS-PAGE gradient gel and then Western blotted using a cocktail of monoclonal antibodies. Monomeric and dimeric envelope protein are indicated.

Chemical cross-linking is particularly useful when used in conjunction with velocity gradient sedimentation. An example of this is shown in Figure 3. The monomer and dimer fractions from a sucrose velocity gradient (Figure 1) were pooled separately, cross-linked with EGS, and analyzed by SDS-PAGE and western blot. Under reducing conditions, the envelope protein from the dimer fractions ran partially as a dimer without any added cross-linker, and increasingly as a dimer as cross-linker was added. The envelope protein from the monomer fractions, on the other hand, migrated as a monomer both with and without cross-linker. The molecular weight estimates from these cross-linking results correlate with the sedimentation estimates from the sucrose velocity gradient (Figure 1). The absence of bands at the higher concentrations is common and most likely due to the non-specific formation of high molecular weight aggregates which fail to enter the gel. Another method to analyze the oligomeric structure of a membrane protein is to cross-link the proteins prior to centrifugation. This is a useful technique if the protein dissociates during centrifugation.

SDS-resistance. Some protein oligomers display varying degrees of resistance to SDS-induced dissociation (See Fig.3; also Doms and Helenius, 1986; Yewdell *et al.*, 1988; Pinter *et al.*, 1989). This is particularly true of membrane proteins, since hydrophobic, alpha helical transmembrane segments may constitute inherently SDS-stable structures (Doms and Helenius, 1986). To determine if a protein exhibits resistance to SDS, we solubilize immunoprecipitated protein in buffers that contain different SDS concentrations (from 0.1% to 1.0%). We also vary the temperature of incubation prior to SDS-PAGE (from room temperature to 95°C), and leave out reducing agents. Under less denaturing conditions, some proteins migrate as oligomeric complexes, including certain influenza HA subtypes and the HIV-1 env protein. The protein may not, however, migrate quantitatively as an oligomer, which sometimes minimizes the usefulness of this technique.

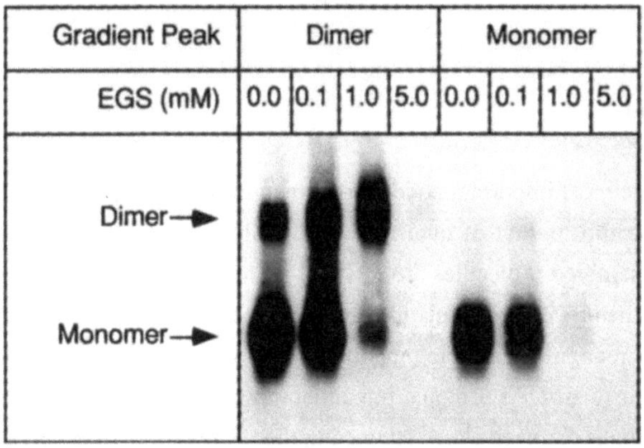

Fig. 3. Chemical cross-linking after centrifugation. Envelope protein was analyzed by velocity gradient sedimentaiton as shown in Figure 1. Fractions containing monomeric and dimeric envelope protein were separately pooled and then cross-linked using EGS (Pierce) at the concentrations indicated. The cross-linked fractions were reduced, boiled, run on a 3%-10% SDS-PAGE gradient gel and then western blotted using a cocktail of monoclonal antibodies.

Antigenic changes. The simplest way to monitor oligomerization is to take advantage of oligomer-specific antibodies, provided they exist (Copeland *et al.*, 1986; Gething *et al.*, 1986; Yewdell *et al.*, 1988). If available, oligomer formation can be monitored simply by performing a pulse chase experiment and comparing the amount of protein immunoprecipitated by the oligomer dependent antibody to that precipitated by a control oligomer independent antibody at each time point. It is important to recognize, however, that assembly could actually occur prior to the formation of any given oligomer dependent epitope. However, since assembly appears to require relatively well folded monomeric subunits, the differences between the real and apparent kinetics of assembly are apt to be minimal.

Quality control

One of the most important concepts to emerge from studies of viral membrane protein folding and assembly is that of quality control. Simply put, this is the process by which misfolded, unassembled molecules are retained in the ER where they are either degraded or rescued. While the precise mechanisms are not clear, there are several possibilities explaining how proteins are recognized as being misfolded and, as a consequence, retained. It is known, for example, that misfolded proteins remain stably associated with molecular chaperones. Since the chaperones themselves are retained in the ER by virtue of ER retention signals, misfolded or incompletely folded proteins to which they are bound may be retained as a consequence. Misfolded proteins also exhibit a tendency to form large, insoluble aggregates that may contain interchain disulfide bonds. Such complexes may not be delivered to or incorporated in transport vesicles. Quality control mechanisms may also exist outside of the ER. For example, a number of mutations in viral membrane proteins have been described that cause them to be retained in the Golgi (Naruse *et al.*, 1986; Garten *et al.*, 1992). Taken together, cells are likely to possess a number of mechanisms to retain misfolded proteins and provide for efficient conformational sorting.

Implications for mutagenesis

Mutations are commonly introduced into viral and cellular membrane proteins to learn something about their structure and function. While mutations can affect a given function directly (for example, by replacing a residue involved in receptor interactions), they can also affect protein function indirectly by inducing general structural alterations. While the consequences of any given mutation on protein structure and function cannot be predicted with certainty, some helpful generalizations can be made to assist in the design of mutational studies. Ectodomain cysteines are almost always involved in inter- or intramolecular disulfide bonds. Elimination of any single cys residue (or the addition of a new cys) leaves an unpaired cys that can participate in aberrant disulfide bonds. Indeed, elimination of single cys residues often leads to the formation of misfolded, disulfide linked complexes, whereas the elimination of a cys pair is often tolerated (Wilcox *et al.*, 1988; Long *et al.*, 1992; Segal *et al.*, 1992). Cys residues involved in interchain disulfide bonds are generally not as important for folding and can often be deleted (Alvarez *et al.*, 1989; Holsinger and Lamb, 1991).

Since addition of N-linked carbohydrate chains is often required for folding (but not necessarily maintenance of native structure), elimination or addition of N-linked sites can have deleterious consequences for folding and assembly. The effects of glycosylation mutations may be cumulative, and may generate a temperature sensitive folding phenotype (Machamer and Rose, 1988b; Roberts *et al.*, 1992). Introduction of new N-linked sites to regions of a protein known or suspected to be surface exposed is better tolerated than the introduction of new sites into regions normally sequestered in the interior of the molecule or involved in subunit-subunit interactions (Schuy *et al.*, 1986; Gallagher *et al.*, 1988; Machamer and Rose, 1988a).

More severe mutations involving insertions or deletions are often less well tolerated. However, swaps between independent folding domains frequently result in proteins that are transported normally. Since the cytoplasmic, transmembrane, and ectodomains fold relatively independently of one another, exchanging these domains often yields useful chimeric molecules.

Summary

Viral membrane proteins have proven to be useful tools with which to dissect the mechanisms by which proteins fold and assemble in the ER, and how these processes relate to protein transport. In addition, studies of viral membrane protein folding, assembly, and transport yield a "biosynthetic framework" that can be used to assess the effects of mutations on protein structure and function. Future studies will have to address a number of unresolved issues, such as the precise roles played by molecular chaperones in the folding and quality control processes, how the ER environment is maintained and regulated, and if quality control mechanisms exist outside of the ER.

References

Alvarez, E., Girones, N., and Davis, R. J. (1989). Intermolecular disulfide bonds are not required for the expression of the dimeric state and functional activity of the transferrin receptor. *EMBO J.* **8,** 31-40.

Anfinsen, C. (1986). Classical protein chemistry in a world of slicing and splicing. *In* "Protein engineering. Applications in science, medicine, and industry" (M. Inouye, and R. Sarma, Eds.), pp. 3-13. Academic, Orlando, Fla.

Bergeron, J. J. M., Brenner, M. B., Thomas, D. Y., and Williams, D. B. (1994). Calnexin: a membrane-bound chaperone of the endoplasmic reticulum. *Trends Biochem.Sci.* **19,** 124-128.

Bergman, L. W., and Kuehl, W. M. (1979). Formation of an intrachain disulfide bond on nascent immunoglobulin light chains. *J. Biol. Chem.* **254,** 8869-8876.

Blond-Elguindi, S., Cwirla, S. E., Dower, W. J., Lipshutz, R. J., Sprang, S. R., Sambrook, J. F., and Gething, M. J. (1993). Affinity panning of a library of peptides displayed on bacteriophages reveals the binding specificity of BiP. *Cell* **75,** 717-718.

Bonifacino, J. S. (1992). Biosynthetic labeling of proteins. *In* "Current Protocols in Molecular Biology" (F. M. Ausubel, R. Brent, R. E. Kingston, D. D. Moore, J. G. Seidman, J. A. Smith, and K. Struhl, Eds.), pp. 10.18.1- 10.18.9. Greene Publishing Assoc., Inc., New York.

Boulay, F., Doms, R. W., Webster, R. G., and Helenius, A. (1988). Posttranslational oligomerization and cooperative acid activation of mixed influenza hemagglutinin trimers. *J. Cell Biol.* **106,** 629-639.

Braakman, I., Helenius, J., and Helenius, A. (1992a). Manipulating disulfide bond formation and protein folding in the endoplasmic reticulum. *EMBO J.* **11,** 1717-1722.

Braakman, I., Helenius, J., and Helenius, A. (1992b). Role of ATP and disulfide bonds during protein folding in the endoplasmic reticulum. *Nature* **356,** 260-262.

Braakman, I., Hoover-Litty, H., Wagner, K. R., and Helenius, A. (1991). Folding of influenza hemagglutinin in the endoplasmic reticulum. *J. Cell Biol.* **114,** 401-411.

Copeland, C. S., Doms, R. W., Bolzau, E. M., Webster, R. G., and Helenius, A. (1986). Assembly of influenza hemagglutinin trimers and its role in intracellular transport. *J. Cell Biol.* **103,** 1179-1191.

deSilva, A., Braakman, I., and Helenius, A. (1993). Post-translational folding of VSV G protein in the endoplasmic reticulum: Involvement of noncovalent and covalent complexes. *J. Cell Biol.* **120,** 647-655.

Doms, R. W. (1990). Oligomerization and protein transport. *Meth. Enzymol.* **191,** 841-854.

Doms, R. W. (1993). Protein conformational changes in virus-cell fusion. *Methods Enzymol.* **221,** 61-72.

Doms, R. W., and Helenius, A. (1986). Quaternary structure of the influenza virus hemagglutinin after acid treatment. *J. Virol.* **60,** 833-839.

Doms, R. W., Helenius, A., and Balch, W. (1987). Role for ATP in the assembly and transport of VSV G protein trimers. *J. Cell Biol.* **105,** 1957-1969.

Doms, R. W., Lamb, R., Rose, J. K., and Helenius, A. (1993). Folding and assembly of viral membrane proteins. *Virology* **193,** 545-562.

Doms, R. W., Ruusala, A., Machamer, C., Helenius, J., Helenius, A., and Rose, J. K. (1988). Differential effects of mutations in three domains on folding, trimerization and intracellular transport of VSV G protein trimers. *J. Cell Biol.* **107,** 89-99.

Earl, P. L., Doms, R. W., and Moss, B. (1990). Oligomeric structure of the human immunodeficiency virus type 1 envelope glycoprotein. *Proc. Natl. Acad. Sci. USA* **87,** 648-652.

Earl, P. L., Moss, B., and Doms, R. W. (1991). Folding, interaction with GRP78-BiP, assembly, and transport of the human immunodeficiency virus type 1 envelope protein. *J. Virol.* **65,** 2047-2055.

Einfeld, D., and Hunter, E. (1988). Oligomeric structure of a prototype retrovirus glycoprotein. *Proc. Natl. Acad. Sci. USA* **85,** 8688-8692.

Gallagher, P., Henneberry, J., Wilson, I., Sambrook, J., and Gething, M.-J. (1988). Addition of carbohydrate side chains at novel sites on influenza hemagglutinin can modulate the folding, transport, and activity of the molecule. *J. Cell Biol.* **107,** 2059-2073.

Gallagher, P. J., Henneberry, J. M., Sambrook, J. F., and Gething, M.-J. H. (1992). Glycosylation requirements for intracellular transport and function of the hemagglutinin of influenza virus. *J. Virol.* **66,** 7136-7145.

Garten, W., Will, C., Buckard, K., Kuroda, K., Ortmann, D., Munk, K., Scholtissek, C., Schnittler, H., Drenckhahn, D., and Klenk, H.-D. (1992). Structure and assembly of hemagglutinin mutants of fowl plaque virus with impaired surface transport. *J. Virol.* **66,** 1495-1505.

Gething, M.-J., McCammon, K., and Sambrook, J. (1986). Expression of wild-type and mutant forms of influenza hemagglutinin: the role of folding in intracellular transport.

Cell **46,** 939-950.

Gibson, R., Schlesinger, S., and Kornfeld, S. (1979). The nonglycosylated glycoprotein of vesicular stomatitis virus is temperature sensitive and undegoes intracellular aggregation at elevated temperatures. *J. Biol. Chem.* **254,** 3600-3607.

Guan, J.-L., Machamer, C. E., and Rose, J. K. (1985). Glycosylation allows cell-surface transport of an anchored secretory protein. *Cell* **42,** 489-496.

Hammond, C., and Helenius, A. (1993). A chaperone with a sweet tooth. *Curr. Biol.* **3,** 884-886.

Holsinger, L. J., and Lamb, R. A. (1991). Influenza virus M2 integral membrane protein is a homotetramer stabilized by formation of disulfide bonds. *Virology* **183,** 32-43.

Hwang, C., Sinskey, A. J., and Lodish, H. F. (1992). Oxidized redox state of glutathione in the endoplasmic reticulum. *Science* **257,** 1496-1502.

Long, D., Wilcox, W. C., Abrams, W. R., Cohen, G. H., and Eisenberg, R. J. (1992). The disulfide bond structure of glycoprotein D of Herpes simplex virus types 1 and 3. *J. Virol.* **66,** 6668-6685.

Machamer, C. E., Doms, R. W., Bole, D. G., Helenius, A., and Rose, J. K. (1990). Heavy chain binding protein recognizes incompletely disulfide-bonded forms of vesicular stomatitis virus G protein. *J. Biol. Chem.* **265,** 6879-6883.

Machamer, C. E., Florkiewicz, R. Z., and Rose, J. K. (1985). A single N-linked oligosaccharide at either of the two normal sites is sufficient for transport of vesicular stomatitis virus G protein to the cell surface. *Mol. Cell Biol.* **5,** 3074-3083.

Machamer, C. E., and Rose, J. K. (1988a). Influence of new glycosylation sites on expression of the vesicular stomatitis virus G protein at the plasma membrane. *J. Biol. Chem.* **263,** 5948-5954.

Machamer, C. E., and Rose, J. K. (1988b). Vesicular stomatitis virus G proteins with altered glycosylation sites display temperature-sensitive intracellular transport and are subject to aberrant intermolecular disulfide bonds. *J. Biol. Chem.* **263,** 5955-5960.

McEwen, C. R. (1967). Tables for estimating sedimentation through linear concentration gradients of sucrose solution. *Anal. Biochem.* **20,** 114-149.

Naruse, H., Scholtissek, C., and Klenk, H.-D. (1986). Temperature-sensitive mutants of fowl plague virus defective in the intracellular transport of the hemagglutinin. *Virus Res.* **5,** 293-305.

Ng, D. T. W., Hiebert, S. W., and Lamb, R. A. (1990). Different roles of individual N-linked oligosaccharide chains in folding, assembly, and transport of the simian virus 5 hemagglutinin-neuraminidase. *Molec. Cell Biol.* **10,** 1989-2001.

Persson, R., and Pettersson, R. F. (1991). Formation and intracellular transport of a heterodimeric viral spike protein complex. *J. Cell Biol.* **112,** 257-266.

Peters, T., and Davidson, L. K. (1982). The biosynthesis of rat serum albumin. In vivo studies on the formation of the disulfide bonds. *J. Biol. Chem.* **257,** 8847-8853.

Pinter, A., Honnen, W. J., Tilley, S. A., Bona, C., Zaghouani, H., Gorny, M. K., and Zolla-Pazner, S. (1989). Oligomeric structure of gp41, the transmembrane protein of human immunodeficiency virus type 1. *J. Virol.* **63,** 2674-2679.

Pinto, L. H., Holsinger, L. J., and Lamb, R. A (1992). Influenza virus M2 protein has ion channel activity. *Cell* **69,** 517-528.

Roberts, C., Garten, W., and Klenk, H.-D. (1993). The role of conserved glycosylation in the maturation and transport of the influenza hemagglutinin. *J. Virol.* **67,** 3048-3060.

Schuy, W., Will, C., Kuroda, K., Scholtissek, C., Garten, W., and Klenk, H.-D. (1986). Mutations blocking the transport of the influenza virus hemagglutinin between the rough endoplasmic reticulum and the Golgi apparatus. *EMBO J.* **5,** 2831-2836.

Segal, M. S., Bye, J. M., Sambrook, J. F., and Gething, M.-J. H. (1992). Disulfide bond formation during the folding of influenza virus hemagglutinin. *J. Cell Biol.* **118,** 227-244.

Wilcox, W. C., Long, D., Sodora, D. L., Eisenberg, R. J., and Cohen, G. H. (1988). The contribution of cysteine residues to antigenicity and extent of processing of herpes simplex virus type 1 glycoprotein D. *J. Virol.* **62,** 1941-1947.

Wilson-Rawls, J., Deutscher, S. L., and Wold, W. S. M. (1994). The signal-anchor domain of adenovirus E3-6.7K, a type III integral membrane protein, can direct adenovirus E3-

gp19K, a type I integral membrane protein, into the membrane of the endoplasmic reticulum. *Virology* **201,** 66-76.

Yewdell, J. W., Yellen, A., and Bächi, T. (1988). Monoclonal antibodies localize events in the folding, assembly, and intracellular transport of the influenza hemagglutinin glycoprotein. *Cell* **52,** 843-852.

Membrane Fusion Induced by the HIV env Glycoprotein: Purification of CD4 for Reconstitution Studies

Charles Larsen[1], Arun Patel[2] and Joe Bentz
Department of Bioscience and Biotechnology
Drexel University
Philadelphia, PA 19104
U. S. A.

Introduction

Membrane fusion is the union of two bilayers resulting in a redistribution of aqueous contents and bilayer components. Fusion is a critical event in biological systems, being a required step during intracellular trafficking, cellular endocytosis and exocytosis, zygote formation, cellular attack (e.g., enveloped virus fusion) and several specialized processes. Despite progress in understanding membrane fusion regulation in a variety of biological systems, the mechanism of the fusion event itself remains unclear. Most model membranes are stable and do not fuse without an external trigger. Biological membrane fusion requires a specific mediator, which is presumed to be a protein.

Fusion proteins are integral membrane proteins that destabilize membrane bilayers by directly mediating lipid mixing and bilayer reorganization. The only well described fusion proteins are in lipid enveloped viruses. Although promising discoveries are occurring in the fields of intracellular membrane trafficking and zygote formation (as detailed in this conference report and in Blobel et al. (1992) and Söllner et al. (1993)), most reports in those fields describe the regulatory elements controlling membrane fusion rather than the central event itself.

Viral fusion may not be representative of all membrane fusion, but the lack of well described membrane fusion proteins in other systems leads many to use viral fusion proteins as models for fusion in other biological processes. The influenza virus fusion protein, hemagglutinin (HA), is the best characterized fusion protein. It is therefore often used as a model for other fusion proteins (White, 1992). Recent reports describing novel conformational changes in HA structure (Bullough et al., 1994; Carr and Kim, 1993) will probably also stimulate new research.

---

[1]Current address and correspondence: Center for Blood Research Laboratories, 800 Huntington Avenue, Boston, MA 02115 U.S.A.
[2]Department of Protein Biochemistry, Mail Code UE0432, SmithKline Beecham Pharmaceuticals, Inc., P.O. Box 1539, King of Prussia, PA 19406 U. S. A.

NATO ASI Series, Vol. H 91
Trafficking of Intracellular Membranes
Edited by M.C. Pedroso de Lima N. Düzgüneş and D. Hoekstra

Studying the fusion mechanisms of other viral fusion proteins may help determine the features of a more generally applicable membrane fusion mechanism. The human immunodeficiency virus (HIV) fusion protein is the envelope glycoprotein gp160 (gp120/41). Despite intense study and much speculation about the HIV infection pathway, little is known about the membrane fusion mechanism used by HIV to gain entry into host cells (Larsen et al., 1992; Ellens and Larsen, 1993). In an attempt to understand better the CD4–dependent fusion mechanism of HIV, we purified native human transmembrane CD4 (tmCD4) for reconstitution into model membranes. Here we review several key questions in the field of HIV fusion, discuss the necessity of using purified CD4-containing model membranes to answer several of those questions and briefly describe the purification of CD4[3]. Finally, a potential contents delivery fusion assay using CD4-containing liposomes is discussed. (See Bentz (1993a), Larsen et al. (1992) and the chapters on virus fusion and HIV in these Proceedings by M.C.P. de Lima, S. Nir, R. Doms and N. Düzgünes for more general reviews.)

The HIV Fusion Mechanism: Background, Questions and Models

HIV env glycoprotein structure and function. This topic is extensively covered in these Proceedings by R. Doms and in our previous reviews (Ellens and Larsen, 1993; Larsen et al., 1992). We wish to stress only a few points regarding this subject. First, it is important to note that gp120/41, like several other viral fusion proteins, shares some common features with HA. These include its composition as two separate subunits, a transmembrane-anchored subunit (gp41) containing at least a portion of the fusion reaction domain, associated with an external subunit (gp120) containing the receptor binding domain, and a quaternary structure as a multimer of heterodimers. There are some important differences with HA, however. First, unlike HA, gp120 and gp41 are associated non-covalently. Second, the transmembrane subunit gp41 has an unusually long cytoplasmic domain, which may play some regulatory role in viral entry (Dubay et al., 1992). Finally, a more complete mutational analysis of the HIV gp120/41 shows that, while mutations in the N-terminus of gp41 (the so-called "fusion peptide") affect fusion without affecting binding to CD4, other mutations in both gp120 and gp41 have similar effects (Larsen et al., 1992; Ellens and Larsen, 1993; Owens et al., 1994; Cao et al., 1994).

Taken together, these facts suggest the likelihood that conformational changes within several regions of both gp120 and gp41 are involved in the HIV fusion reaction. An earlier report that HIV fusion may occur in the total absence of gp120 (Perez et al., 1992) was recently challenged (Marcon and Sodroski, 1994). Although the three-dimensional structural determination

[3]A detailed technical description of the CD4 purification has been submitted to the journal *AIDS*.

of gp120/41 may not be achieved soon, several groups are using monoclonal antibodies to probe these proteins (Moore et al., 1993b, 1994). Any new information regarding envelope glycoprotein secondary or tertiary structure will be extremely useful, in conjunction with biophysical fusion information, for defining the architecture of the HIV fusion site (Bentz, 1991; Bentz et al., 1993).

Direct fusion assays are required. Membrane fusion is an absolute requirement for at least two important stages of HIV pathogenesis: viral capsid entry (the first step of viral infection after binding to the target cell surface) and syncytium formation (a cytopathic effect in which an infected cell fuses with an uninfected cell). Most experiments designed to investigate HIV fusion rely on infectivity or syncytia assays as an indirect measure of the fusion event. Commonly used infectivity assays are rate dependent on events such as viral replication and/or antigen production, which take place well after virus entry. Syncytium formation is a more direct assay of gp120/41-induced fusion, but the detection of syncytium formation appears to require additional steps beyond membrane fusion (Dimitrov et al., 1991). When well controlled, these measurements can identify specific viral or cellular molecules required for virus infection, but determination of how these molecules interact to trigger fusion requires measuring the event on its own time scale (Bentz et al., 1991).

Furthermore, several lines of evidence suggest either that these assays are sensitive to processes other than fusion or that virus entry and syncytia may utilize different fusion mechanisms. For example, some investigators report mutations in the virus envelope glycoprotein (Helseth et al., 1990; Kowalski et al., 1991) and target cell CD4 (Camerini and Seed, 1990; Corbeau et al., 1993) that dramatically affect syncytium formation without affecting infectivity. Given the limitations of these assays, it is not surprising that contradictory reports have also appeared (e.g., Broder and Berger, 1993).

Such differences might represent variation in efficiencies of the same fusion mechanism, or they may reflect different fusion mechanisms. Alternatively, the detection of syncytium formation may require additional steps beyond fusion. These differences argue for more direct fusion assays to answer mechanistic questions, and they also demonstrate that syncytia assays cannot be used to study the whole range of virus isolates. Non-syncytia producing isolates may provide critical insights into fusion-dependent virus tropism and cell entry.

Examples of published direct entry studies include EM of early virus-cell interactions (Goto et al., 1988; Grewe et al., 1990; Pauza and Price, 1988; Stein et al., 1987), lipid-mixing fusion assays of virus with liposomes (Larsen et al., 1990, 1993) and cells (Sinangil et al., 1988), and a ^{32}P-labeled virus binding and cellular uptake assay (Pauza and Price, 1988). Paradoxically, most of these direct fusion assays challenge several common views about immunodeficiency virus entry. For example, Pauza and Price (1988), Goto et al. (1988) and Grewe et al. (1990) found that HIV entry is not restricted to the plasma membrane. Larsen et al. (1990, 1993) showed that HIV-1 and SIV fuse (although often inefficiently) with a variety of liposomes lacking CD4 or any other protein, and that this fusion is faster and more extensive at values below pH 7.5. However, direct fusion assays must also be interpreted carefully, particularly when they only measure lipid mixing (Düzgünes and Bentz, 1988; Stamatatos and Düzgünes, 1993).

Key questions in HIV fusion. Three key questions in HIV fusion are of interest to us:

1. What are the target membranes for fusion (CD4-dependent and CD4-independent)?
2. Is an auxiliary receptor required in addition to the binding (e.g., CD4 or galactosylceramide) receptor?
3. What is the architecture of the HIV fusion site?

Indirect assays of HIV fusion have led to conflicting results and hypotheses with respect to the fusion mechanism. Although CD4-expressing cells appear to be the most common targets for HIV infection, it is clear that HIV can fuse with and infect cells that do not express CD4 (Larsen et al., 1992, 1993). However, no HIV-1 strain is known to be capable of inducing syncytia in CD4-negative cells. CD4 is not the only receptor used by HIV for binding to target membranes. Galactosylceramide (gal cer) appears to be the most promising candidate as an alternative receptor, but there may be several others. Interestingly, model membranes containing gal cer may be particularly useful for studying the CD4-independent fusion mechanism (Long et al., 1994).

Recently, the hypothesis that, in addition to CD4, (an)other auxiliary membrane molecule(s) may be involved in HIV fusion has gained increasing attention (Broder et al., 1993; Golding et al., 1994; Harrington and Geballe, 1993; Moore et al., 1993a; Stefano et al., 1993). The concept of a second (or auxiliary) receptor in CD4-dependent HIV entry has been hypothesized since 1986 (Klatzmann and Gluckman, 1986) as an explanation for the fact that some CD4-positive cells are not susceptible to infection by immunodeficiency viruses. Since our last review (Ellens and Larsen, 1993), several new auxiliary receptor candidates have been proposed (Callebaut et al., 1994; Ebenbichler et al., 1993; Henderson and Qureshi, 1993; Sato et al., 1994), but no generally accepted second receptor has been identified to date (see articles to which Callebaut et al., 1994 is a response).

Several alternative explanations to the secondary "fusion receptor" hypothesis exist for the inability to establish an HIV-1 infection in certain CD4-positive cells (Ellens and Larsen, 1993). One possibility is that the CD4 in these cells is structurally altered at some site(s) critical for HIV-1 entry (other than at the gp120 binding site). The role of CD4 in post-binding events is often overlooked as an explanation for the inability to establish an HIV-1 infection in CD4-positive cells. We, along with others have proposed beginning models for studying the fusion event mechanism (Allan, 1993; Ellens and Larsen, 1993; Moore et al., 1993a). Our model proposes that gp120/41 mediated fusion is triggered by a fusogenic (gp120/41) dissociation intermediate, which can form spontaneously at a relatively slow rate. This allows for a relatively inefficient CD4-independent fusion process. The rate of gp120/41 dissociation can be enhanced by gp120 binding to CD4 and/or possibly other receptor molecules, resulting in an increased density of fusogenic dissociation intermediates and hence an enhanced fusion efficiency. Implicit in this proposal is that specific primary and auxiliary receptors are not absolutely required for the fusion reaction, but function to enhance the efficiency of the event.

The third key question regarding the HIV fusion mechanism is the architecture of the fusion site. Determination of the architecture of a fusion site is a prerequisite for delineating the mechanism of a fusion protein (Bentz et al., 1990; Bentz, 1993b). This requires knowledge of whether fusion requires more than one gp120/41 oligomer, which can be studied by measuring fusion efficiency as a function of fusion protein surface density (Bentz, 1991). Such an approach was successful with influenza hemagglutinin (HA)-expressing cells (Ellens et al., 1990; Bentz et al., 1990). The most straightforward method of adapting this approach to HIV fusion would require reconstitution of CD4 (or an alternative primary receptor) with or without putative auxiliary receptors (as needed) into model membranes (such as liposomes).

Model membranes allow rigorous kinetic characterization of the fusion process (see Bentz (1993a) and reviews in these Proceedings by S. Nir), enable the target composition to be manipulated to identify the dependence of viral fusion on target membrane lipids and incorporated receptors (Alford et al., 1994; and see review in these Proceedings by N. Düzgünes), and allow for the direct quantitative comparison of the fusion characteristics of different viruses. To obtain clear and direct biochemical information regarding this complex process, it will prove useful to study the fusion characteristics of HIV with model membranes containing the CD4 receptor. Liposomes containing CD4 could be used to study directly the necessity and sufficiency of CD4 in the HIV fusion reaction. If purified tmCD4 were the only protein in the membrane, the target membrane would be reduced to its basic components (lipids and functional CD4).

Our primary goal is to examine rigorously the gp120/41-induced CD4-dependent membrane fusion event using a well-defined model membrane system. In theory, we could use reconstituted gp120/41 in model membranes, but attempts by others to produce such "virosomes" from HIV have failed (Cornet et al., 1992). Therefore, we attempted to produce liposomes containing only CD4 in the lipid bilayer. This requires purification of functionally active transmembrane CD4. To achieve our goal, we had three specific aims: to purify functional CD4 from a human cell line; to produce reconstituted CD4-bearing liposomes; and to use those liposomes to develop a direct contents delivery fusion assay.

CD4: Background and Purification

Background. CD4 is an integral membrane glycoprotein, expressed predominantly on the surface of helper T lymphocytes, thymocytes and macrophages. Under normal physiological circumstances it functions as an accessory to the T cell antigen receptor for major histocompatibility complex class II antigen (MHC II) (Chuck et al., 1990; Doyle and Strominger, 1987) and as a complex with $p56^{lck}$ to deliver intracellular activating signals (Rudd et al., 1988). CD4 also functions as the major receptor for HIV-1 infection of human cells and binds to the viral envelope glycoprotein gp120. For an overview of CD4 structure and function, see Sweet et al. (1991).

CD4 extraction and purification. We chose to purify transmembrane CD4 from human cells rather than from non-human cells expressing transfected CD4 (Webb et al., 1989) because we wanted to avoid the possibility that any minor oligosaccharide differences in CD4 might alter its behavior during the HIV fusion reaction. We used a high expression sub-clone (SupT1-18) of a human T lymphoma cell line as our source of human CD4 (isolated by and obtained from Dr. Alemseged Truneh (SmithKline Beecham Pharmaceuticals, King of Prussia, PA, USA) from the cell line Sup-T1 (Smith et al., 1984) (supplied by Dr. James Hoxie, University of Pennsylvania, Philadelphia, PA, USA)). The cells were grown using standard human cell culture techniques, and frozen cell pellets were stored at -80 °C.

Our purification strategy was to solubilize and de-lipidate the tmCD4, bind CD4 from the cell extract to an immunoaffinity resin containing an anti-CD4 antibody, remove contaminating proteins that might bind non-specifically, wash away lipid while the CD4 was still bound to the resin, acid elute the purified tmCD4 in our detergent of choice, immediately neutralize and freeze the product for future use (before or after concentration). We would then reconstitute CD4 in a lipid bilayer using standard detergent dialysis reconstitution techniques (Silvius, 1992). Immunoaffinity chromatography was chosen as the most likely method to achieve a one-step purification. One of us (AP) evaluated a series of antibody matricies for optimal resin capacity and recovery of soluble CD4 (sCD4). All antibodies were conjugated to Sepharose CL4B. One antibody in particular gave the highest recovery, and we used that matrix to purify tmCD4.

An extract of these cells was obtained by shearing the cells in a low osmotic strength buffer containing protease inhibitors and glycerol, followed by membrane solubilization using 1% Triton X-100 (Pierce Chemical Co.). All future steps prior to washing the immunoaffinity resin contained protease inhibitors. Triton X-100 was chosen as the solubilizing detergent because it is a mild, non-denaturing detergent with a low critical micelle concentration. It can therefore solubilize membrane bilayers at relatively low concentrations. Additionally, Triton X-100 is easily exchangeable with other non-ionic detergents.

The solubilized cell lysate was treated with nucleases to reduce viscosity, and after an ultracentrifuge step, the supernatant was 0.45 μm filtered and loaded onto a column containing the immunoaffinity resin. Most of the contaminating proteins and lipids never bound to the resin, as assessed by optical density of the unbound fractions. After a high salt wash step and a saline rinse, the CD4 was eluted in acetic acid containing detergent, glycerol and NaCl. The eluate was immediately buffered, neutralized to approximately pH 7 and frozen. Microgram amounts of protein were recovered from 1-3 x 10^9 cells, as we expected.

The eluate was essentially a single band on SDS-PAGE (55 kd), and it was quantified in a gp120-binding ELISA assay. Recovery was approximately 50% on average. The N-terminal amino acid sequence of the eluate was a single sequence identical to that predicted for CD4 from its DNA sequence (Littman et al., 1988). In a direct binding assay, the purified molecule had the same binding constant to gp120 as the recombinant soluble form. The protein can be purified in the presence of octylglucoside, the detergent that will be used in the reconstitution of CD4 into liposomes.

CD4 Reconstitution and Contents Delivery Assay

CD4 reconstitution. One possible use of purified full-length CD4 is its reconstitution into liposomes. It is particularly useful that we can purify full-length CD4 in the presence of octylglucoside, a detergent that is commonly used for reconstitution of transmembrane proteins into lipid bilayers. Previous reports using partially purified CD4 reconstituted into liposomes (Cudd et al., 1990) or enriched CD4 reconstituted into plasma membrane vesicles (Puri et al., 1992) gave encouraging results for studying specific CD4 liposome-HIV interactions. Liposomes may also prove useful as carriers of antiviral or cytotoxic compounds that could be directed preferentially toward cell-free virus or specifically toward infected cells expressing viral membrane glycoproteins on their surface (Cudd et al., 1990; Nicolau et al., 1990), but that is not within the scope of this report. One of us (CL) is now working in the laboratory of Dr. C. Nicolau (Center for Blood Research Labs (CBRL), Harvard Medical School) on such a project. Preliminary results from that project suggest that such a carrier system may be capable of blocking HIV replication, at least in vitro (Larsen et al., unpublished observations).

Liposomes containing CD4 as their only membrane protein would permit the study of the CD4-dependent HIV-1 fusion reaction in a well-defined system. Specifically, purified membrane-bound CD4 could be used to determine the necessity and sufficiency of CD4, auxiliary receptors and the fusion site architecture for the HIV-1 fusion reaction. Additionally, such liposomes could be used for studying gp120/41 conformational changes upon binding and/or fusion. Purified full-length CD4 will be a useful tool in understanding the process of interaction between HIV and its target membranes as well as for comparison with sCD4 in other immunological binding assays.

Contents delivery fusion assay. We had planned to adapt to HIV an extremely sensitive contents mixing assay originally developed for influenza HA-mediated fusion (Ellens et al., 1990; Bentz et al., 1990). A cartoon of our strategy is presented as Figure 1. The strategy utilizes delivery of the plant toxin gelonin encapsulated in CD4-reconstituted liposomes to mea-

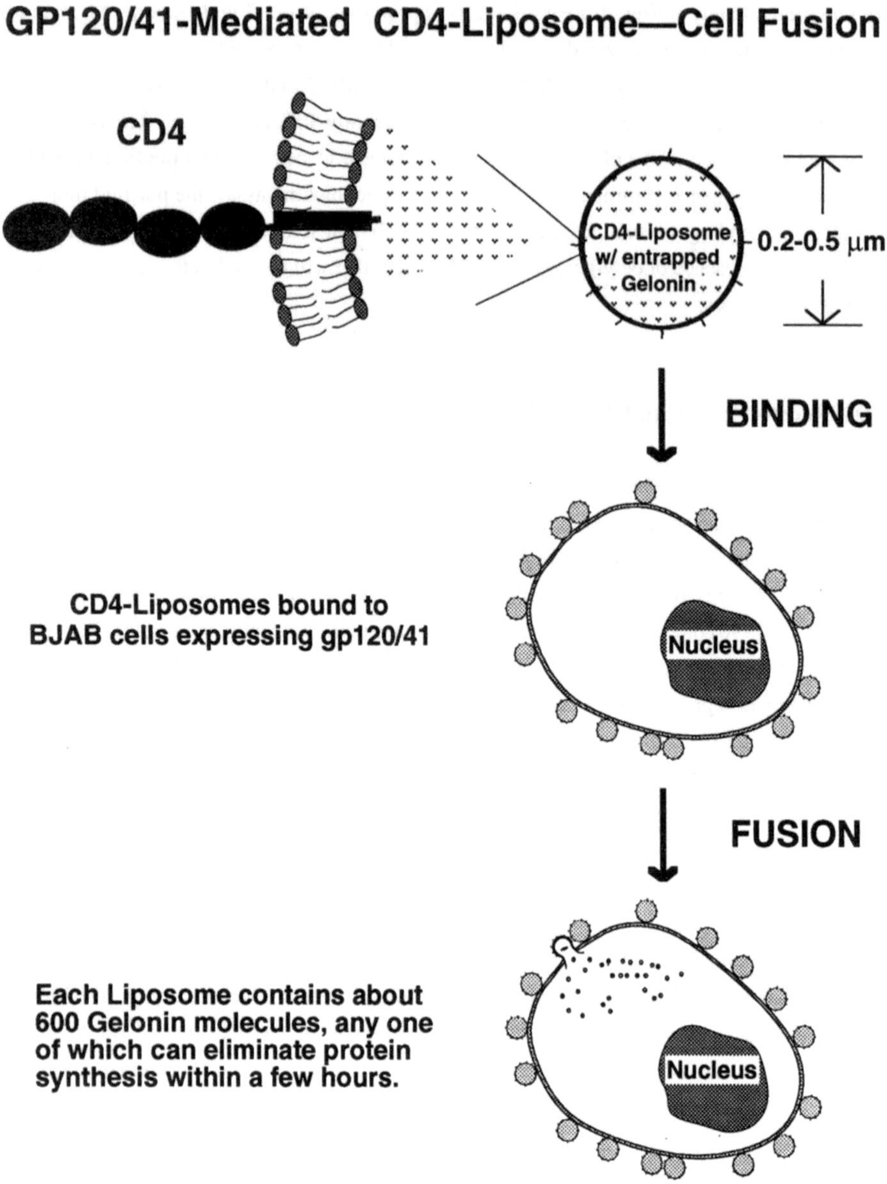

Figure 1. CD4-Liposome-Cell Fusion Strategy. The drawing is roughly to scale. The CD4 ectodomain is approximately 130 Å in length with a diameter of 25 to 30 Å.

sure fusion with viral glycoprotein expressing cell lines. We planned to study the fusion between CD4-reconstituted liposomes and BJAB cells. BJAB is a transformed B cell line that permanently expresses functional HIV gp120/41 under promoter control (Jonak et al., 1993). The assay depicted in Figure 1 measures contents delivery directly (by assaying cell death due to inhibition of protein synthesis caused by the internal presence of gelonin) and not just lipid mixing (potentially "leaky" fusion (Düzgünes and Bentz, 1988)).

Comparison of contents mixing between CD4-reconstituted liposomes and control liposomes would allow direct determination of the relevance of various molecules in gp120/41 induced fusion. After incubation for a fixed time (minutes), sCD4 and/or antibodies against CD4 or gp120 could be added to halt fusion. The cells could then be assayed for protein synthesis (the metabolic step blocked by gelonin) using a radioactive amino acid incorporation assay. The number of individual fusion events per cell could be calculated directly by incubating the cells with various ratios of gelonin-containing to buffer-containing CD4-reconstituted liposomes (Ellens et al., 1990; Bentz et al., 1993b). Testing the effects of other environmental parameters (e.g., neutralizing antibodies, proteolysis, pH, and chemical blocking agents) on the number of fusion events per cell could also be performed.

Acknowledgments

The authors thank Alemseged Truneh for providing the SupT1-18 cell line, Joanne Miller and Harma Ellens for help with cell culture, Brian Hellmig for the OKT4 and gp120, Dean McNulty for the amino acid sequencing, Dwight Moore and Shing Mai for CD4 ELISA advice and materials and Ray Sweet for helpful discussions. This work was partially supported by research grants AI08117 (CL) and GM31506 (JB) from the National Institutes of Health and by in-kind support from SmithKline Beecham Pharmaceuticals, Inc.

References

Alford D, Ellens H, Bentz J (1994) Fusion of influenza virus with sialic acid-bearing target membranes. Biochemistry 33: 1977-1987

Allan JS (1993) Receptor-mediated activation of the viral envelope and viral entry. AIDS 7 (S1): S43-S50

Bentz J In: Adv Membr Fluidity, vol 5. Aloia RC, Curtain CC, Gordon LM (eds) (1991) Membrane fusion: viral fusion proteins and lipid intermediates. Alan R. Liss, Inc., New York; pp. 259-287

Bentz J (1993a) Viral Fusion Mechanisms. CRC Press Boca Raton, FL

Bentz J In: Viral Fusion Mechanisms. Bentz J (ed) (1993b) Membrane fusion intermediates and the kinetics of membrane fusion. CRC Press Boca Raton; pp. 453-474

Bentz J, Ellens H, Alford D (1990) An architecture for the fusion site of influenza hemagglutinin. FEBS Lett 276: 1-5

Bentz J, Ellens H, Alford D In: Viral Fusion Mechanisms. Bentz J (ed) (1993) Architecture of the influenza hemagglutinin fusion site. CRC Press Boca Raton; pp. 163-199

Blobel CP, Wolfsberg TG, Turck CW, Myles DG, Primakoff P, White JM (1992) A potential fusion peptide and an integrin ligand domain in a protein active in sperm-egg fusion. Nature 356: 248-252

Broder CC, Berger EA (1993) CD4 molecules with a diversity of mutations encompassing the CDR3 region efficiently support human immunodeficiency virus type 1 envelope glycoprotein-mediated cell fusion. J Virol 67: 913-926

Broder CC, Dimitrov DS, Blumenthal R, Berger EA (1993) The block to HIV-1 envelope glycoprotein-mediated membrane fusion in animal cells expressing human CD4 can be overcome by a human cell component(s). Virology 193: 483-491

Bullough PA, Hughson FM, Skehel JJ, Wiley DC (1994) Structure of influenza haemagglutinin at the pH of membrane fusion. Nature 371: 37-43

Callebaut C, Jacotot E, Krust B, Hovanessian AG (1994) CD26 antigen and HIV fusion? Response. Science 264: 1162-1165

Camerini D; Seed B (1990) A CD4 domain important for HIV-mediated syncytium formation lies outside the virus binding site. Cell 60: 747-754.

Cao J, Vasir B, Sodroski JG (1994) Changes in the cytopathic effects of human immunodeficiency virus type 1 associated with a single amino acid alteration in the ectodomain of the gp41 transmembrane glycoprotein. J Virol 68: 4662-4668

Carr CM, Kim PS (1993) A spring-loaded mechanism for the conformational change of influenza hemagglutinin. Cell 73: 823-832

Chuck RS, Cantor CR, Tse DB (1990) CD4 T-cell antigen receptor complexes on human leukemia T cells. Proc Natl Acad Sci USA 87: 5021-5025

Corbeau P, Benkirane M, Weil R, David C, Emiliani S, Olive D, Mawas C, Serre A, Devaux C (1993) Ig CDR3-like region of the CD4 molecule is involved in HIV-induced syncytia formation but not in viral entry. J Immunol 150: 290-301

Cornet B, Decroly E, Ruysschaert J-M, Vandenbranden M In: Adv Membr Fluidity, vol 6, Membrane Interactions of HIV. Aloia RC, Curtain CC (eds) (1992) Reconstitution of human immunodeficiency virus envelope. Wiley-Liss New York; pp. 377-390

Cudd A, Noonan CA, Tosi P-F, Melnick JL, Nicolau C (1990) Specific interaction of CD4-bearing liposomes with HIV-infected cells. J Acquir Immune Defic Synd 3: 109-114

Dimitrov DS, Golding H, Blumenthal R (1991) Initial stages of HIV-1 envelope glycoprotein-mediated cell fusion monitored by a new assay based on redistribution of fluorescent dyes.

AIDS Res Hum Retrovir 7: 799-805

Doyle C, Strominger JL (1987) Interaction between CD4 and class II MHC molecules mediates cell adhesion. Nature 330: 256-259

Dubay JW, Roberts SJ, Hahn BH, Hunter E (1992) Truncation of the human immunodeficiency virus type 1 transmembrane glycoprotein cytoplasmic domain blocks virus infectivity. J Virol 66: 6616-6625

Düzgünes N, Bentz J In: Spectroscopic Membrane Probes, Vol. I. Loew LM (ed) (1988) Fluorescence assays for membrane fusion. CRC Press Boca Raton; pp. 117-159

Ebenbichler CF, Röder C, Vornhagen R, Ratner L, Dierich MP (1993) Cell surface proteins binding to recombinant soluble HIV-1 and HIV-2 transmembrane proteins. AIDS 7:489-495

Ellens H, Bentz J, Mason D, Zhang F, White JM (1990) Fusion of influenza hemagglutinin-expressing fibroblasts with glycophorin-bearing liposomes: role of hemagglutinin surface density. Biochemistry 29: 9697-9707

Ellens H, Larsen C In: Viral Fusion Mechanisms. Bentz J (ed) (1993) CD4-induced change in gp120/41 conformation and its potential relationship to fusion. CRC Press Boca Raton; pp. 291-312

Golding H, Manischewitz J, Vujcic L, Blumenthal R, Dimitrov DS (1994) The phorbol ester phorbol myristate acetate inhibits human immunodeficiency virus type 1 envelope-mediated fusion by modulating an accessory component(s) in CD4-expressing cells. J Virol 68: 1962-1969

Goto T, Harada S, Yamamoto N, Nakai M (1988) Entry of human immunodeficiency virus (HIV) into MT-2, human T cell leukemia virus carrier cell line. Archives of Virology 102: 29-38

Grewe C, Beck A, Gelderblom, HR (1990) HIV: early virus-cell interactions. J Acquir Immune Defic Syndr 3: 965-974

Harrington RD, Geballe AP (1993) Cofactor requirement for human immunodeficiency virus type 1 entry into a CD4-expressing human cell line. J Virol 67: 5939-5947

Henderson LA, Qureshi MN (1993) A peptide inhibitor of human immunodeficiency virus infection binds to novel human cell surface polypeptides. J Biol Chem 268: 15291-15297

Jonak ZL, Clark RK, Matour D, Trulli S, Craig R, Henri E, Lee E, Greig R, Debouck C (1993) A human lymphoid cell line with functional human immunodeficiency virus type 1 evelope. AIDS Res Hum Retroviruses 9: 23-31

Klatzmann D, Gluckman JC (1986) HIV infection: facts and hypotheses. Immunology Today 7: 291-296

Kowalski M, Bergeron L, Dorfman T, Haseltine W, Sodroski J (1991) Alteration of human immunodeficiency virus type 1 cytopathic effect by a mutation affecting the transmembrane envelope glycoprotein. J Virol 65: 281-291

Larsen CE, Alford DR, Young LJT, McGraw TP, Düzgünes N (1990) Fusion of simian immunodeficiency virus with liposomes and erythrocyte ghost membranes: effects of lipid composition, pH and calcium. J Gen Virol 71: 1947-1955

Larsen C, Ellens H, Bentz J In: Adv Membr Fluidity, vol 6, Membrane Interactions of HIV. Aloia RC, Curtain CC (eds) (1992) Membrane fusion induced by the human immunodeficiency virus env glycoprotein. Wiley-Liss New York; pp. 143-166

Larsen CE, Nir S, Alford DR, Jennings M, Lee K-D, Düzgünes N (1993) Human immunodeficiency virus type 1 (HIV-1) fusion with model membranes: kinetic analysis and the role of lipid composition, pH and divalent cations. Biochim Biophys Acta 1147: 223-236

Littman DR, Maddon PJ, Axel R (1988) Corrected CD4 sequence. Cell 55: 541

Long D, Berson JF, Cook DG, Doms RW (1994) Characterization of human immunodeficiency virus type 1 gp120 binding to liposomes containing galactosylceramide. J Virol 68: 5890-5898

Marcon L, Sodroski J (1994) Gp120-independent fusion mediated by the human immunodeficiency virus type 1 gp41 envelope glycoprotein: a reassessment. J Virol 68: 1977-1982

Moore JP, Jameson BA, Weiss RA, Sattentau Q In: Viral Fusion Mechanisms. Bentz J (ed) (1993a) The HIV-cell fusion reaction. CRC Press Boca Raton; pp. 233-289

Moore J, Sattentau Q, Jameson B, Sodroski J (1993b) Monoclonal antibodies to HIV-1 gp120: a request. AIDS Res Hum Retrovir 9: 695

Moore JP, Sattentau QJ, Wyatt R, Sodroski J (1994) Probing the structure of the human immunodeficiency virus surface glycoprotein gp120 with a panel of monoclonal antibodies. J Virol 68: 469-484

Nicolau C, Tosi, P-F, Arvinte T, Mouneimne Y, Cudd A, Sneed L, Madoulet C, Schulz B, Barhoumi R In: Horizons in Membrane Biotechnology. (1990) CD4 inserted in red blood cell membranes or reconstituted in liposome bilayers as a potential therapeutic agent against AIDS. Wiley-Liss New York; pp. 147-177

Owens RJ, Burke C, Rose JK (1994) Mutations in the membrane-spanning domain of the human immunodeficiency virus envelope glycoprotein that affect fusion activity. J Virol 68: 570-574

Pauza CD, Price TM (1988) Human immunodeficiency virus infection of T cells and monocytes proceeds via receptor-mediated endocytosis. J Cell Biol 107: 959-968

Perez LG, O'Donnell MA, Stephens EB (1992) The transmembrane glycoprotein of human immunodeficiency virus type 1 induces syncytium formation in the absence of the receptor binding glycoprotein. J Virol 66: 4134-4143

Puri A, Dimitrov DS, Golding H, Blumenthal R (1992) Interactions of CD4+ plasma membrane vesicles with HIV-1 and HIV-1 envelope glycoprotein-expressing cells. J Acquir Immune Defic Syndr 5: 915-920

Rudd CE, Trevillyan JM, Dasgupta JD, Wong LL, Schlossman SF (1988) The CD4 receptor is complexed in detergent lysates to a protein-tyrosine kinase (pp56) from human T lymphocytes. Proc Natl Acad Sci USA 85: 5190-5196

Sato AI, Balamuth FB, Ugen KE, Williams W, Weiner DB (1994) Identification of CD7 glycoprotein as an accessory molecule in HIV-1-mediated syncytium formation and cell-free infection. J Immunol 152: 5142-5152

Silvius JR (1992) Solubilization and functional reconstitution of biomembrane components. Annu Rev Biophys Biomol Struct 21: 323-348

Sinangil F, Loyter A, Volsky DJ (1988) Quantitative measurement of fusion between human immunodeficiency virus and cultured cells using membrane fluorescence dequenching. FEBS Lett 239: 88-92

Smith SD, Shatsky M, Cohen PS, Warnke R, Link MP, Glader BE (1984) Monoclonal antibody and enzymatic profiles of human malignant T-lymphoid cells and derived cell lines. Cancer Res 44: 5657-5661

Söllner T, Whiteheart SW, Brunner M, Erdjument-Bromage H, Geromanos S, Tempst P, Rothman JE (1993) SNAP receptors implicated in vesicle targeting and fusion. Nature 362: 318-324

Stamatatos L, Düzgünes N (1993) Simian immunodeficiency virus (SIV_{mac}251) membrane lipid mixing with human CD4+ and CD4- cell lines *in vitro* does not necessarily result in internalization of the viral core proteins and productive infection. J Gen Virol 74: 1043-1054

Stefano KA, Collman R, Kolson D, Hoxie J, Nathanson N, Gonzalez-Scarano F (1993) Replication of a macrophage-tropic strain of human immunodeficiency virus type 1 (HIV-1) in a hybrid cell line, CEMx174, suggests that cellular accessory molecules are required for HIV-1 entry. J Virol 67: 6707-6715

Stein BS, Gowda SD, Lifson JD, Penhallow RC, Bensch, KG, Engleman EG (1987) pH-Independent HIV entry into CD4-positive T cells via virus envelope fusion to the plasma membrane. Cell 49: 659-668

Sweet RW, Truneh A, Hendrickson WA (1991) CD4: its structure, role in immune function and AIDS pathogenesis, and potential as a pharmacological target. Curr Opinion Biotech 2: 622-633

Webb NR, Madoulet C, Tosi P-F, Broussard DR, Sneed L, Nicolau C, Summers MD (1989) Cell surface expression and purification of human CD4 produced in baculovirus-infected cells. Proc Natl Acad Sci USA 86: 7731-7735

White JM (1992) Membrane fusion. Science 258: 917-924

CYCLING OF RAB PROTEINS: ROLE OF RAB GDI IN THE REVERSIBLE MEMBRANE ASSOCIATION OF RAB GTP-ASES

Oliver Ullrich and Marino Zerial
European Molecular Biology Laboratory
Postfach 10.2209
Meyerhofstrasse 1
D-69012 Heidelberg
Germany

Summary

Small GTPases of the Rab family are specific regulators of intracellular vesicular traffic. Mutational analysis of Rab proteins suggests that it is the membrane-bound and GTP-bound form of the molecule that is active in vesicle docking or fusion. Since Rab proteins are also present in a cytosolic pool in the cell, it has been postulated that they recycle via a soluble intermediate to serve multiple rounds of vesicular traffic. Here we discuss our evidence for the involvement of Rab GDI (GDP-Dissociation Inhibitor) in the cycling of Rab proteins between the cytosol and the membrane. We find that Rab GDI is able to both dissociate Rab proteins from and deliver them to the membrane. Rab GDI-mediated membrane association of Rab proteins is a multistep mechanism which includes a nucleotide exchange reaction.

Introduction

The Rab family of small GTPases is implicated in intracellular vesicle transport such as exocytosis, endocytosis and transcytosis (Pfeffer, 1992; Zerial and Stenmark, 1993). The first two members of this class of proteins were identifed in yeast. Ypt1p (Gallwitz et al., 1983) and Sec4p (Salminen and Novick, 1987) were shown to regulate transport from the endoplasmic reticulum (ER) to the Golgi apparatus and from the Golgi apparatus to the cell surface, respectively (Segev et al., 1988; Salminen and

NATO ASI Series, Vol. H 91
Trafficking of Intracellular Membranes
Edited by M.C. Pedroso de Lima N. Düzgüneş and D. Hoekstra

Novick, 1987). The large number of Rab GTPases identified in mammalian cells (~30 different members) and localized to various intracellular compartments reflects the complexity of the trafficking network in the cell, with each transport step being regulated by at least one Rab protein. In addition, specialized transport pathways in differentiated cells require cell type specific Rab proteins. For instance, cells which display regulated secretion, specifically express the Rab3A and Rab3C (Fischer von Mollard et al., 1990; 1994) and polarized epithelial cells have also been shown to require specific Rab proteins (Lütcke et al., 1993).

Like other members of the Ras superfamily, Rab proteins exist in two distinct conformations, depending on the bound nucleotide. The molecular switch-function of Rab proteins is applied to control vesicle docking or fusion. The nucleotide state of Rab proteins is mainly regulated by interacting proteins, since the intrinsic rates of GTP hydrolysis and nucleotide dissociation are very low. Rab GDI is a factor that inhibits exchange of GDP for GTP, whereas GEF (guanine-nucleotide-exchange factor) stimulates this process. A third regulatory protein is GAP (GTPase-activating protein) which promotes GTP hydrolysis.

It has been proposed that the cycle of GTP binding and hydrolysis provides directionality to the vesicle docking/fusion process (Bourne, 1988; Pfeffer, 1992; Zerial and Stenmark, 1993). According to the current model, a specific Rab protein binds to the donor compartment, a process which is somehow coupled to the exchange of GDP for GTP. The GTP-bound form is incorporated into transport vesicles that bud from the donor membrane. Following docking and fusion of the vesicle with the target membrane, GTP is hydrolysed. The view that GTP hydrolysis is not required prior to fusion is suggested by *in vivo* experiments with a mutant of Rab5, Rab5 Q79L, that is mainly in the GTP-bound form (Li and Stahl, 1993; Stenmark et al., 1994). Overexpression of this mutant resulted in stimulation rather than inhibition of membrane fusion. It thus appears that GTP hydrolysis may function to inactivate the protein, so that the GDP-bound form of the Rab protein is removed from the target membrane and transported as a cytosolic intermediate back to the donor membrane to serve another round of vesicular transport.

Rab proteins exist in two pools in the cell, one membrane-bound and the other soluble. Cytosolic Rab proteins are complexed to Rab GDI (Araki et al., 1990; Regazzi et al., 1992; Soldati et al., 1993; Ullrich et al., 1993), a factor first identified as a protein that inhibits the dissociation of GDP from Rab3A (Sasaki et al., 1990). Rab GDI was shown to inhibit membrane binding of Rab3A and to dissociate Rab3A from synaptic membranes and vesicles (Araki et al., 1990). In addition, Rab GDI has also been found to be active on Sec4p and Rab11 (Sasaki et al., 1991; Ueda et al., 1991). The geranylgeranylation of the COOH-terminal cysteine motif of Rab3A was shown to be an essential requirement for the interaction with Rab GDI (Araki et al., 1991).

In our studies we have used a permeabilized cell system to investigate the role of Rab GDI in the postulated cycling of Rab proteins between the cytosol and the membrane. Since Rab proteins display distinct cysteine motifs in the C-terminus we first tested the specificity of Rab GDI for the removal of various members of the Rab protein family from the membrane. Second, we have studied the mechanism whereby the cytosolic form of Rab proteins (complexed to Rab GDI) re-associates with the membrane. Finally, we adressed the issue at which stage the proposed GDP/GTP-exchange reaction occurs.

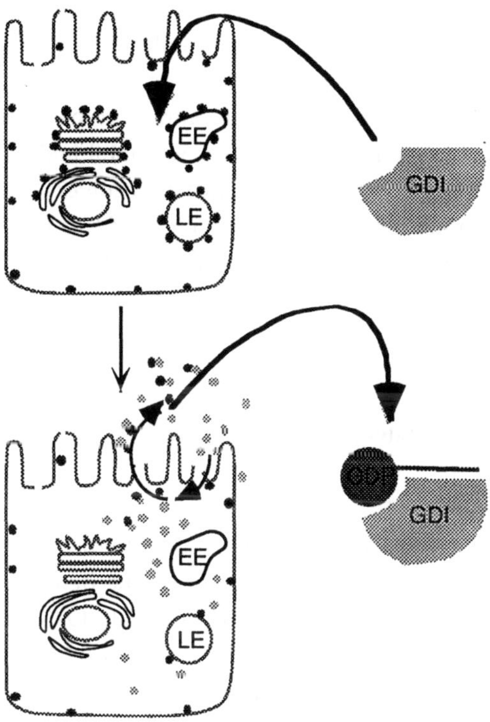

Fig. 1 Permeabilized cell system to study the role of Rab GDI on membrane-bound Rab proteins. Incubation of MDCK cells with SLO leads to the formation of pores in the plasma membrane. Rab GDI diffuses into the permeabilized cells and solubilizes Rab proteins that are bound to various compartments. The cytosolic complex of Rab GDI and the GDP-form of Rab proteins is washed out and the membrane-associated and soluble Rab proteins are detected by western blot analysis.

Rab GDI as a solubilizing factor for Rab proteins

We have used Madin-Darby canine kidney (MDCK) cells permeabilized with the bacterial toxin streptolysin O to test the general activity of Rab GDI in modulating the membrane association of Rab proteins (Ullrich et al., 1993) (Fig. 1). When permeabilized cells were incubated with Rab GDI purified from bovine brain, we found that Rab GDI removed all endogenous Rab proteins we have tested from the membrane. We further investigated the effect of Rab GDI on the membrane association of exogenous Rab proteins synthesized *in vitro* in a rabbit reticulocyte lysate translation system. In the absence of Rab GDI, [^{35}S]methionine-labelled *in vitro* translated Rab proteins bound to the membrane of permeabilized cells but not of unpermeabilized cells. Binding was a temperature-dependent process and required geranylgeranylation of the Rab proteins. In contrast, in the presence of Rab GDI membrane association of exogenous Rab proteins was inhibited. Rab GDI appears to be specific for Rab proteins since it was not active on members of the Rho and Ras families of small GTP-binding proteins. Maximal effect of Rab GDI on the removal of Rab proteins from the membrane was at 0.8 µM, a concentration close to the estimated cytosolic concentration of Rab GDI in MDCK cells (1.2 µM).

Rab GDI mediates the association of Rab5 with the membrane

Removal of Rab proteins from the membrane by Rab GDI leads to the formation of stable complexes of Rab GDI and Rab proteins, that also can be detected in the cytosol of cells (Araki et al., 1990; Regazzi et al., 1992; Soldati et al., 1993; Ullrich et al., 1993). We therefore tested in our permeabilized cell system how the cytosolic form of Rab proteins reassociates with the membrane (Ullrich et al., 1994). Our approach was to reconstitute a GDP-Rab5-Rab GDI complex from purified components *in vitro*. Since geranylgeranylation is required for interaction with Rab GDI (Araki et al., 1991), modified Rab5 was purified from the membranes of insect cells infected with a recombinant Rab5-baculovirus, by CHAPS extraction and Mono Q anion-exchange chromatography. His_6-tagged Rab GDI was expressed in *E. coli* and purified by Ni^{2+}-NTA-agarose and Mono Q anion-exchange chromatography. The GDP-Rab5-Rab GDI complex was obtained by incubating 1.5 µM GDP-Rab5 with 3 µM Rab GDI at 30°C for 10 min, followed by gel filtration to remove uncomplexed Rab GDI. The purified Rab5-Rab GDI complex was then added to the permeabilized MDCK cells to test whether Rab5-Rab GDI is competent to bind to the membrane.

Rab5 was found to associate with permeabilized but not with intact cells in a concentration- and time-dependent manner, reaching saturation at 50 nM of exogenous complex after 30 min incubation time (Fig. 2). In contrast, there was no significant increase in the amount of Rab GDI bound to the membrane, indicating that upon binding of Rab5 to the membrane Rab GDI is released into the cytosol. Membrane association of Rab5 was temperature-dependent but did not require cytosol and ATP (Fig. 2). In the absence of Rab GDI or detergent, Rab5 was insoluble and could not bind to the membrane. We further investigated whether exogenous Rab5 was correctly targeted and functionally active in permeabilized cells. Rab5 is localized to the plasma membrane and early endosomes and functions in the early endoctic pathway (Chavrier et al., 1990; Bucci et al., 1992). Using confocal immunofluorescence microscopy we found that exogenous Rab5 co-localized with transferrin-receptor, a marker of plasma membrane and early endosome, but not with the late endosome/lysosome marker lamp-1 (Ullrich et al., 1994). These results indicate that purified Rab5 was properly targeted in permeabilized cells. When permeabilized cells were incubated for increasing periods of time (5, 15, and 30 min), we could observe a time-dependent increase in the size of early endosomes. The same effect was previously shown for cells transiently overexpressing Rab5 (Bucci et al., 1992) and demonstrates that exogenous Rab5 was functionally active in the permeabilized cell system.

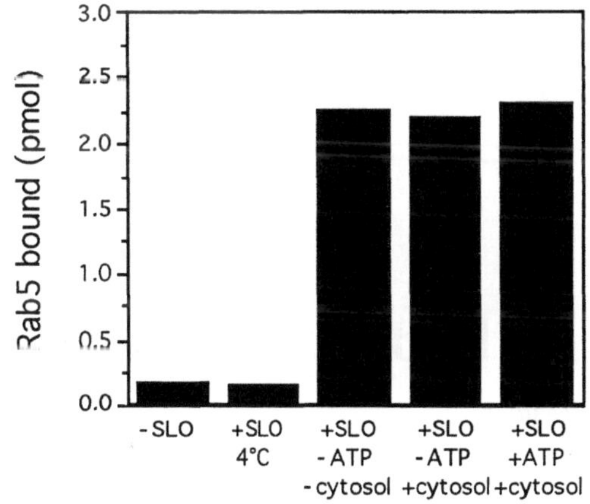

Fig. 2 Binding properties of Rab5-Rab GDI complex to membranes of SLO-permeabilized cells. Permeabilized (+SLO) and unpermeabilized (-SLO) cells were incubated with 50 nM Rab5-Rab GDI complex for 15 min and the amount of membrane-bound Rab5 was quantified by western blot analysis. No significant binding of exogenous Rab5 could be detected to the membrane of unpermeabilized cells at 37°C and to permeabilized cells incubated at 4°C. In contrast, binding was efficient to the membrane of permeabilized cells and did not require cytosol and ATP.

Membrane association of Rab5 is accompanied by nucleotide exchange

Current models propose that membrane attachment of Rab proteins is coupled to the exchange of GDP for GTP, mediated by a guanine-nucleotide-exchange factor (Bourne et al., 1988; Takai et al., 1992; Pfeffer, 1992; Zerial and Stenmark, 1993). Having reconstituted the membrane association of Rab5 *in vitro*, we determined the guanine-nucleotide form of Rab5 which is associated with the membrane (Ullrich et al., 1994). When permeabilized cells were incubated with [^{3}H]GDP-Rab5-Rab GDI complex (Rab5 pre-loaded with [^{3}H]GDP) for different periods of time, increasing amounts of Rab5 in the GDP-bound form could be detected on the membrane, reaching a maximum after 15 min and decreasing to background level after 60 min (Fig. 3). These results indicate that Rab5 first associates in the GDP-bound form with the membrane but that this process is followed by GDP-release. We then tested whether GTP is exchanged for GDP bound to Rab5. Permeabilized cells were incubated with GDP-Rab5-Rab GDI complex in the presence of [^{35}S]GTPγS for different periods of times, and the amount of radioactivity bound to Rab5 on the membrane was determined. A time-dependent increase of [^{35}S]GTPγS-Rab5 could be detected on the membrane, indicating that membrane associated GDP-Rab5 is subsequently converted into the GTP-bound form.

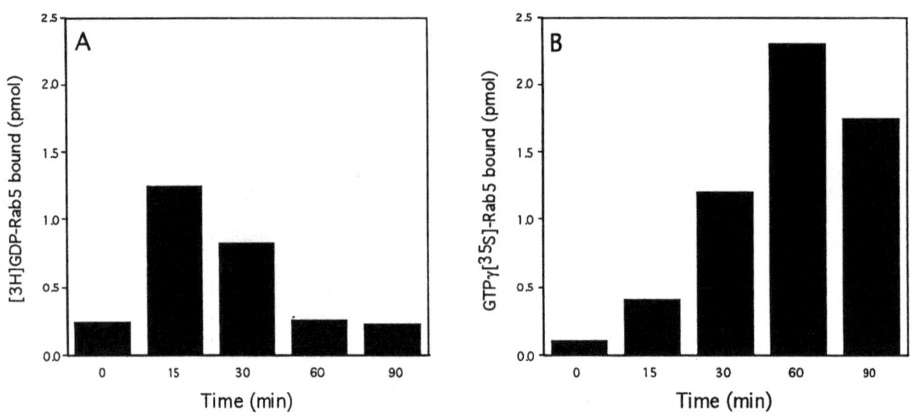

Fig. 3 Membrane association of Rab5 is accompanied by nucleotide exchange. (A) Time course of [^{3}H]GDP-Rab5 association with the membrane. Permeabilized cells were incubated with the complex of [^{3}H]GDP-Rab5-Rab GDI for the indicated times and the amount of [^{3}H]GDP-Rab5 bound to the membrane was measured. (B) Time course of [^{35}S]GTPγS binding to Rab5 on the membrane. Unlabelled Rab5-Rab GDI was incubated with permeabilized cells in the presence of [^{35}S]GTPγS for different periods of time and the membrane-associated [^{35}S]GTPγS-Rab5 was quantified after immunoprecipitation.

Conclusions and perspectives

Rab GDI fulfils all requirements for a factor that plays a central role in recycling of Rab proteins. First, Rab GDI specifically removes Rab proteins from the membrane of permeabilized cells, leading to the formation of cytosolic heterodimers of Rab GDI and Rab proteins. Second, in complex with Rab5, Rab GDI mediates the reassociation of Rab5 with the membrane. Upon binding of Rab5 to the membrane Rab GDI is released into the cytosol. Initially, membrane association occurs in the GDP-bound form, but this is subsequently converted into the GTP-bound form. Similar results were obtained by Soldati et al. (1994) for Rab9-Rab GDI complex. This group could show that Rab GDI also mediates the membrane attachment of Rab9 to late endosomes, and that this process is accompanied by nucleotide exchange. Furthermore, the role of Rab GDI in recycling of Rab proteins is confirmed by *in vivo* experiments in yeast. Depletion of the yeast Rab GDI homolog Gdi1p leads to loss of the soluble pool of Sec4p (Garrett et al., 1994) and protein transport is inhibited at several stages of the secretory pathway, suggesting that recycling of multiple Rab proteins is affected.

Further work is required to identify factor(s) responsible for membrane attachment and nucleotide exchange. It will be interesting to determine whether these activities are mediated by one or two distinct proteins.

References

Araki S, Kaibuchi K, Sasaki T, Hata Y, Takai Y (1991) Role of the C-terminal region of smg p25A in its interaction with membranes and the GDP/GTP exchange protein. Mol Cell Biol 11:1438-1447.

Araki S, Kikuchi A, Hata Y, Isomura M, Takai Y (1990) Regulation of reversible binding of smg p25A, a ras p21-like GTP-binding protein, to synaptic plasma membranes and vesicles by its specific regulatory protein, GDP dissociation inhibitor. J Biol Chem 265:13007-13015.

Bourne HR (1988) Do GTPases direct membrane traffic in secretion? Cell 53:669-671.

Bucci C, Parton RG, Mather IH, Stunnenberg H, Simons K, Hoflack B, Zerial M (1992) The small GTP-ase rab5 functions as a regulatory factor in the early endocytic pathway. Cell 70:715-728.

Chavrier P, Parton RG, Hauri HP, Simons K, Zerial M (1990) Localization of low molecular weight GTP-binding proteins to exocytic and endocytic compartments. Cell 62:317-329.

Fischer von Mollard G, Khokhlatchev A, Südhof T, Jahn R (1994) Rab3C is a synaptic vesicle protein that dissociates from synaptic vesicles after stimulation of exocytosis. J Biol Chem 269:10971-10974.

Fischer von Mollard G, Mignery GA, Baumert M, Perin MS, Hansson TJ, Burger PM, Jahn R, Südhof T (1990) Rab3 is a small GTP-binding protein exclusively localized to synaptic vesicles. Proc Natl Acad Sci USA 87:1988-1992.

Gallwitz D, Donath C, Sander C (1983) A yeast gene encoding a protein homologous to the human c-has/bas proto-oncogene product. Nature 306:704-707.

Garrett MD, Zahner JE, Cheney CM, Novick PJ (1994) GDI1 encodes a GDP dissociation inhibitor that plays an essential role in the yeast secretory pathway. EMBO J 13:1718-1728.

Li G, Stahl PD (1993) Structure-function relationship of the small GTPase rab5. J Biol Chem 268:24475-24480.

Lütcke A, Jansson S, Parton RG, Chavrier P, Valencia A, Huber LA, Lehtonen E, Zerial M (1993) Rab17, a novel small GTPase, is specific for epithelial cells and is induced during cell polarization. J Cell Biol 121:553-564.

Pfeffer SR (1992) GTP-binding proteins in intracellular transport. Trends Cell Biol 2:41-46.

Regazzi R, Kikuchi A, Takai Y, Wollheim CB (1992) The small GTP-binding proteins in the cytosol of insulin-secreting cells are complexed to GDP dissociation inhibitor proteins. J Biol Chem 267:17512-17519.

Salminen A, Novick P (1987) A ras-like protein is required for a post-Golgi event in yeast secretion. Cell 49:527-538.

Sasaki T, Kaibuchi K, Kabcenell AK, Novick PJ, Takai Y (1991) A mammalian inhibitory GDP/GTP exchange protein (GDP dissociation inhibitor) for smg p25A is active on the yeast SEC4 protein. Mol Cell Biol 11:2909-2912.

Sasaki T, Kikuchi A, Araki S, Hata Y, Isomura M, Kuroda S, Takai Y (1990) Purification and characterization from bovine brain cytosol of a protein that inhibits the dissociation of GDP from and the subsequent binding of GTP to smg p25, a ras p21-like GTP-binding protein. J Biol Chem 265:2333-2337.

Segev N, Mulholland J, Botstein D (1988) The yeast GTP-binding YPT1 protein and a mammalian counterpart are associated with the secretion machinery. Cell 52:915-924.

Soldati T, Riederer MA, Pfeffer SR (1993) Rab GDI: a solubilizing and recycling factor for rab9 protein. Mol Biol Cell 4:425-434.
Soldati T, Shapiro AD, Svejstrup ABD, Pfeffer SR (1994) Membrane targeting of the small GTPase Rab9 is accompanied by nucleotide exchange. Nature 369:76-78.
Stenmark H, Parton RG, Steele-Mortimer O, Lütcke A, Gruenberg J, Zerial M (1994) Inhibition of rab5 GTPase activity stimulates membrane fusion in endocytosis. EMBO J 13:1287-1296.
Takai Y, Kaibuchi K, Kikuchi A, Kawata M (1992) Small GTP-binding proteins. Int Rev Cytol 133:187-231.
Ueda T, Takeyama Y, Ohmori T, Ohyanagi H, Saitoh Y, Takai Y (1991) Purification and characterization from rat liver cytosol of a GDP dissociation inhibitor (GDI) for liver 24K G, a ras p21-like GTP-binding protein, with properties similar to those of smg p25A GDI. Biochem 30:909-917.
Ullrich O, Horiuchi H, Bucci C, Zerial M (1994) Membrane association of Rab5 mediated by GDP-dissociation inhibitor and accompanied by GDP/GTP exchange. Nature 368:157-160.
Ullrich O, Stenmark H, Alexandrov K, Huber LA, Kaibuchi K, Sasaki T, Takai Y, Zerial M (1993) Rab GDI as a general regulator for the membrane association of rab proteins. J Biol Chem 268:18143-18150.
Zerial M, Stenmark H (1993) Rab GTPases in vesicular transport. Curr Opin Cell Biol 5:613-620.

Microtubule Dependent Transport and Fusion of Phagosomes with the Endocytic Pathway

Janis K. Burkhardt, Ariel Blocker , Andrea Jahraus and Gareth Griffiths
European Molecular Biology Laboratory
Postfach 10.2209
69012 Heidelberg
Germany

The endocytic pathway in general

Cells internalize material from the extracellular fluid by several different mechanisms (for a recent review on the diversity of mechanisms see Watts and Marsh, 1992). The two best characterized mechanisms are the clathrin-dependent pathway that functions for most receptor-mediated uptake as well as a variable proportion of the bulk fluid uptake, and phagocytosis. For a number of years our group has focused on the organization and function of the "classical" clathrin mediated endocytic pathway, both in fibroblasts and in polarized MDCK cells (see Griffiths and Gruenberg, 1991). Collectively, our results argue that, following the initial clathrin coated pit/vesicle uptake, material destined for degradation will traverse four distinct cellular organelles, the early endosome, the multivesicular body-like endosomal carrier vesicle (ECV), the late endosome (or prelysosomal compartment) and the terminal lysosome (Fig. 1). In our view the early and late endosomes represent true cellular compartments that are pre-existing in the cell and whose passage to daughter cells during mitosis is, we propose, essential for cell viability. In contrast, we propose that the ECV and the terminal lysosomes are transient vesicles (although these may be very long lived) which bud off the early and late endosomes, respectively. In line with this prediction we put forward the hypothesis that the ECV and dense lysosomes are not essential for daughter cell viability since they can be formed by budding from pre-existing compartments. While this view of the structures we now refer to as lysosomes may seem heretical, an increasing list of data can be put forward to argue that it is the late endosomes or prelysosomes that are the key functional degradation compartment in the cell (see Griffiths, 1990).

NATO ASI Series, Vol. H 91
Trafficking of Intracellular Membranes
Edited by M.C. Pedroso de Lima N. Düzgüneş and D. Hoekstra

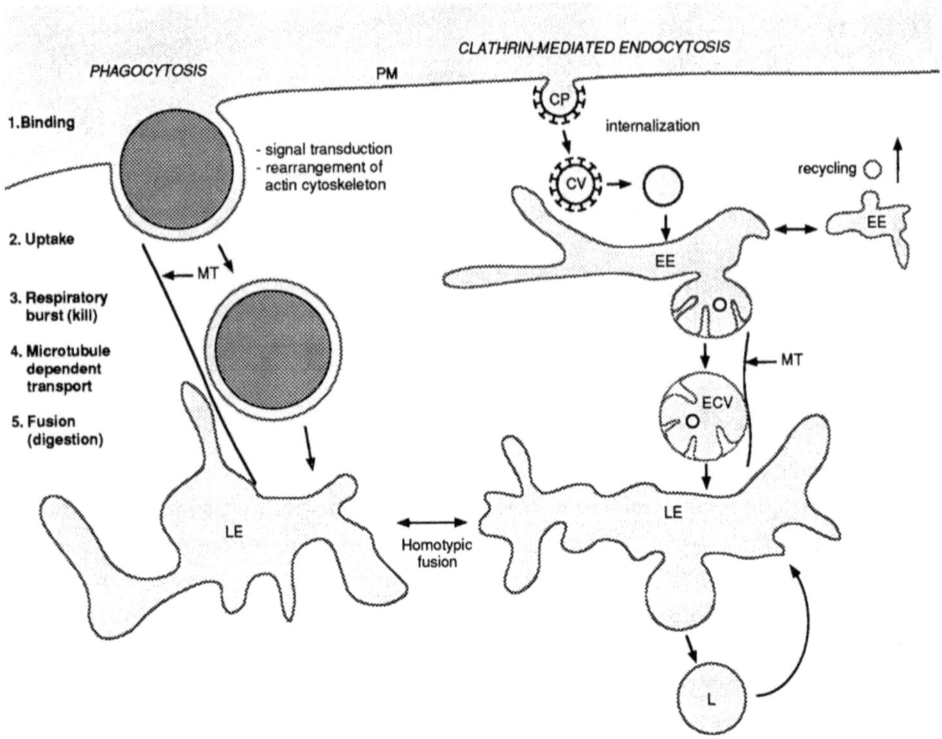

Fig 1. Schematic representation of our view of the convergence of the endocytic and phagocytic pathways. For the endocytic pathway, we argue that early endosomes (EE) and late endosomes (LE) are dynamic, but pre-existing compartments that can fuse in a homotypic fashion. ECV represents the endosomal carrier vesicle, which we believe can bud off the early endosome and move along microtubules towards their minus ends in a dynein dependent fashion (see Aniento et al 1993) to the perinuclear region of the cell where they fuse with late endosomes. We propose that the terminal endocytic compartment, the lysosomes, are transient vesicles that can bud off the bud off the late endosome and can fuse back with that compartment thereby replenishing the concentration of acid hydrolases when required. For the phagocytic route, we propose that the phagosome fuses preferentially with the late endosome, a process that is dependent on microtubules. Though the phagosome moves bidirectionally on microtubules, it has a net inward movement. This, together with microtubule mediated movement of lysosomes, facilitates mixing.

Phagocytosis

Phagocytosis is the mechanism by which cells take up relatively large particles (> 0.3 μm) such as bacteria, yeast or protozoa. While many if not all cells appear to possess some capacity for this process, it is a highly developed function of so-called professional phagocytes such as macrophages and neutrophils. The process can be functionally divided into five main steps (Fig. 1, for review see Silverstein et al. 1989). Following binding to receptors on the cell surface in a multivalent fashion, the particle is enclosed by a domain of the plasma membrane, a process that involves complex signal transduction events as well as major local rearrangement of the actin cytoskeleton (Greenberg et al. 1993). As soon as the phagosome separates from the plasma membrane a series of biochemical reactions, referred to as the respiratory burst, is switched on. In this transient process a series of cytoplasmic and membrane-associated enzymes such as the NADPH-dependent oxidase function to secrete a toxic concentration of free radicals such as O^{-}_{2} and OH^{-} into the lumen of the phagosome (Morel et al. 1991). The function of this process is to kill the ingested micro-organism, a process which some highly evolved micro-organisms manage to evade.

Following the respiratory burst the phagosome fuses with late compartment(s) of the endocytic pathway, a process whereby the ingested micro-organism is exposed to the acid environment and to lysosomal hydrolases. According to conventional dogma the phagosome fuses with the lysosome to form the so-called phagolysosome. However, the data from our work would argue rather that phagosomes fuse with the late endosome/prelysosomal compartment (Rabinowitz et al., 1992). Further, the available data argue strongly that the formation of phagolysosomes depends critically upon the involvement of microtubules (Jahraus et al., 1993; Desjardins et al., 1994)

Our group is currently involved in studying these late events in the formation of phagolysosomes. We will now focus on these late events, describing our experimental strategy and summarizing the present stage of our knowledge.

Use of latex beads as phagocytic markers

Latex beads (> 0.3 μm) have been extensively used as phagocytic markers. Using a modification of an earlier approach developed by Wetzel and Korn (1969) we purified latex bead containing phagosomes/phagolysosomes, essentially to homogeneity (Desjardins et al., 1994). In this procedure J774 macrophage cells are allowed to internalize 1 μm latex beads for a defined period. Subsequently, the cells are homogenized and the post-nuclear supernatant is adjusted to 40% sucrose and loaded onto a discontinuous sucrose gradient (with steps of 40, 25 and 10% sucrose). Following centrifugation at the latex bead-enclosing membranes float to the interface between 25% and 10% sucrose. The crucial point is that, in the absence of beads, essentially all the cellular material has a density significantly lower than that of the beads.

We have undertaken an extensive biochemical characterization of this latex bead fraction using 2D gel electrophoresis at different times of internalization up to 48 hrs (Desjardins et al., 1994; Desjardins et al., submitted). One significant finding has been the fact that while the biochemical composition of this compartment changes significantly with time, no two time points up to 48 hrs appear to have the same composition. This surprising observation suggests that the latex bead compartment never reaches a steady state. Another is that transfer of material from lysosomes into the phagosomes appears to occur by multiple, partial mixing events, with some lysosomal proteins being transferred more rapidly than others. This finding is consistent with previous work (Wang and Goren, 1987), and suggests that phagosome fusion with late endocytic organelles should not be viewed as a one-time movement of the particle into the lysosome, but as a gradual exchange of contents between the two compartments. Although the physiological meaning of these data not yet clear, this biochemical map of phagosomes at different times provides the framework for our current *in vitro* studies to follow 1) phagosome-endosome fusion and 2) phagosome binding and motility along microtubules.

<u>In vitro fusion assay of phagosomes with endosomes</u>

One of us (AJ) has recently set up an *in vitro* assay to study the interactions of latex bead phagosomes/phagolysosomes with the organelles of the endocytic pathway. The basis for this assay is the work of Jean Gruenberg who has followed the homotypic fusion *in vitro* of early endosomes and late endosomes with themselves as well as the vectoral fusion of ECV with late endosomes (Gruenberg et al., 1989; Gruenberg and Clague, 1992).

In our assay avidin is covalently bound to 1 μm latex beads that are then internalized for different times into J774 macrophages. In a different set of the same cells biotin conjugated to horseradish peroxidase (HRP) is internalized via fluid phase endocytosis. This HRP marker can be selectively internalized into different stages of the endocytic pathway depending on the condition (see Gruenberg and Howell, 1989). For the assay the avidin-latex bead membrane fraction is purified from the donor cell while, for the acceptor cell (selectively filled with biotin-HRP) the total post nuclear supernatant is prepared. The two fractions are then mixed at 37°C in the presence of cytosol and ATP for the fusion reaction. Subsequently, the latex beads are again separated on the step gradient and the amount of bound biotin HRP is quantified.

The present state of this assay is that a significant and specific fusion signal can be obtained between phagolysosomes (1 hr internalization of latex followed by 1 hr chase) and endocytic organelles from the acceptor cells where the biotin HRP had been internalized for 40 min, a condition in which the marker is distributed in early endosomes, ECV and late endosomes. This fusion event is dependent both on a threshold concentration of exogenous cytosol and on ATP. Preliminary data suggest that the cytosol from J774 macrophages gives a significantly higher signal when compared to cytosol prepared from fibroblast cells.

Binding of phagosomes to microtubules and microtubule-dependent motility.

Work from our laboratory and others has shown that the phagosomal/lysosomal system within macrophages is highly dynamic. For example, Knapp and Swanson (1990) showed that the morphology of the lysosomal compartment in bone marrow macrophages is very plastic, responding to the demands of the phagocytic load placed on the cells. We have observed that both phagosomes and late endocytic organelles are highly motile within the cell. In J774 macrophages, late endosomes and lysosomes containing colloidal gold and phagosomes containing latex beads move actively within the cytoplasm, encountering one another multiple times and coming apart again (Desjardins et al 1994). We believe that these movements are important for digestion of phagocytosed material, as they likely provide the basis for efficient mixing of incoming phagocytosed material with the digestive enzymes stored in late endosomes and lysosomes. These movements may be particularly important in light of the multiple fusion-fission events which appear to be involved in transferring material between these compartments. In support of this view, there is evidence that certain cellular pathogens evade destruction in phagosomes by inhibiting the intracellular movements of degradative compartments (D'Arcy Hart, et al. 1987).

Several pieces of evidence indicate that at least some of the observed movements of phagosomes and late endocytic structures are due to microtubule-mediated motility. The first clues come from simply observing these movements. Using the system mentioned above to visualize lysosomes and late endosomes in living cells, we observe that both sets of organelles move along linear tracks in the cell. If these cells are treated with the microtubule-depolymerizing drug nocodazole, directed particle movement ceases, including the movement of phagosomes and lysosomes (Burkhardt et al., manuscript in preparation). Similarly, Swanson et al. (1987) showed that microtubule-depolymerizing agents cause the disruption of the tubular lysosome system. Indeed, the radial extension of tubular lysosomes was shown to be mediated by the microtubule motor protein kinesin (Hollenbeck and Swanson, 1990, Swanson, et al, 1992).

These microtubule-mediated movements are important for the process of phagocytosis in general, as indicated by the finding that treatment of cells with the microtubule-depolymerizing drug nocodazole substantially slows the fusion of incoming phagosomes with lysosomes and late endosomes. We have observed this effect using two very different assays to measure the mixing of lysosomal contents with the phagosome compartment. The first measures the access of sucrose accumulated in lysosomes ("sucrosomes") with phagosomes containing latex beads covalently coupled to invertase (Jahraus, et al. 1994). Fusion between phagosomes and sucrosomes is observed as a reduction in the osmotic swelling of the sucrosomes due to action of the invertase on the otherwise indigestible sucrose. If the experiment is performed in the presence of nocodazole, little loss of sucrosome volume is observed, indicating that phagosome-lysosome fusion is reduced in the absence of microtubules. Similar results were obtained using an assay which measures the transfer of lysosomal membrane proteins into the latex bead compartment (Desjardins, et al. 1994). Latex beads were fed to macophages for various times, and the transfer of LAMP1, a resident membrane protein of late endocytic organelles, into the latex bead phagosome was quantified using immuno-electron microscopy. In comparison to control cells, where the LAMP1 label on phagosomes increased by 6-fold, cells incubated with nocodazole showed little LAMP1 transfer to phagosomes (label increased by only 1.5 fold). In both assays, the inhibition of phagosome fusion with late endocytic compartments was reversed when the nocodazole was removed and microtubules were permitted to reform. Note that in both assays, inhibition by nocodazole was not complete; fusion of phagosomes with late endocytic compartments continued at low levels. Though it is possible that this residual fusion is the result of a few highly stable microtubules remaining in the cells, we think this is unlikely. Rather, we think that this fusion is the result of random encounters between organelles. According to this view, the role of microtubules is to facilitate interaction between phagosomes and late endocytic compartments.

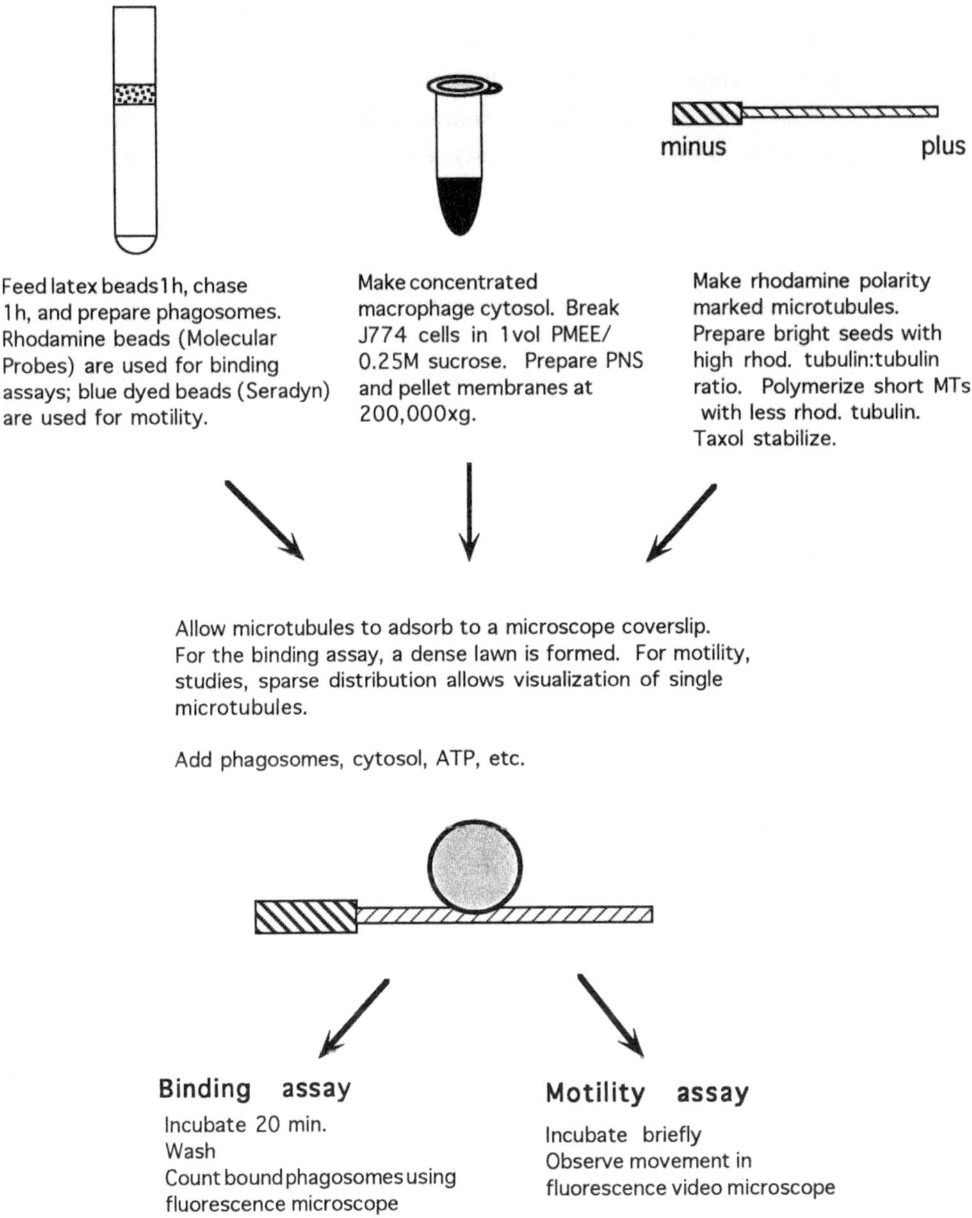

Fig. 2. General outline for our assays measuring phagosome binding to microtubules and microtubule-dependent phagosome motility.

In an effort to dissect the machinery responsible for phagosome interaction with microtubules, we (JKB and AB) have established a collaborative project with Drs. F. Severin and A. Hyman (EMBL). This collaboration has resulted in the development of in vitro assays to measure phagosome-microtubule binding and microtubule-based phagosome movement. The two assays are closely related, and are outlined in Fig. 2.

For the binding assay, phagosomes containing rhodamine-latex beads are purified from tissue culture macrophages. The ability of the phagosomes to bind to microtubules is then tested using a modification of the assay developed by Sorger et al. (submitted) to measure kinetochore - microtubule interactions. Microtubules, dimly labelled with rhodamine tubulin, are stabilized with taxol and adsorbed to a coverslip in a small assay chamber to form a lawn, and the phagosomes are added and allowed to bind. After washing away the unbound phagosomes, the bound phagosomes are counted in the fluorescence microscope. We find that binding of phagosomes to microtubules is minimal in the absence of cytosol, but when 3mg/ml cytosol is added, numerous phagosomes bind to the lawn of microtubules (Burkhardt et al., manuscript in preparation). In addition to cytosol, proteins of the phagosome membrane are required for binding; uninternalized latex beads do not bind, and heat treatment or proteolysis of the phagosome abolishes binding. Phagosome-microtubule interaction is differentially sensitive to digestion with a panel of proteases, suggesting that specific phagosome membrane proteins are involved.

One of us (AB) developed a related assay to assess the movement of purified phagosomes on microtubules in the fluorescence video microcope. This assay combines a previously described method for studying the movement of secretory granules on microtubules (Burkhardt et al. 1993) with the use of polarity marked rhodamine microtubules (Howard and Hyman, 1993). Many aspects of this motility assay are similar to the binding assay described above, facilitating comparison of the two types of phagosome-microtubule interactions. However, there are several key differences. First, in place of rhodamine beads, latex beads colored with blue dye are used; these fluoresce only dimly in the rhodamine channel, making it possible to visualize both the bead and the rhodamine-microtubule on which it is moving. The second major difference is that the optimal concentration of cytosol for motility is much higher than that which is required in the binding assay. With care, it is possible to prepare macrophage cytosol in concentrations up to 30-50 mg/ml; and these preparations work best in the motility assay.

The preliminary findings from the motility assay indicate that isolated phagosomes do move on microtubules, and that this movement requires ATP and concentrated cytosol. The results so far show that phagosomes move bidirectionally on microtubules, with preferential movement toward the microtubule minus end. Since microtubules in cells are organized with their plus ends at the cell surface and their minus ends at the cell center (Bergen, et al. 1980) this preferential minus-end directed movement would result in accumulation of phagosomes at the cell center. This is in good agreement with our observations of phagosome movements in intact cells, where we observe movement of individual particles toward and away from the cell periphery for many hours after internalization, but a net accumulation near the nucleus over time.

The differences in our results so far using the assays for binding and motility, in particular the different cytosol requirements, suggest that these assays are detecting two different types of phagosome-microtubule interactions. These are likely to be mediated by members of the two families of cytosolic microtubule binding proteins, motor proteins and MAPs. Both the plus-end motor kinesin and the minus-end motor cytoplasmic dynein are detectable in our macrophage cytosol as well as in the phagosome fraction. In addition, we detect other members of the kinesin superfamily. Though we have yet to conclusively demonstrate this point, it is probable that the cannonical kinesin and cytoplasmic dynein are responsible for the mediating phagosome motility in our assay. In contrast, it is likely that MAP proteins mediate the majority of interactions we observe in the binding assay. Motor-mediated binding is likely to be of low affinity, while the interactions we observe in the binding assay withstand substantial dilution and shear forces during washing. In our view, it is likely that MAPs and motor proteins work together to localize phagosomes and lysosomes within the cell and to facilitate their interactions. High affinity interactions may be required to initiate binding of phagosomes to microtubules, and to anchor stationary phagosomes along tracks where they are likely to encounter late endosomes and lysosomes. When these bound organelles are triggered to move, either by the binding of motor proteins or by activation of pre-bound motor proteins, the binding factors would release the organelle and the motor proteins would mediate motility. Clearly, such a system requires regulation, either through the direct interaction of binding proteins and motor proteins, or through accessory activating factors. A general scheme depicting the machinery proposed to mediate phagosome-microtubule interactions is shown in Fig. 3.

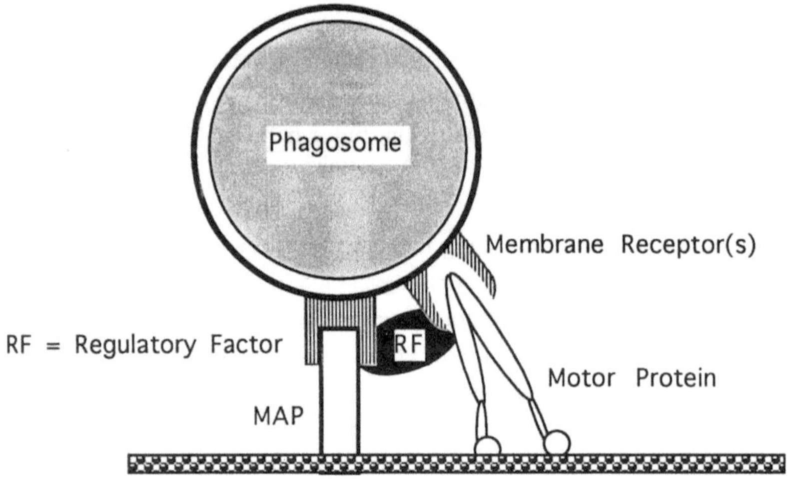

Fig. 3. Simplified representation of the cellular machinery currently thought to be involved in mediating phagosome binding and motility on microtubules. Shown here is a microtubule binding protein (MAP), which anchors non-motile phagosomes on microtubules and perhaps also initiates the binding required for subsequent motilty. A motor protein drives movement on the microtubule when the binding protein releases. It is likely that phagosomes interact with at least two different motors, to mediate movement toward and away from the cell center. We show here two different membrane receptor proteins, one for the MAP and one for one motor protein. Clearly, models involving 1, 2, or 3 receptors can be envisaged. Finally, a regulatory factor is required to transmit information between the MAP and the motor protein. Data from several groups suggests that one or more of these cytosolic accessory factors is an absolute requirement to form an active motor protein on a membranous organelle (see Schroer 1991; Schroer and Sheetz 1991; Burkhardt, et al. 1993).

Acknowlegements
The authors thank J. Pickles for help with the manuscript and P. Reidinger for help with illustrations. J.K.B. is a fellow of the Human Frontier Science Program.

References

Aniento, F, Emans, N, Griffiths, G, and Gruenberg, J (1993) Cytoplasmic dynein-dependent vesicular transport from early to late endosomes. J. Cell Biol. 123:1373-1387.

Bergen, L, Kuriyama, R, and Borisy, GG (1980) Polarity of microtubules nucleated by centrosomes and chromosomes of CHO cells in vitro. J. cell Biol. 84:151-159.

Burkhardt, JK, McIlvain, JM, Sheetz, MP, and Argon, Y (1993) Lytic granules from cytotoxic T cells exhibit kinesin-dependent motility on microtubules in vitro. J. Cell Sci. 104:151-162.

D'Arcy Hart, P, Young, MR, Gordon, AH, and Sullivan, KH (1987) Inhibition of phagosome-lysosome fusion in macrophages by certain mycobacteria can be explained by inhibition of lysosomal movements observed after phagocytosis. J. Exp. Med. 166:933-946.

Desjardins, M, Huber, L, Parton, R, and Griffiths, G (1994) Biogenesis of phagolysosomes proceeds through a sequential series of interactions with the endocytic apparatus. J. Cell Biol. 124:677-688.

Desjardins, M, Celis, JE, van Meer, G, Dieplinger, H, Griffiths, G and Huber, L Molecular characterization of phagolysosomes in professional and non-professional phagocytes. (submitted)

Greenberg, S, Chang, P, and Silverstein, SC (1993) Tyrosine phosphorylation is required for Fc receptor-mediated phagocytosis in mouse macrophages. J. Exp. Med. 177: 529-534.

Griffiths, G (1990) The compartments of the endocytic pathway. in *Proc 2nd European Workshop on Endocytosis*. P. Cortoy ed., pp73-83 Springer Verlag, Heidelberg.

Griffiths, G and Gruenberg, J (1991) The arguments for pre-existing early and late endosomes. Trends Cell Biol. 1:5-9.

Gruenberg, J, Griffiths, G, and Howell, KE (1989) Characterization of an early endosome and putative endocytic carrier vesicles in vivo and with an assay of vesicle fusion in vitro. J. Cell Biol. 108:1301-1316.

Gruenberg, J, and Clague, MJ (1992) Regulation of intracellular membrane transport. Curr. Opin. Cell Biol. 4:593-599.

Hollenbeck, PJ and Swanson, JA (1990) Radial extension of macrophage tubular lysosomes supported by kinesin. Nature 346:864-866.

Howard, J, and Hyman, AA (1993) Preparation of marked microtubules for the assay of the polarity of microtubule-based motors by fluorescence microscopy. in *Motility Assays for Motor Proteins*, Scholey, JM, ed. Academic Press, San Diego.

Jahraus, A, Storrie, B, Griffiths, G and Desjardins, M (1994) Evidence for retrograde traffic between terminal lysosomes and the prelysosomal/late endosome compartment. J. Cell Sci. 107:145-157.

Knapp, PE, and Swanson, JA (1990) Plasticity of the tubular lysosomal compartment in macrophages. J. Cell Sci. 95:433-439.

Morel, F, Doussiere, J, and Vignais, PV (1991) The superoxide-generating oxidase of phagocytic cells: Physiological molecular and pathological aspects. Eur. J. Biochem. 201:523-546.

Rabinowitz, S, Horstmann, H, Gordon, S and Griffiths, G (1992) Immunocytochemical characterization of the endocytic and phagolysosomal compartments in peritoneal macrophages. J Cell Biol. 116:95-112.

Schroer, TA (1991) Association of motor proteins with membranes. Curr. Opinion Cell Biol. 3:133-137.

Schroer, TA, Sheetz, MP (1991) Functions of microtubule-based motors. Ann. Rev. Physiol. 53:629-652.

Silverstein, SC, Greenberg, S, DiVirgilio, F, and Steinberg, TH (1989) Phagocytosis. in *Fundamental Immunology*. Paul, WE, ed. Raven Press, New York.

Sorger, PK, Severin, FF, and Hyman, AA Factors required for the binding of reassembled yeast kinetochores to microtubules in vitro. Submitted.

Swanson, J, Bushnell, A, and Silverstein, SC (1987) Tubular lysosome morphology and distribution within macrophages depend on the integrity of cytoplasmic microtubules. Proc. Nat. Acad. Sci. 84:1921-1925.

Swanson, JA, Locke, A, Ansel, P, and Hollenbeck, PJ (1992) Radial movement of lysosomes along microtubules in permeabilized macrophages. J. Cell Sci. 103:201-209.

Wang, Y and Goren, MB (1987) Differential and sequential delivery of fluorescent lysosomal probes into phagosomes in mouse peritoneal macrophages. J. Cell Biol. 104:1749-1754.

Watts, C and Marsh, M (1992) Endocytosis: what goes in and how. J. cell Sci. 103:1-8.

Wetzel, MG and Korn, ED (1969) Phagocytosis of latex beads by Acanthamoeba castellanii (NEFF). III Isolation of the phagocytic vesicles and their membranes. J. cell Biol. 43:90-104.

GTPases: Key regulatory components of the endocytic pathway.

M. Alejandro Barbieri, Maria Isabel Colombo, Guangpu Li, Luis Segundo Mayorga* and Philip Stahl.

From the Department of Cell Biology and Physiology, Washington University School of Medicine, St. Louis, Missouri 63110 and from Instituto de Histologia y Embriologia (CONICET-Universidad Nacional de Cuyo), Mendoza,Argentina.*

KEY WORDS/ ABSTRACT: ENDOCYTOSIS/ RAB5/ ADP-RIBOSYLATION FACTOR/ HETEROTRIMERIC G PROTEINS/ NSF/ SNARES/ PLA2/ ENDOSOME FUSION

GTP-binding proteins or GTPases are versatile cyclic molecular switches (Gilman, A.G. 1987 ; Bourne et al., 1990, 1992). In the past few years there has been an explosion of interest in unraveling the role of GTPases in membrane traffic (Balch et al., 1990; Bourne, H.R., 1988). GTP-binding proteins are classified into two broad families: ras-like monomeric GTP-binding proteins and the heterotrimeric G proteins. Evidences from genetics, immunolocalization, and functional assays has established that two subfamilies of monomeric GTP-binding proteins, the Rab and ARF subfamilies, are required for membrane traffic (Pfeffer, S.R., 1992; Rothman, J.E. and Orci, L., 1992). More recently, attention has turned to the role of heterotrimeric G proteins in membrane traffic (Balch et al., 1992; Barr et al., 1992). This review will focus solely on the recent data that suggest that heterotrimeric G proteins and monomeric GTP-binding proteins are involved in endosome-endosome fusion and in endocytic transport.

In vitro endosome-endosome fusion: Characteristics

The complex pattern of intracellular membrane vesicle traffic is responsible not only for delivery of newly synthesized proteins to their correct intracellular locations but also for the appropriate distribution of macromolecules which have been internalized either by fluid-phase uptake or by receptor-mediated endocytosis. Recent experimental advances have led to the reconstitution of several membrane vesicle fusion events important for transport along the endocytic pathway (Davey et al., 1985; Gruenberg et al., 1986; Braell et al., 1987; Diaz et al., 1988; Woodman, P.G. and Warren, G., 1988) and

NATO ASI Series, Vol. H 91
Trafficking of Intracellular Membranes
Edited by M.C. Pedroso de Lima N. Düzgüneş and D. Hoekstra

the subsequent sorting of internalized probes (Wessling-Resnick, M. and Braell, W.A. 1990). Other reconstitution strategies have focussed on vesicular transport between Golgi cisternae (Balch et al., 1984), transfer from the endoplasmic reticulum to the Golgi (Beckers et al., 1987), and the recycling from prelysosomes to the trans-Golgi network (Goda, Y. and Pfeffer, S.R., 1986). Cell-free reconstitution of membrane trafficking also provides a powerful tool for analysis of gene products identified by genetic means to be involved in secretion (Novick et al ., 1981; Baker et al., 1988; Bacon et al., 1988; Backer et al.,1990). Ultimately, characterization of such cell-free assay systems will permit the identification of the biochemical elements which mediate and regulate membrane traffic events *in vivo*.

Fusion among early endocytic vesicles has been reconstituted in cell-free systems by several groups (Gruenberg et al., 1986; Braell et al., 1987; Diaz et al., 1988; Woodman, P.G. and Warren, G., 1988). The fusion assay employed in our studies utilizes two probes: (1) a mouse monoclonal antibody raised against dinitrophenol (anti-DNP IgG) that has been derivatized with mannose to make it a high-affinity ligand for the macrophage mannose receptor and (2) dinitrophenol-derivatized β–glucuronidase, a glycoprotein recognized by the mannose receptor. In this assay, two separate populations of J774-E clone macrophages are allowed to internalize the probes for 5 min. at 37°C. Cells are homogenized and endosomal fractions are obtained by differential centrifugation. These membrane fractions are mixed in a reaction mixture containing salts, gel filtered-cytosol, and ATP. Fusion between endosomes loaded with different probes results in the formation of an immune complex that can be immunoprecipitated and quantitated by taking advantage of the enzymatic activity of β-glucuronidase. Fusion, as assessed by immune complex formation, is cytosol, time and temperature dependent, reaching a plateau after a 30 min. incubation at 37°C. The rate of fusion also increases with the total amount of vesicles present in the system, and it is inhibited by dilution. Fusion depends on the presence of ATP, which is supplied by an ATP-regenerating system. Adenylimidodiphosphate, a non-hydrolyzable analog of ATP, does not support fusion, suggesting that ATP hydrolysis is required. Endosome fusion requires the presence of 50-70 mM salts. The requirement for cytosol is saturable at about 1.2 mg/ml.

GTP-binding proteins have been implicated in transport along the endocytic and exocytic pathways (Diaz et. al., 1988; Malhotra et al., 1988; Beckers, J.C.M. and Balch, W.E., 1989). Non hydrolyzable analogs of GTP, such as GTPγS, have been shown to block transport from the endosplasmic reticulum (ER) to the Golgi (Beckers et al., 1989), among Golgi stacks (Melancon et al., 1987) and between endosomes (Mayorga et al.,

1989). Curiously, when the GTPγS effect was carefully examined in the endosome fusion assay, the inhibitory effect of GTPγS was observed only at high concentrations of cytosol; when GTPγS was tested at low concentrations of cytosol, a stimulatory effect was observed (Mayorga et al., 1989). The stimulatory effect of GTPγS on endosome fusion was dependent on salts, cytosol, ATP and was effectively blocked by anti-NSF antibody. It is possible that the stimulatory effect of GTPγS at low cytosol concentrations and the inhibitory effect at high cytosol concentrations are related to the same mechanism. An excess of factors bound irreversibly to the membranes may inhibit fusion by steric hindrance or may promote futile reactions that consume some additional factors present in limiting concentration (Mayorga et al., 1989). Alternatively, two or more GTP-binding proteins may be active in endosome fusion: one mediating the binding of cytosolic proteins and another participating in some regulatory step that is blocked by the non-hydrolyzable analog.

Rab5 in endocytosis

Like others GTPases, rab5 proteins are thought to act as a regulatory molecule that recognizes target proteins through a nucleotide-dependent conformational change (Bourne et al., 1988). One possible candidate for a target is a rabphilin-3A-like molecule that binds specifically to the GTP-bound form of rab3a (Shirataki et al., 1993). Association of rab proteins to the membrane depends on C-terminal isoprenylation which is essential for function. The membrane association is regulated by rab-GDP dissociation inhibitor (rab GDI) (Araki et al., 1990; Regazzi et al., 1992; Soldati et al., 1993) and is accompanied by exchange of bound GDP with GTP (Ullrich et al., 1994; Soldati et al., 1994). Nucleotide exchange is probably catalyzed by a GDP dissociation stimulator (GDS) (Burstein, E.S. and Macara, I.G., 1992; Burton et al., 1993). Conversely, the switch from GTP- to GDP-bound state occurs through hydrolysis of GTP, a process stimulated by one or more GTPase activating proteins (GAPs) (Walworth et al., 1992; Strom et al., 1993; Walworth et al., 1989; Becker et al., 1991). While mutational studies have indicated that cycling between the GTP- and GDP-bound forms is critical for the function of rab proteins, the exact role of the GTP hydrolysis step is not clear (Goud et al.,1991; Li et al., 1993; Simons et al., 1993). GTP hydrolysis might be required as a kinetic proofreading device prior to membrane fusion (Bourne, H.R., 1988), as supported by the inhibitory effect of non-hydrolysable GTP-analogue in various cell-free system (Mayorga et al., 1989; Bomsel et al., 1990). However, intracellular membranes contain different GTPases (Pfeffer, S.R., 1992) and, so far, it has not been possible to specifically affect rab protein function with GTPγS. Therefore, the alternative possibility still exists that rab proteins might use GTP hydrolysis as a means of switching from an 'active' to an 'inactive' conformation (Takai et al., 1993; Novick et al., 1993).

We have used the *in vitro* endosome-endosome fusion and overexpression of rab 5 in cultured BHK-21cells to study the role of GTPase activity of rab5, a rate limiting factor in the endocytic pathway. Over-expression of rab5:WT resulted in a 3-fold stimulation of horseradish peroxidase uptake in BHK-21 cells. Deletion of the entire C-terminal tetrapeptide motif abolished rab5 activity. In addition to the previously reported N133I mutation (Bucci et al., 1992; Li et al., 1993), the S34N mutation also resulted in a guanine nucleotide binding defective form that was a dominant inhibitor of endogenous rab5 activity. To explore the importance of C-terminal domain of rab5, second-site mutations in rab5:S34N and rab5:N133I were introduced. When the double mutants were tested in the HRP uptake assay, they were inactive demonstrating the C-terminal isoprenylation is necessary for S34N and N133I to inhibit fluid-phase endocytosis. The Q79L mutation (the ras equivalent Q61L decreases intrinsic and GAP-activated GTPase activity) had no effect on rab5 activity suggesting that rab5:GTP is required prior to membrane fusion, whereas the GTP hydrolysis by rab5 occurs after membrane fusion and functions to inactivate the protein. We also examined the effect of rab5:WT and mutants in the *in vitro* endosome-endosome fusion. Endosome fusion was markedly activated by rab5 addition but only when rab5 was in the GTP or GTPγS-form, rab5:GDPβS was inactive (Barbieri et al., 1994). Rab5:GTPγS further enhanced endosome fusion when compared to rab5:GTP, indicating that GTP hydrolysis is not required for vesicle docking and fusion. This finding confirmed previous results *in vivo*, where we found that rab5:Q79L, was a potent stimulator of endocytosis. The stimulatory effect of rab5:GTPγS is time and temperature dependent, reaching a plateau after a 30 min incubation at 37°C, and was dependent of ATP and was effectively blocked by anti-NSF antibody. In the absence of cytosol, stimulation of fusion by rab5:GTPγS was dependent on the amount of rab5 added, suggesting that factors required for fusion are already present on the membrane. This was, in part, confirmed when the membranes were washed with 0.2 M KCl resulting in a partial loss of rab5 stimulation. These results suggest that some of the proteins necessary for the fusion events are probably membrane-associated peripheral proteins.

When the two GTP-binding defective rab5 mutants (rab5:N133I and rab5:S34N) were tested in *in vitro* endosome fusion both were inhibitory. The inhibitory effect was absolutely dependent of the isoprenyl group since nonprenylated rab5 mutants or double mutants (rab5:S34N/ΔC4 and rab5:N133I/ΔC4) were inactive. This finding also confirmed earlier work that the double mutants were unable to block HRP uptake into BHK-21 cells (Li et al., 1994)

ADP-ribosylation factor (ARF) and endosome fusion.

ADP-ribosylation factors (ARFs) comprise a family of small GTP-binding proteins that includes more than 15 structurally related gene products (Clark et al., 1993). ARF was first identified as the cofactor required for the ADP-ribosylation of Gαs by cholera toxin (Kahn and Gilman, 1984; Kahn and Gilman, 1986). In the last few years ARF proteins have been implicated in a number of membrane traffic events. The initial evidence for a role of ARF in vesicular transport came from genetic studies in yeast where deletion of the ARF1 genes resulted in a secretory defect (Stearns et al., 1990). Using several *in vitro* assays that reconstitute transport between different compartments, it has been shown that ARF is an essential component required for transport through the secretory pathway (Balch et al., 1992; Kahn et al., 1992). In addition, ARF is also involved in nuclear membrane fusion in *Xenopus* egg extracts (Boman et al., 1992). We provided evidence that ARF also modulates endosome-endosome fusion (Lenhard et al., 1992). The addition of recombinant ARF1 to the *in vitro* assay resulted in a GTPγS-dependent inhibition of fusion. Furthermore, myristoylation at the NH2-terminal glycine and GTP binding appear to be critical for ARF activity during an early prefusion step required for endosomal fusion.

Previous studies have also demonstrated that ARF is associated with Golgi membranes (Stearns et al., 1990; Donaldson et al; 1991) and Golgi-derived COP-coated vesicles (Serafini et al., 1991). It has been shown that ARF plays an essential role in regulating binding of cytosolic coat proteins to the membranes (Donaldson et al., 1992; Palmer et al., 1993). In a recent report we have also presented evidence suggesting that both heterotrimeric G proteins (see below) and ARF regulate priming of endosomal membranes for fusion (Lenhard et al., 1994). Thus, G proteins and ARF may regulate endocytosis by mediating the binding of cytosolic factor(s) required for fusion to the endosomal membrane.

Heterotrimeric G proteins and endosome-endosome fusion

Classically, heterotrimeric G proteins are thought to coupled to transmembrane receptors and transduce extracellular signals to intracellular effectors (Gilman, A., 1987). Heterotrimeric G proteins are a trimeric complex of α, β, and γ subunits. When a transmembrane receptor is activated by an extracellular signal (light, binding of a ligand etc.) the receptor interacts with the G protein and stimulates the exchange of GDP for GTP onto the Gα subunits. Binding of GTP causes the G protein to dissociate into a and $\beta\gamma$ subunits. There is considerable evidence that both α and $\beta\gamma$ subunits are able to activate a broad range of intracellular effectors (Gilman, A., 1987; Clapham, D. and Neer, E., 1993). As mentioned above trimeric G proteins have always been believed to function at the plasma membrane but the finding that there were intracellular pools of heterotrimeric G proteins (Codina et al, 1988; Ali et al., 1989) suggests that the G proteins might have other functions. A growing body of evidence now indicates that G proteins also regulate intracellular transport events (Bomsel and Mostov, 1992; Nuoffer and Balch, 1994).

An approach to demonstrate that G proteins are involved in membrane traffic is to show that various agents that affect G protein function alter a specific vesicular transport step. We presented the first evidence indicating that one or more heterotrimeric G proteins regulate *in vitro* fusion between endosomes. Our conclusions are based on the effects observed upon addition of a peptide that resembles activated receptors and upon addition of purified G$\beta\gamma$ subunits (Colombo et al., 1992) to the fusion assay. Excess of free G$\beta\gamma$ subunits in general antagonizes the activation of a G protein by complexing and sequestering the Gα subunit. Addition of purified G$\beta\gamma$ subunits inhibited endosome-endosome fusion suggesting that an activated Gα subunit was involved in the process. However, it is also possible that the free G$\beta\gamma$ subunits may interact directely with a downstream effector. Indeed, the effects of G$\beta\gamma$ subunits on the downstream effectors may be either synergistic or antagonistic.

Recently, we have presented further evidence that confirms the involvement of G proteins in endosome fusion. In order to identify the G protein involved in the fusion process we tested a large number of agents that specifically interact with Gαs in the *in vitro* assay. These include

(i) peptides corresponding to the cytoplasmic domain of G protein-coupled receptors that mimic the interaction of the receptors with the G proteins,

(ii) anti-G proteins antibodies that impede G protein function, and

(iii) cholera toxin, a bacterial toxin that specifically ADP-ribosylates Gas resulting in a permanently activated G protein. The use of these reagents allowed us to identify Gs as one of the G proteins regulating the process of endosome recognition and/or fusion (Colombo et al., 1994a). Moreover, our results also raise the possibility that signal transduction through cytoplasmic domains of receptors may participate in the regulation of endocytic trafficking.

The involvement of multiple GTPases has also been implicated in clathrin-coated vesicle formation. Using an *in vitro* assay that reconstitute early stages of the endocytic process, S. Smith and collaborators (Carter et al., 1993), have shown that distinct classes of GTP-binding proteins are differentially involved in coated pit assembly, invagination, and coated vesicle budding.

In addition, we have presented the first evidence indicating a role for heterotrimeric G proteins as regulators of endocytosis *in vivo*. As a tool to study the role of G proteins *in vivo* we have used the membrane permeant agent aluminum fluoride (AlF). Fluoroaluminate complexes (AlF) activates trimeric G proteins by mimicking the γ-phosphoryl group of GTP when GDP is present in the guanine binding site of the Gα subunit. Consequently, AlF acts as a potent and reversible activator of heterotrimeric G proteins. Our results indicate that in intact cells AlF inhibits fusion of early endosomes with an intracellular proteolytic compartment. Also AlF affected the efficiency of mixing of sequentially internalized endocytic probes (Colombo et al., 1994b). Kahn and collaborators have shown that most of the small GTP-binding proteins from the Rab and ARF families are not activated by AlF (Kahn, R., 1992). Therefore the target of fluoroaluminate may be a trimeric G protein that participates in the endocytic process.

In summary, we have provided several lines of evidence for the participation of heterotrimeric G proteins in fusion of early endosomes and in the regulation of endocytic trafficking. Findings from our laboratory and others, provide new insights into the processes involved in receptor-mediated endocytosis, but clearly, much remains to be learned about the role of G proteins in these membrane trafficking events.

Conclusions:

It is now clear that multiple GTP-binding proteins are involved both in endocytosis and exocytosis (reviewed by Nuoffer and Balch, 1994). It seems that each of these GTPases regulates a series of distinct biochemical steps required for vesicular transport and fusion events. A growing consensus points to the possibility that rab proteins may play an important role in vesicle targeting/fusion, while ARF and trimeric G proteins appears to form part of the machinery involved in coat assembly and budding. Each of these GTPases may act in a coordinated fashion with the accessory proteins that modulate the GTPase cycle and with a specific set of upstream and downstream effectors, such as NSF, SNAPs and phospholipases involved in the fusion process (see Figure 1). Therefore, GTPases confer vectoriality and fidelity to the transport process, and play a critical role in the maintenance of organelle structure and function.

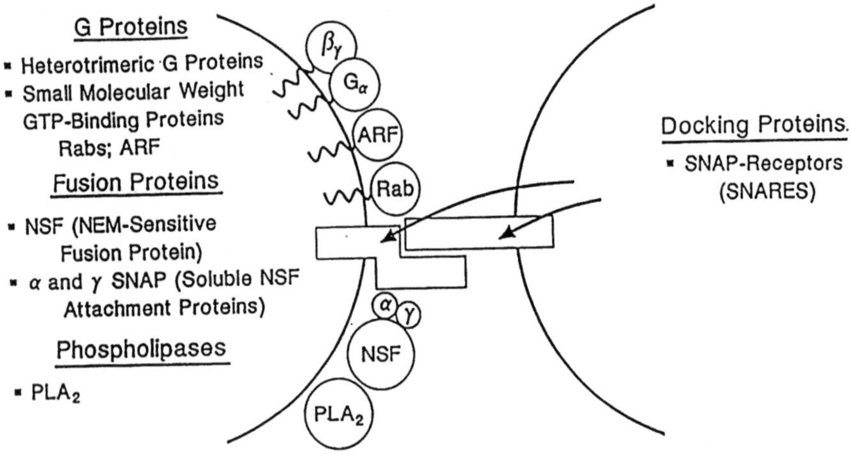

Figure 1. The diagram depicts the possible involved components in endosome fusion

References

- Ali, N., Milligan, G., Evans, H. (1989) Distribution of G-proteins in rat liver plasma-membrane domains and endocytic pathways. *Biochem. J.* **261**: 905-910
- Araki, S., Kikuchi, A., Hata, Y., Isomura, M. and Takai, Y. (1990) Regulation of reversible binding of smgp25, a ras p21-like GTP-binding protein, to synaptic plasma membrane and vesicles by its specific regulation protein, GDP dissociation factor *J. Biol. Chem.* **265**: 13007-10470
- Bacon, R.A., Salminen, A.K., Ruohola, H., Novick, P. and F erro-Novick, S. (1988) The GTP-binding protein, YPT1 is required for transport *in vitro* : the Golgi apparatus is defective in YPT1 mutants *J. Cell. Biol.* **109** : 1015-1022
- Backer, D., Westehube, L., Schehmean, R., Botstein, D. and Segev, N. (1990) GTP-binding ypt1 protein and calcium function independently in a cell-free protein transport reaction *Proc. Natl. Acad. Sci. U.S.A.* **87**: 355-3359
- Baker, D., Hicke, L., Rexach, M., Schleyer, M. and Schekman, R. (1988) Reconstitution of SEC gene product-dependent inter compartmental protein transport *Cell* **54**: 335-344
- Balch, W.E., Dumphy, W.G., Braell, W.A. and Rothman, J.E. (1984) Reconstitution of the transport of protein between successive compartments of the Golgi measured by the coupled incorporation of N-acetylglucosamine *Cell* **39**: 405-416
- Balch, W.E., Kahn, R.A. and Schwninger, R. (1992) ADP-ribosylation factor is required for vesicular trafficking between the endoplasmic reticulum and the cis-Golgi compartment *J. Biol. Chem.* **267**: 13053-13061
- Balch, W.E. (1990) Small GTP-binding proteins in vesicular transport *Trends Biochem. Sci.* **15**: 4473-477
- Balch, W.E. (1992) Molecular dissection of early stages of eukaryotic secretory pathway *Current Biol.* **2** : 157-160
- Barbieri, M.A., Li, G., Colombo, M.I., and Stahl, P. (1994) Rab5, an early acting endosomal GTPase, supports in vitro endosome fusion without GTP hydrolysis *J. Biol. Chem.***269**: 18720-18722
- Barr, F.A., Leyte, A. and Huntter, W.B. (1992) Multiple trimeric G proteins on the trans Golgi network exert stimulatory and inhibitory effects on the secretory vesicles formation *Trends Cell Biol.* **2**: 91-94
- Becker, J, Tan, T.J., Trepte, H.H. and Gallwitz, D. (1991) Mutational analysis of the putative effector domain of the GTP-binding ypt1 protein in yeast suggest specific regulation by a novel GAP activity *EMBO J.* **10**: 785-792
- Beckers, C.J.M., Keller, D.S. and Balch, W.E. (1987) Preparation of semi-intact Chinese hamster ovary cells for reconstitution of endoplasmic reticulum to Golgi transport in a cell-free system *Cell* **50**: 523-534
- Beckers, C.J.M. and Balch, W.E. (1989) Calcium and GTP essential components in vesicular trafficking between the endoplasmic reticulum and Golgi apparatus *J. Cell Biol.* **108**: 1245-11256
- Beckers, C.J. M., Block, M.R., Glick, B.S., Rothman, J.E. and Balch, W.E. (1989) Vesicular transport between the endoplasmic reticulum and the Golgi stack requires the NEM-sensitive fusion protein *Nature* **339**: 397-398
- Boman, A.L., Taylor, R.C., Melancon, P. and Wilson, K L. (1992) A role for ADP-ribosylation factor in nuclear vesicles dynamic *Nature* **358**: 512-514
- Bomsel, M., Parton, R., Kuznestsov, S.A., Schoer,T.A. and Gruenberg, J. (1990) Microtubule-and motor-dependent fusion *in vitro* between apical and basolateral endocytic vesicles from MDCK cells *Cell* **62**: 719-731
- Bourne, H.R., Sanders, D.A. and McCormick, F. (1990) The GTPase superfamily: structure and molecular mechanism *Nature* **349**: 117-127
- Bourne, H.R. and Steyer, L. (1992) G proteins *Nature* **355**: 541-543

- Bourne, H.R. (1988) Do GTPases direct membranes traffic in secretion? *Cell* **53**: 669-671
- Braell, W.A. (1987) Fusion between endocytic vesicles in a cell-free system *Proc. Natl. Acad. Sci U.S.A.* **84**: 1137-1141.
- Bucci, C., Parton, R.G., Mather, I.M. Stunneberg, H., Simons, K., Hoflack, B. and Zerial, M. (1992) The small GTPase rab5 functions as a regulatory factor in the early endocytic pathways *Cell* **70**: 715-728.
- Burstein, E.S. and Macara, I.G. (1992) Characterization of Guanine nucleotide-releasing factor and GTP activating protein that are specific for the ras-related protein rab3A. *Proc. Natl. Acad. Sci. U.S.A.* **89:** 1154-1158.
- Burton, J., Roberts, D., Montaldi, M., Novick, P. and Camilli, P.D. (1993) A mammalian guanine-nucleotide releasing factor enhances function of yeast secretory protein SEC4 *Nature* **361**: 464-467.
- Carter, L.L., Redelmeir, T.E., Wollenweber, L.A. and Schmid, S.L. (1993) Multiple GTP-binding proteins participate in clathrin coated vesicles-mediated endocytosis *J. Cell. Biol.* **120**: 37-45.
- Clapham, D. and Neer, E. (1993) New role for Gproteins βγ-dimers in trans membrane signalling *Nature* **365**: 403-406.
- Clark, J., Moore, L., Krasinskas, A., Way, J., Battey, J., Tamkun, J., and Kahn, R. (1993) Selective amplification of additional members of the ADP-ribosylation factor family:cloning of additional human and drosophila ARF like *Proc. Natl. Acad. Sci. USA.* **90**: 8952-8956.
- Codina, J. , Kimura, S., Krauss-Friedman, N. and Highan, S. (1988) Demonstration of the presence of G protein in hepatic microsomes *Biochem. Biophys. Res. Comm.*
- Colombo, M. I., Mayorga, L.S., Casey, P. and Stahl, P.D. (1992) Evidence of a role for heterotrimeric GTP-binding proteins in endosome fusion *Science* **255**: 1695-1697.
- Colombo, M.I., Mayorga, L.S., Nishimoto, I., Ross, E.M. and Stahl., P.D. (1994a) Gs regulation of endosome fusion suggests a role for signal transduction pathways in endocytosis *J. Biol. Chem.* **269**: 14919-14923.
- Colombo, M. I., Lenhard, J. M., Mayorga, L. S., Beron, W., Hall, H., and Stahl, P. D. (1994b) Inhibition of endocytic transport by aluminum fluoride implicates GTPases as regulators of endocytosis *Mol. Membr. Biol.* **11**: 93-100.
- Davey, J., Hurtley, S .M. and Warren, G. (1985) Reconstitution of endocytic fusion in cell-free system *Cell* **43**: 643-652.
- Diaz, R., Mayorga, L. and Stahl, P. (1988) *In vitro* fusion of endosomal following receptor-mediated endocytosis *J . Biol. Chem.* **263**: 6093-6100.
- Donaldson, J.G., Cassel, D., Kahn, R.A. and Klausner, R.D. (1992) ADP-ribosylation factor a small GTP-binding protein is required for binding of coatomer protein β-COP to Golgi membrane *Proc. Natl. Acad. Sci. USA* **89**: 6408-6412.
- Donaldson, J.G., Lippincott-Schwartz, J. and Klausner, R.D. (1991) Guanine nucleotides modulates the effects of brefeldin A in semi-permeable cells: regulation of the association of a 110 KD peripheral membrane protein with the Golgi apparatus *J. Cell Biol.* **112**: 579-588.
- Gilman, A.G. (1987) G proteins: transducers of receptor-generated signals *Ann. Rev. Biochem.* **56**: 615-649.
- Goda, Y. and Pfeffer, S.R. (1986) Cell free system to study vesicular transport along the secretory and endocytic pathways *Cell* **55**: 309-320.
- Goud, B. and MacCaffrey, M (1991) Small GTP-binding proteins and their role in transport *Curr. Opin. Cell. Biol.* **3** : 626-633.
- Gruenberg, J.E. and Howell, K.E. (1986) Reconstitution of vesicle fusions occurring in endocytosis with a cell-free system *EMBO J.* **5**: 3091-3101.

- Kahn, R.A. and Gilman, A.G. (1984) ADP-ribosylation of Gs promotes the dissociation of its α and βετα subunits *J. Biol. Chem.* **259**: 6235-6240.
- Kahn, R.A. and Gilman, A.G. (1986) The protein cofactor necessary for ADP-ribosylation of Gs by cholera toxin is itself a GTP binding protein *J. Biol. Chem.* **261**: 7906-7911.
- Kahn, R.A., Randazzo, P., Serafini, T., Weiss, O., Rulka, C., Clark, J., Amherdt, M., Roller, P., Orci, L. and Rothman, J. E. (1992) The amino terminus of ADP-ribosylation factor is a critical inhibitor determinant of ARF activates and is a potent and specific protein transport *J. Biol. Chem.* **267**: 13039-13046.
- Lenhard, J.M., Kahn, R.A. and Stahl, P.D. (1992) Evidence for ADP-ribosylation factor as a regulator of *in vitro* endosome-endosome fusion *J. Biol. Chem.* **267**: 13047-13052.
- Lenhard, J.M., Colombo, M.I. and Stahl, P.D. (1994) Heterotrimeric GTP-binding proteins and ARF regulate priming of endosomal membranes for fusion *Arch. Biochem. Biophys.* **312**:
- Li, G. and Stahl, P.D. (1993) Structure-function relationship of the small GTPase rab5 *J. Biol. Chem.* **268** : 24475-24480.
- Li, G., Barbieri, M.A, Colombo, M.I. and Stahl, P. (1994) Structure features of the GTP-binding defective rab5 mutants required for their inhibitory activity on endosome fusion *J. Biol. Chem.* **269**: 14631-14635.
- Malhotra, V., Orci, L., Glick, B.S., Block, M.R. and Rothman, J.E. (1989) Role of N-ethylmaleimide-sensitive transport component in promoting fusion of transport vesicles with cisternae of the Golgi stack *Cell* **54**: 221-227.
- Mayorga, L.S., Diaz, R., Colombo, M.I. and Stahl P.D. (1989) GTPγS stimulation of endosome fusion suggests a role of GTP binding protein in the priming of vesicles prior to fusion *Cell Reg.* **1**: 113-124.
- Mayorga, L.S., Diaz, R. and Stahl, P.D. (1989) Regulatory role for GTP-binding proteins in endocytosis *Science* **244**: 1475-1477.
- Melancon, P., Glick, B.S., Malhorta, V., Weidman, P.J., Serafini, T. Gleson N.L., Orci, L. and Rothman, J.E. (1987) A role of GTP-binding proteins in vesicular transport through the Golgi complex *Cell* **51**: 1053-1062.
- Nuoffer, C. and Balch, W.E. (1994) A GDP-bound of rab 1 inhibits protein export from the endoplasmic reticulum and transport between Golgi compartments *Ann. Rev. Biochem* **63**: 949-990.
- Novick, P., Ferro, S. and Schekman, R. (1981) Order of events in the yeast secretory pathways *Cell* **25**:4461-469
- Novick, P and Brennwald, P. (1993) Friend and family: the role of the rab 6 GTPases in vesicular traffic *Cell* **49** :527-538.
- Palmer, D.J., Helms, J.B., Beckers, C.J.M., Orci, L. and Rothman, J.E. (1993) Binding of coatomer to Golgi membranes requires ADP-ribosylation factor *J. Biol. Chem.* **268**: 12083-12089.
- Pfeffer, S.R. (1992) Targetting of proteins to the lysosome *Trends Cell Biol.* **2** : 41-46.
- Regazzi, R., Kikucechi, A., Takai, Y. and Wolheim,C.B. (1992) The small GTP-binding protein in the cytosol of insulin-secreting cells are coupled to GDP dissociation inhibitor proteins *J. Biol. Chem.* **267**: 17512-17519.
- Rothman, J.E. and Orci, L. (1992) Molecular dissection of the secretory pathway *Nature* **355** : 409-415.
- Serafini, T., Orci, L., Amherdt, M., Brunner, M., Kahn, R.A. and Rothman, J.E. (1991) ADP-ribosylation is a subunit of the coat of Golgi-derived COP-coated vesicles: a novel role for a GTP-binding protein *Cell* **67**: 239-253.
- Shiorataki, H., Sakoda,T., Yamagushi.T., Miyazaki, M., Kaibuchi,.K., Kishida, S., Wada, K. and Takai Y. (1993) Rabphilin rab3A, a putative target protein for p25A/rab3A p25 small GTP-binding protein related to synaptotagmin *Mol. Cell. Biol.* **13**: 2061-2068.

- Simons, K. and Zerial, M. (1993) Rab proteins and the road maps for intracellular transport *Neuron* **11** : 789-799.
- Soldati, T., Riedeer, M.A. and Pfeffer, S.R. (1993) Rab GDI, a solubilization and recycling factor for rab 9 proteins *Mol. Biol. Cell.* **4**: 4225-434.
-Soldati, T., Shapiro, A.D., Divac Svejsttrup, D.B. and Pfeffer, S.R. (1994) Membrane targeting of small GTPase rab9 is accompanied by nucleotide exchange *Nature* **369**: 76-78.
- Stearns, T., Willingham, M.C., Botstein, D. and Kahn, R.A. (1990) ADP-ribosylation factor is functionally and physically associated in the Golgi complex *Proc. Natl. Acad. Sci. USA* **87**: 1238-1242.
- Strom, M., Vollmer, P., Tan, T.J. and Gallwitz, D. (1993) A yeast GTPase-binding protein that interacts specifically with member of the ytp1/rab family inhibitor *Nature* **361**: 731-739.
- Takai, Y., Vollmer, P., Tan, T.J. and Gallwitz, D (1993) Small GTP-binding proteins *Int. Rev. Cytol.***133** : 187-231
- Ullrich, O., Horiuchi, H., Bucci, C. and Zerial, M. (1994) Membrane association rab5, mediated by GDP-dissociation inhibitor and accompanied by GDP/GTP exchange *Nature* **368**: 1157-160.
- Woodman, P.G. and Warren, G. (1988) Fusion of endocytic vesicles in a cell-free system *Eur. J. Biochem.* **173**: 101-108.
- Wessling-Resnick, M. and Braell, W.A. (1990) Characterization of the mechanism of endocytic vesicle fusion *in vitro* *J. Biol. Chem.* **265:** 116751-16759
- Walworth, N.C., Brennwald, P., Kabcenell, A.K., Garrett, M. and Novick, P. (1992) Hydrolysis of GTP by SEC4 protein plays an important role in vesicular transport and is stimulated by a GTPases activating protein in saccharomyces cerevisiae *Mol. Cell Biol.* **12** : 2017-2028.
-Walworth, N.C., Goud, B., Kabcenell, A.K. and Novick, P. (1989) Mutational analysis of SEC4 suggests a cyclical mechanism for the regulation of vesicular traffic *EMBO J.* **8** : 1685-1693.

GLYCOLIPIDS OF CELLULAR SURFACES: TOPOLOGY OF METABOLISM, FUNCTION AND PATHOBIOCHEMISTRY OF GLYCOLIPID BINDING PROTEINS

Konrad Sandhoff and Gerhild van Echten-Deckert
Institut für Organische Chemie und Biochemie der Universität Bonn
Gerhard-Domagk-Straße 1
D-53121 Bonn
Germany

I. Introduction

Glycosphingolipids (GSL) form cell and differentiation specific patterns in the outer leaflet of the plasma bilayer membrane. Gangliosides are sialic acid containing glycosphingolipids (GSL), which are highly enriched in nervous tissue, in which their more complex derivatives –namely, di,- tri,- and tetrasialogangliosides– are particulary prevalent. Except GM4, all gangliosides are derived from lactosylceramide (LacCer) and contain glucosylceramide (GlcCer) as a backbone in their molecule (Fig. 1). As components of the cell surface GSL function as binding sites for toxins, viruses, and bacteria (Karlsson, 1989). Moreover they are essential for cellular growth (Hanada *et al.*, 1992; Spiegel, 1993), stabilize cellular membranes and keep them impermeable to protons even at low pH (Patton *et al.*, 1992). GSL are modulators of growth factor receptors and are involved in cell adhesion processes as ligands for selectins (for review see Hakomori and Igarashi, 1993). A variety of intracellular sphingolipid metabolites were found to function as lipid second messengers (Merrill, 1991; Kolesnick, 1992; Olivera and Spiegel, 1992) suggesting their role in signal transduction (Okazaki *et al.*, 1989).

NATO ASI Series, Vol. H 91
Trafficking of Intracellular Membranes
Edited by M.C. Pedroso de Lima N. Düzgüneş and D. Hoekstra

GP1c (NeuAcα2 → 8NeuAcα2 → 3Galβ1 → 3GalNAcβ1 → 4(NeuAcα2 → 8NeuAcα2 → 8NeuAcα2 → 3)Galβ1 → 4Glcβ1 → 1Cer):

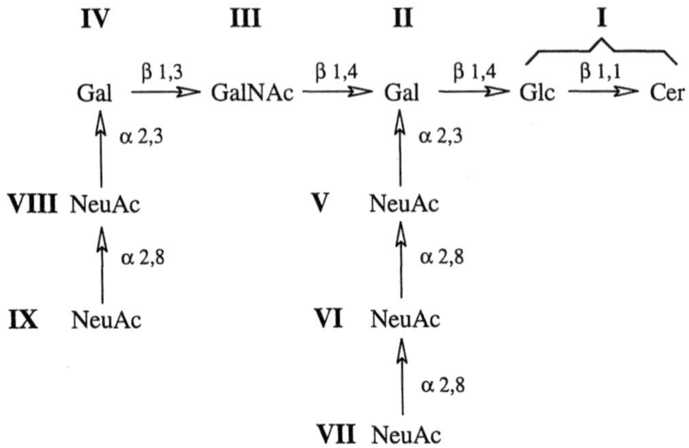

II - inner galactose
IV - outer galactose

GlcCer	- glucosylceramide, **I**
LacCer	- lactosylceramide, **I, II**
GM3	- **I, II, V**
GD3	- **I, II, V, VI**
GT3	- **I, II, V - VII**
GA2	- **I - III**
GM2	- **I - III, V**
GD2	- **I - III, V, VI**
GT2	- **I - III, V - VII**
GA1	- **I - IV**
GM1a	- **I - V**
GD1b	- **I - VI**
GT1c	- **I - VII**
GM1b	- **I - IV, VIII**
GD1a	- **I - V, VIII**
GT1b	- **I - VI, VIII**
GQ1c	- **I - VIII**
GD1c	- **I - IV, VIII, IX**
GT1a	- **I - V, VIII, IX**
GQ1b	- **I - VI, VIII, IX**
GP1c	- **I - IX**

FIG. 1: Structure of the ganglioside GP1c. The structures of glycosphingolipids (GSL) derived from LacCer are part of the GP1c structure, containing the sugar residues as shown. The terminology used for gangliosides is that recommended by Svennerholm (1963).

II. Biosynthesis of glycosphingolipids: Cellular compartmentalization and topology

GSL biosynthesis starts in the endoplasmic reticulum (ER) with the formation of ceramide, which then is glycosylated stepwise in the Golgi apparatus. The mechanism by which the growing molecules are transported from the ER through the Golgi cisternae to the plasma membrane has not yet been defined (for review, see Schwarzmann and Sandhoff, 1990). Although a vesicle-bound exocytic membrane flow has been considered, the involvement of glycolipid binding and/or transfer proteines cannot be excluded at this time (for review, see van Echten and Sandhoff, 1993).

The enzymes catalyzing the formation of ceramide via 3-dehydrosphinganine, sphinganine, and dihydroceramide are bound to the cytosolic face of the ER membrane (Mandon *et al.*, 1992). The 4-*trans* double bond of the sphingoid base is introduced after acylation of sphinganine at the level of dihydroceramide (Rother *et al.*, 1992). The localization and topology of the enzyme catalyzing the desaturation of dihydroceramide has not yet been studied.

All the glycosylation steps presented in Fig. 2 are localized in the Golgi apparatus. Information about the localization of the individual glycosyltransferases within the Golgi stack was derived from metabolic studies in cultured cells using transport inhibitors (Miller-Prodraza and Fishman, 1984; Saito *et al.*, 1984; Hogan *et al.*, 1988; van Echten and Sandhoff, 1989; van Echten *et al.*, 1990). The most impressive uncoupling of ganglioside biosynthesis was observed in primary cultured neurons in the presence of brefeldin A (van Echten *et al.*, 1990). The results suggested that GM3 and GD3 are synthesized in early Golgi compartments, whereas complex gangliosides such as GM1a, GD1a, GD1b, GT1b, and GQ1b are formed in a late Golgi compartment or even in the *trans* Golgi network, beyond the brefeldin A-induced block. These findings were later confirmed in Chinese hamster ovary (CHO) cells (Young *et al.*, 1990).

These indirect pieces of evidence obtained using drugs in cultured cells were supported by Golgi subfractionation studies in rat liver (Trinchera and Ghidoni, 1989; Trinchera *et al.*, 1990) as well as in primary cultured neurons (Iber *et al.*, 1992a). However, data obtained in different cell systems should be interpreted with caution, especially when unnatural and truncated sphingolipid precursors such as NBD-C_6-ceramide [*N*{-6-(7-nitrobenzo-2-oxa-1,3-diazol-4-yl)-aminohexanoyl}-D-*erythro* sphingosine] are used that undergo transbilayer movement (flip-flop) in membranes (reviewed by Sandhoff and van Echten, 1993).

With respect to the topology of the glycosyltransferases involved in GSL biosynthesis, many studies were done during the last decade. Glucosyltransferase catalyzing the formation of GlcCer is accessible from the cytosolic side of the Golgi (Trinchera *et al.*, 1991). Its precise localization is, however, not yet clear and could be on the Golgi and/or in a pre-Golgi compartment (Coste *et al.*, 1985, 1986; Futerman and Pagano, 1991 Trinchera *et al.*, 1991; Jeckel *et al.*, 1992). The topology of galactosyltransferase I (LacCer-synthase) seems to be clear now. The finding that mutant CHO cells, with an intact enzyme but lacking the translocator for UDP-Gal into the Golgi lumen, have strongly reduced levels of LacCer and other galactose containing GSL (Briles *et al.*, 1977; Deutscher and Hirschberg, 1986), suggested a luminal topology for galactosyltransferase I. Similar results were obtained in a polarized MDCK-cell mutant (Brändli *et al.*, 1988). Very recently the luminal topology of LacCer-synthase has been demonstrated by Lannert et al. (1994) using semi-intact cells as well as intact rat liver Golgi membranes and a truncated GlcCer as a substrate. However, these results are contradictory to the findings of Trinchera et al. (1991) who found exactly the oppposite, namely a cytosolic topology for LacCer synthesis in isolated rat liver Golgi membranes. Thus we think that at the present time one cannot completely exclude the existence of both a cytosolic and a luminal enzyme catalyzing LacCer biosynthesis. As concluded from mutant cells which were not completely depleted of galactose containing GSL as well as from experiments in which a truncated substrate has been used, LacCer biosynthesis largely occurs in the lumen of Golgi vesicles. However, under certain conditions there might exist the possibility that a related enzyme facing the cytosol, also catalyzes, though to a much smaller extent, biosynthesis of LacCer.

Very recently a stereospecific inhibitor of LacCer-synthase has been described (Zacharias *et al.*, 1994). Clarification of the mechanism of inhibition could give us more information on this key biosynthetic enzyme, and possibly on its topology.

The sequential addition of subsequent monosaccharide and sialic acid residues to the growing oligosaccharide chain, yielding GM3, GD3, and more complex gangliosides, is catalyzed by membrane-bound glycosyl-transferases that are restricted to the luminal face of the Golgi membranes (Carey and Hirschberg, 1981; Fleischer, 1981; Yusuf *et al.*, 1983; Fig. 2). However, it is still an open question how GlcCer, the starting material for the synthesis of complex gangliosides, is transferred from the cytosolic to the luminal side of the Golgi membranes. Similarly it is unclear, how ceramide formed at the cytosolic surface of ER membranes reaches the luminal side of early Golgi membranes for sphingomyelin biosynthesis (Futerman *et al.*, 1990; Jeckel *et al.*, 1990).

As illustrated in Fig. 2 only a few glycosyltransferases are involved in the elongation of the oligasaccharide chains of LacCer, GM3, GD3, and GT3, thereby generating the manifold members of the different ganglioside series. Brady and co-workers suggested very early that the same glycosyltransferases might be involved in mono- and disialoganglioside synthesis (Cumar *et al.*, 1972; Pacuszka *et al.*, 1978). Years later, our group proved for the first time that almost all glycosyltransferases involved in ganglioside biosynthesis are rather unspecific for their acceptors. Competition experiments performed in rat liver Golgi demonstrated that these enzymes catalyze the transfer of the same sugar molecule to analogous glycolipid acceptors that differ in the number of sialic acid residues bound to the inner galactose of the oligosaccharide chain (Pohlentz *et al.*, 1988; Iber and Sandhoff, 1989; Iber *et al.*, 1989, 1991, 1992b); these enzymes also use neoglycolipids with changed hydrophobic anchors (Pohlentz *et al.*, 1992). For example, sialyltransferase IV *in vitro* generates GM1b from GA1, GD1a from GM1a, GT1b from GD1b, and even GM3 from LacCer (Iber *et al.*, 1991).

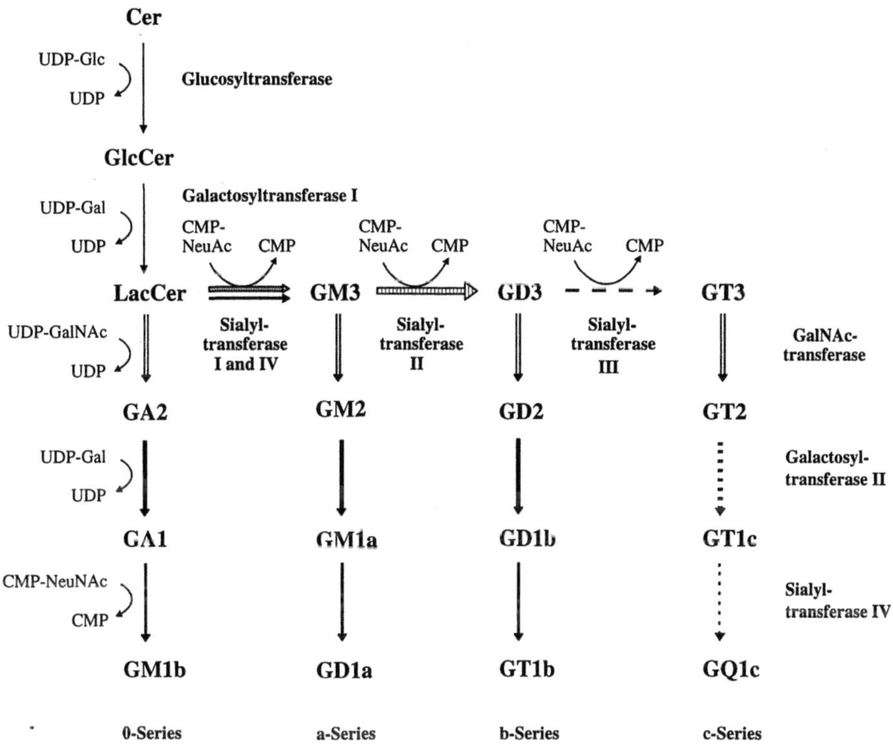

FIG. 2: General scheme for ganglioside biosynthesis starting with ceramide (Cer). All steps are catalyzed by glycosyltransferases of Golgi membranes. All glycosylation steps (except the formation of GlcCer) have a luminal topology.

Although many studies on GSL biosynthesis have been published during the last two decades, many questions concerning the topology of some enzymes involved in sphingolipid biosynthesis (e.g., dihydroceramide desaturase) and the mechanism of intracellular transfer of sphingolipids remain unanswered.

III. Lysosomal degradation of glycosphingolipids: Topology and mechanism

After endocytosis, possibly also via noncoated invaginations, GSL reach the same endosomes through which ligands internalized via coated pits are channeled, and accumulate in a time-dependent manner in multivesicular bodies (Tran *et al.,* 1987). Endosomes are important sorting centers from which materials may be directed to the lysosomes, to the Golgi, or back to the plasma membrane (Kok *et al.,* 1991). Whether the components of the plasma membrane finally become part of the lysosomal membrane is not clear. We think this assumption is quite unlikely, considering the massive luminal glycocalyx coat of lysosomal membranes that makes selective degradation by lysosomal enzymes impossible. Alternatively, the occurrence of multivesicular bodies suggested the idea that parts of the endosomal membranes –possibly those enriched in components derived from the plasma membrane–bud off into the endosomal lumen and thus form intraendosomal vesicles. These vesicles could be delivered by successive processes of membrane fission and fusion directly into the lysosol for final degradation of their components (Fürst and Sandhoff, 1992 and reviewed recently by Sandhoff and Klein, 1994).

In the lysosomes, GSL are degraded stepwise by the action of specific acid exohydrolases (Fig. 3). Degradation of GSL with short hydrophilic head groups also requires the assistance of small glycoprotein cofactors called sphingolipid activator proteins (SAPs) to attack their lipid substrates (Fig. 3). Some of these SAPs have been identified as GSL binding proteins. They complex membrane-bound GSL, lift them from the plane of the membrane, and present them as substrates to the lysosomal hydrolases (Fürst and Sandhoff, 1992).

The five SAPs known to date are encoded on two genes. One gene carries the genetic information for the GM2-activator and the second for the *sap*-precursor (prosaposin), which is processed to four homologous proteins–saposins A-D (*sap*-A to D; Fürst *et al.,* 1988;O'Brien *et al.,* 1988; Nakano *et al.,* 1989). The function, specificity, physicochemical properties, mechanism of action, and biosynthesis of SAPs have been reviewed in detail by Fürst and Sandhoff (1992) and therefore are not discussed further here.

IV. Pathobiochemistry: Inherited enzyme and activator protein defiencies

As already mentioned, final degradation of sphingolipids occurs in the lysosome. Here, they are degraded in a stepwise manner, starting at the hydrophilic end of the molecules (Fig. 3). As indicated in Fig. 3, almost every one of the degrading hydrolases can be deficient,

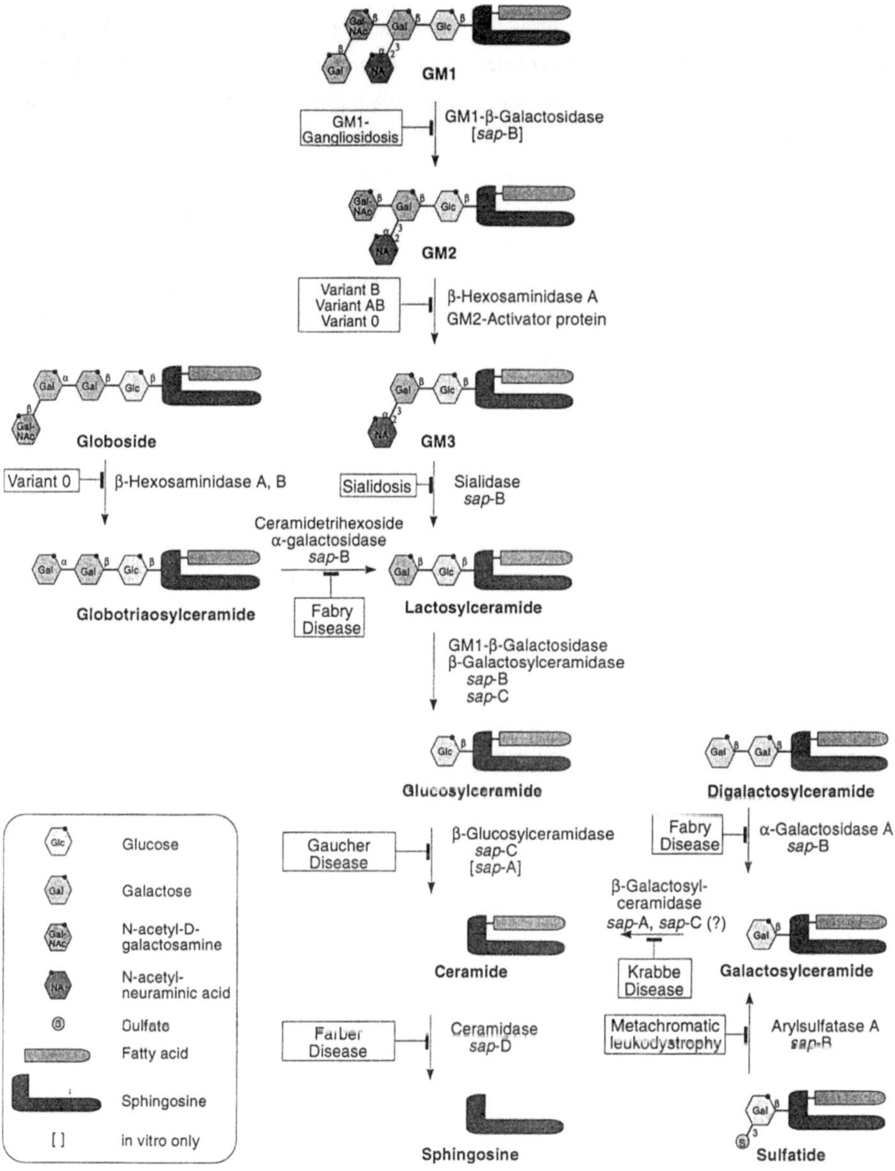

FIG. 3: Degradation scheme of sphingolipids. Known enzymes and sphingolipid activator proteins (SAPs) involved in each degradation step, as well as known diseases caused by defects of either one are given.
Variant B: infantile GM2-gangliosidosis (Tay-Sachs disease), hexosaminidase α-subunit deficiency.
Variant 0: GM2-gangliosidosis (Sandhoff disease), hexosaminidase β-subunit deficiency.
Variant AB: infantile GM2-gangliosidosis, GM2-activator deficiency.

thus leading to a human lipid storage disease. The inherited deficiency of one of these ubiquitously occurring enzymes causes the lysosomal storage of its substrates. The diseases resulting from these defects are rather heterogeneous from the biochemical as well as from the clinical point of view (for review, see Moser *et al.*, 1989; Barranger and Ginns, 1989; O'Brien, 1989; Sandhoff et al.; 1989).

The analysis of sphingolipid storage diseases without detectable hydrolase deficiency resulted in the identification of several point mutations in the GM2-activator gene (reviewed by Sandhoff and Klein, 1994) and in the *sap*-precursor gene. Mutations affecting the *sap*-C domain (Gaucher factor) resulted in a variant form of Gaucher's disease, mutations affecting the *sap*-B domain (sulfatide activator) resulted in variant forms of metachromatic leukodystrophy, whereas a mutation in the start codon ATG of the *sap*-precursor resulted in a defect of several SAPs and in a simultaneous storage of ceramide and sphingolipids with short carbohydrate head groups, e.g. glucosylceramide, lactosylceramide, galactosylceramide, sulfatids and GM3 in the patient's tissue. Under the electron microscope most of the storage appeared as multivesicular bodies, which resemble most likely intraendosomal and intralysosomal vesicles, supporting the view on the topology of endocytosis and intralysosomal digestion mentioned above (Fürst and Sandhoff, 1992 and reviewed recently by Sandhoff and Klein, 1994).

Acknowledgments: We thank Judith Weisgerber for typing the manuscript, Bernd Liessem, Gunther Dierks and Michael Reber for arranging the figures.

References

Barranger JA, Ginns EI (1989) Glucosylceramide lipidosis: Gaucher disease. In: The Metabolic Basis of Inherited Disease (Eds: Scriver CR, Beaudet AL, Sly WS, Valle D) McGraw-Hill, New York, 6th Ed, Vol II: 1677-1698

Brändli AW, Hansson GC, Rodriguez-Boulan E, Simons K (1988) A polarized epithelial cell mutant deficient in translocation of UDP-galactose into the Golgi complex. J Biol Chem 263: 16283-16290

O'Brien JS (1989) β-Galactosidase deficiency (GM1, gangliosidosis, galactosialidosis, and Morquio syndrome type B); ganglioside sialidase deficiency (mucolipidosis IV). In: The Methabolic Basis of Inherited Disease (Eds: Scriver CR, Beaudet AL, Sly WS, Valle D) McGraw-Hill, New York, 6th Ed, Vol II: 1797-1806

O'Brien JS, Kretz KA, Dewji N, Wenger DA, Esch F, Fluharty AL (1988) Coding of two sphingolipid activator proteins (SAP-1 and SAP-2) by same genetic locus. Science 241: 1098-1101

Briles EB, Li E, Kornfeld S (1977) Isolation of wheat germ agglutinin-resistant clones of chinese hamster ovary cells deficient in membrane sialic acid and galactose. J Biol Chem 252: 1107-1116

Carey DJ, Hirschberg CB (1981) Topography of sialoglycoproteins and sialyltransferases in mouse and rat liver Golgi. J Biol Chem 256: 989-993

Coste H, Martel M-B, Azzar G, Got R (1985) UDP glucose-ceramide glycosyltransferase from porcine submaxillary glands is associated with the Golgi apparatus. Biochim Biophys Acta 814: 1-7

Coste H, Martel M-B, Got R (1986) Topology of glucosylceramide synthesis in Golgi membranes from porcine submaxillary glands. Biochim Biophys Acta 858: 6-12

Cumar FA, Talman JF, Brady RO (1972) The biosynthesis of a disialoganglioside by galactosyltransferase from rat brain tissue. J Biol Chem 247: 2322-2327

Deutscher SL, Hirschberg CB (1986) Mechanism of galactosylation in the Golgi apparatus. J Biol Chem 261: 96-100

Fleischer B (1981) Orientation of glycoprotein galactosyltransferase and sialyltransferase enzymes in vesicles derived from rat liver Golgi apparatus. J Cell Biol 89: 246-255

Fürst W, Machleidt W, Sandhoff K (1988) The precursor of sulfatide activator protein is processed to three different proteins. Hoppe Seyler's Z Physiol Chem 369: 317-328

Fürst W, Sandhoff K (1992) Activator proteins and topology of lysosomal sphingolipid catabolism. Biochim Biophys Acta 1126: 1-16

Futerman AH, Stieger B, Hubbard AL, Pagano RE (1990) Sphingomyelin synthesis in rat liver occurs predominantly at the *cis* and medial cisternae of the Golgi apparatus. J Biol Chem 265: 8650-8657

Futerman AH, Pagano RE (1991) Determination of the intracellular sites and topology of glucosylceramide synthesis in rat liver. Biochem J 280: 295-302

Hakomori S, Igarashi Y (1993) Gangliosides and glycosphingolipids as modulators of cell growth, adhesion and transmembrane signaling. Adv Lipid Res 25: 147-162

Hanada K, Nishijima M, Kiso M, Hasegawa A, Fujita S, Ogawa T, Akamatsu Y (1992) Sphingolipids are essential for the growth of chinese hamster ovary cells. J Biol Chem 267: 23527-23533

Hogan MV, Saito M, Rosenberg A (1988) Influence of monensin on ganglioside anabolism and neurite stability in cultured chick neurons. J Neurosci Res 20: 390-394

Iber H, Sandhoff K (1989) Identity of GD1c, GT1a and GQ1b synthase in Golgi vesicles from rat liver. FEBS Lett 248: 18-22

Iber H, Kaufmann R, Pohlentz G, Schwarzmann G, Sandhoff K (1989) Identity of GA1-, GM1a,- and GD1b synthase in Golgi vesicles from rat liver. FEBS Lett 248: 18-22

Iber H, van Echten G, Sandhoff K (1991) Substrate specificity of $\alpha 2 \rightarrow 3$ sialyltransferases in ganglioside biosynthesis of rat liver Golgi. Eur J Biochem 195: 115-120

Iber H, van Echten G, Sandhoff K (1992a) Fractionation of primary cultured neurons: distribution of sialyltransferases involved in ganglioside biosynthesis. J Neurochem 58: 1533-1537

Iber H, Zacharias C, Sandhoff K (1992b) The c-series gangliosides GT3, GT2, and GP1c are formed in rat liver Golgi by the same set of glycosyltransferases that catalyze the biosynthesis of asialo-, a-, and b-series gangliosides. Glycobiology 2: 137-142

Jeckel D, Karrenbauer A, Birk R, Schmidt RR, Wieland F (1990) Sphingomyelin is synthesized in the *cis* Golgi. FEBS Lett 261: 155-157

Jeckel D, Karrenbauer A, Burger KNJ, van Meer G, Wieland F (1992) Glycosylceramide is synthesized at the cytosolic surface of various Golgi subfractions. J Cell Biol 117: 259-267

Karlsson K-A (1989) Animal glycosphingolipids as membrane attachment sites for bacteria. Annu Rev Biochem 58: 309-350

Kok JW, Babia T, Hoekstra D (1991) Sorting of sphingolipids in the endocytic pathway of HT29 cells. J Cell Biol 114: 231-239

Kolesnick R (1992) Ceramide: a novel second messenger. Trends Cell Biol 2: 232-236

Lannert H, Bünning C, Jeckel D, Wieland FT (1994) Lactosylceramide is synthesized in the lumen oft the Golgi apparatus. FEBS Lett 342: 91-96

Mandon E, Ehses I, Rother J, van Echten G, Sandhoff K (1992) Subcellular localization and membrane topology of serine palmitoyltransferase, 3-dehydrosphinganine reductase, and sphinganine N-acyltransferase in mouse liver. J Biol Chem 267: 11144-11148

Merrill AH (1991) Cell regulation by sphingosine and more complex sphingolipids. J Bioenerg Biomembr 23: 83-104

Miller-Prodraza H, Fishman PH (1984) Effect of drugs and temperature on biosynthesis and transport of glycosphingolipids in cultured neurotumor cells. Biochim Biophys Acta 804: 44-51

Moser HW, Moser AB, Chen WW, Schram AW (1989) Ceramidase deficiency: Farber lipogranulomatosis. In: The Metabolic Basis of Inherited Disease (Eds: Scriver CR, Beaudet AL, Sly WS, Valle D) McGraw-Hill, New York, 6th Ed, Vol II: 1645-1654

Nakano T, Sandhoff K, Stümper J, Christomanou H, Suzuki K (1989) Structure of full-length cDNA coding for sulfatide activator, a co-β-glucosidase and two other homologous proteins: Two alternate forms of the sulfatide activator. J Biochem (Tokyo) 105: 152-154

Okazaki T, Bell RM, Hannun YA (1989) Sphingomyelin turnover induced by vitamin D3 in HL-60 cells. J Biol Chem 264: 19076-19080

Olivera A, Spiegel S (1992) Ganglioside GM1 and sphingolipid breakdown products in cell proliferation and signal transduction pathways. Glycoconjugate J 9: 110-117

Pacuszka T, Duffard RO, Nishimura RN, Brady RO, Fishman PH (1978) Biosynthesis of bovine thyroid gangliosides. J Biol Chem 253: 5839-5846

Patton JL, Srinivasan B, Dickson RC, Lester RL (1992) Phenotypes of sphingolipid-dependent strains of saccharomyces cerevisiae. J Bacteriology 174: 7180-7184

Pohlentz G, Klein D, Schwarzmann G, Schmitz D, Sandhoff K (1988) Both GA2, GM2, and GD2 synthases and GM1b, GD1a, and GT1b synthases are single enzymes in Golgi vesicles from rat liver. Proc Natl Acad Sci USA 85: 7044-7048

Pohlentz G, Schlemm S, Egge H (1992) 1-Deoxy-1-phosphatidylethanolamino-lactitol-type neoglycolipids serve as acceptors for sialyltransferases from rat liver Golgi vesicles. Eur J Biochem 203; 387-392

Rother J, van Echten G, Schwarzmann G, Sandhoff K (1992) Biosynthesis of sphingolipids: dihydroceramide and not sphinganine is desaturated by cultured cells. Biochem Biophys Res Commun 189: 14-20

Saito M, Saito M, Rosenberg A (1984) Action of monensin, a monovalent cationophore, on cultured human fibroblasts: evidence that it induces high cellular accumulation of glucosyl- and lactosylceramide (gluco- and lactocerbroside). Biochemistry 23: 1043-1046

Sandhoff K, Conzelmann E, Neufeld EF, Kaback MM, Suzuki K (1989) The GM2 gangliosidosis. In: The Methabolic Basis of Inherited Disease (Eds: Scriver CR, Beaudet AL, Sly WS, Valle D) McGraw-Hill, New York, 6th Ed, Vol II: 1808-1839

Sandhoff K, van Echten G (1993) Ganglioside metabolism. Topology and regulation. Adv Lipid Res 26: 119-142

Sandhoff K, Klein A (1994) Intracellular trafficking of glycosphingolipids: role of sphingolipid activator proteins in the topology of endocytosis and lysosomal digestion. FEBS Lett: in press

Schwarzmann G, Sandhoff K (1990) Metabolism and intracellular transport of glycosphingolipids. Biochemistry 29: 10865-10871

Spiegel S (1993) Sphingosin and sphingosin 1-phosphate in cellular proliferation: relationship with protein kinase C and phosphatidic acid. J Lip Mediators 8: 169-175

Svennerholm L (1963) Chromatographic separation of human brain gangliosides. J Neurochem 10: 613-623

Tran D, Carpentier J-L, Sawano I, Gorden P, Orci L (1987) Ligands internalized through coated or noncoated invaginations follow a common intracellular pathway. Proc Natl Acad Sci USA 84: 7957-7961

Trinchera M, Ghidoni R (1989) The glycosphingolipid sialyltransferases are localized in different sub-Golgi compartments in rat liver. J Biol Chem 264: 15766-15769

Trinchera M, Pirovano B, Ghidoni R (1990) Sub-Golgi distribution in rat liver of CMP-NeuAc:GM3- and CMP-NeuAc:GT1b $\alpha 2 \rightarrow 8$ sialyltransferases and comparison with the distribution of the other glycosyltransferase activities involved in ganglioside biosynthesis. J Biol Chem 265: 18242-18247

Trinchera M, Fabbri M, Ghidoni R (1991) Topography of glycosyltransferases involved in the initial glycosylations of gangliosides. J Biol Chem 266: 20907-20912

van Echten G, Sandhoff K (1989) Modulation of ganglioside biosynthesis in primary cultured neurons. J Neurochem 52: 207-214

van Echten G, Iber H, Stotz H, Takatsuki A, Sandhoff K (1990) Uncoupling of ganglioside biosynthesis by brefeldin A. Eur J Cell Biol 51: 135-139

van Echten G, Sandhoff K (1993) Ganglioside metabolism. Enzymology, topology and regulation. J Biol Chem 268: 5341-5344

Young WW Jr, Lutz MS, Mills SE, Lechler-Osborn S (1990) Use of brefeldin A to define sites of glycosphingolipid synthesis: GA2/GM2/GD2 synthase is *trans* to the BFA block. Proc Natl Acad Sci USA 87: 6838-6842

Yusuf HKM, Pohlentz G, Sandhoff K (1983) Tunicamycin inhibits ganglioside biosynthesis in rat liver Golgi apparatus by blocking sugar nucleotide transport across the membrane vesicles. Proc Natl Acad Sci USA 80: 7075-7079

Zacharias C, van Echten-Deckert G, Plewe M, Schmidt RR, Sandhoff K (1994) A truncated epoxy-glucosylceramide uncouples glycosphingolipid biosynthesis by decreasing lactosylceramide synthase activity. J Biol Chem 269: 13313-13317

Niemann-Pick Disease Type A and B- Natural History Of Lysosomal Sphingomyelinase

Klaus Ferlinz und Konrad Sandhoff
Institut für Organische Chemie und Biochemie
Universität Bonn
Gerhard-Domagk-Str. 1
D-53121 Bonn
Germany

1. Niemann-Pick Disease

Historical review

In 1914 the German physician Albert Niemann reported upon a child affected with hepatosplenomegaly, lymphadenopathy and impairment of the central nervous system who died before 2 years of age [Niemann, 1914]. Histological studies on pathological cells performed by Ludwig Pick revealed the occurrence of characteristic foamy cells, similar but not identical with those found in Gaucher disease [Pick, 1927]. In 1934, E. Klenk discovered a massive accumulation of sphingomyelin in tissue of patients [Klenk, 1935]. Due to the broad heterogeneity of clinical and pathological manifestations, classification of Niemann-Pick disease in three distinct types A, B and C was proposed by Crocker in 1961 [Crocker, 1961]. Types D-F, showing different pathobiological properties, have been added during the following years. In 1965, R.O. Brady and coworkers finally demonstrated a profound decrease in acid sphingomyelinase activity in affected cells [Brady et al., 1966].

Classification of Niemann-Pick disease

It soon became clear that only types A and B of Crocker are due to a primary genetic defect in acid sphingomyelinase and that types C and D are genetically distinct diseases. Here, abnormal and retarded intracellular translocation of exogenous cholesterol from the lysosomes into cytoplasmic compartments is heavily impaired. The molecular defect in patients with type C and D which also exhibit alterations in sphingomyelin metabolism by secondary mechanisms is not yet understood [Vanier et al., 1991]. Leukocytes, liver, spleen and brain of type C, D and E cells have shown normal or even elevated levels of acid sphingomyelinase

NATO ASI Series, Vol. H 91
Trafficking of Intracellular Membranes
Edited by M.C. Pedroso de Lima N. Düzgüneş and D. Hoekstra

activity, whereas partial deficiency could be demonstrated in cultured fibroblasts [see Spence and Callahan, 1989, for review].

This report will focus only on Niemann-Pick type A and B. They are both autosomal recessive disorders of sphingomyelin catabolism caused by a primary deficiency of lysosomal acid sphingomyelinase and have been reported in man, dogs and cats [Brady et al., 1961; Bunza et al., 1979; Wenger et al., 1980]. Somatic cell fusion studies demonstrated that type A and B result from allelic mutations in the acid sphingomyelinase gene [Besley et al., 1980].

Niemann-Pick type A is characterized by a progressive and severe neurodegenerative course manifested by failure to thrive, psychomotor retardation and demise before the first 2 years of life. Niemann-Pick disease type B is a moderate and non-neuropathic form primarily characterized by visceral deposition of sphingomyelin in lysosomes leading to hepatosplenomegaly and pulmonary involvement. Mildly affected patients may survive into the fifth- or sixth decade of life before clinical onset of the disease [see Spence and Callahan, 1989 for review]. It is worth mentioning that some clinical reports of atypical forms of Niemann-Pick disease have been discussed. Cases of slowly progressing neurovisceral variants with protracted course as well as a visceral variant with minimal neurological lesions, phenotypically representing the Niemann-Pick disease type B but biochemically very close to type A have been reported. They also contribute to the enormous heterogeneity of Niemann-Pick disease [Elleder, 1989].

2. Sphingomyelin and Sphingomyelinases

Metabolism of sphingomyelin

Sphingomyelin is composed of a long-chain base, usually sphingosine (2-amino-4-octadecene-1,3-diol with two assymetric carbon atoms at positions 2 and 3) or to a minor extent sphinganine, a long-chain fatty acid (saturated fatty acids thus as stearic acid predominate), and phosphorylcholine. Only one of the four possible diastereoisomers, 2S,3R-D(+)-erythro-sphingomyelin is found in significant amounts as a component of mammalian vacuolar membranes [deDuve, 1975; Blouin et al., 1977]. Several investigations on lipid composition in liver and spleen indicated an average of 7 to 15 weight percent sphingomyelin of total phospholipids [Vanier, 1983; Besley and Elleder, 1986], whereas erythrocytes, peripheral nerve tissue and brain have usually higher levels, ranging from 20 to 30% [Koval and Pagano, 1991].

Biosynthesis of endogenous sphingomyelin is believed to occur predominantly on the luminal side of the cis-Golgi-compartment [Futerman et al., 1990; Jeckel,

1990]. The most important pathway of sphingomyelin synthesis involves a direct transfer of phosphorylcholine from phosphatidylcholine (PC) to ceramide independently from CDP-choline (cytidine-5'-diphosphocholine) synthesis [Diringer and Koch, 1973; Ullman and Radin, 1974; Bernert and Ullman, 1981; Marggraf et al., 1982; Voelker and Kennedy, 1982].

The retrograde ability of forming PC via sphingomyelin-degradation, demonstrated even in sphingomyelinase-deficient cells of Niemann-Pick type A, might be evidence for a biochemical balance between both phospholipids [Spence et al., 1983].

Besides this predominant pathway of sphingomyelin biosynthesis several minor ones have been discussed. A pathway operating in tissues with a high phosphatidylethanolamine (PE) content, such as brain, involves transfer of phosphorylethanolamine from PE to ceramide [Bennert, 1981; Malgat, 1986] with subsequent methylation of the amino-alcohol residue. Negligible sphingomyelin synthesis detected in mitochondria, obviously uses sphingosine as acceptor for CDP-choline which is further modified to sphingomyelin by acylation with acyl-CoA [Fujino and Negishi, 1968].

Final degradation of sphingomyelin is initiated by an acid sphingomyelinase in the lysosomes. Additionally, several acidic and neutral sphingomyelinases have been discussed [Rao and Spence, 1976; Vanha-Pertulla, 1988; Chatterjee and Gosh, 1989] (see 2c.).

Purification and characterization of acid sphingomyelinase

Purification of lysosomal acid sphingomyelinase, hydrolyzing sphingomyelin to ceramide and phosphorylcholine (Figure 1), was achieved by several groups. However, different purification procedures starting from human placenta, spleen, brain or urine yielded enzyme preparations of various protein size, number of subunits as well as different specific activities [Yamanaka and Suzuki, 1982; Jones et al., 1981; Weitz et al., 1983].

Owing to the hydrophobic character, low abundance and poor immunogenicity of this membrane associated protein, early investigations have not led to a satisfying and reproducible characterization. Thus, molecular mass, specific activity, substrate specificity, stimulators and inhibitors of the enzyme remained controversial for quite a long time. In 1987 Quintern and coworkers finally succeeded in purification of acid sphingomyelinase to homogeneity from human urine [Quintern et al., 1987; Quintern and Sandhoff, 1991].

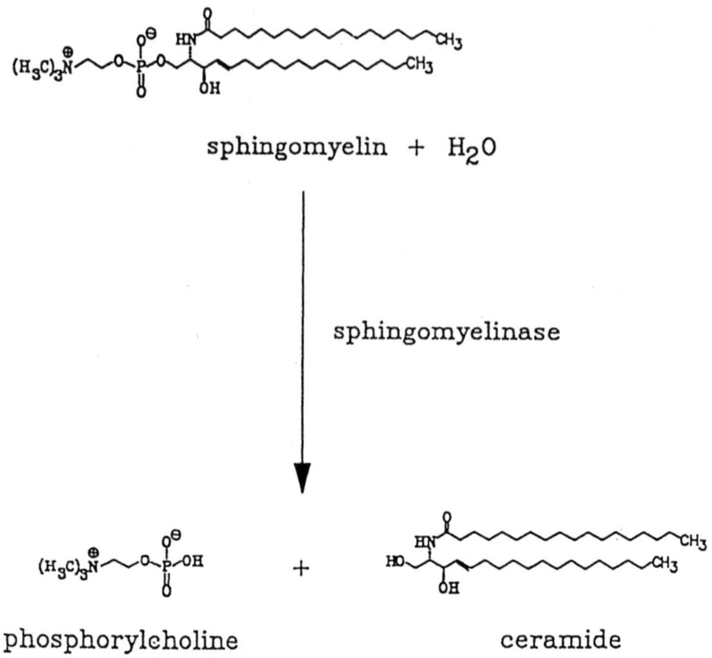

Figure 1: **Hydrolysis of sphingomyelin by acid sphingomyelinase**

The observation that urine of patients seriously affected by peritonitis or traumata express drastically elevated levels of sphingomyelinase activity has facilitated the isolation enormously [Quintern et al., 1989a].

Purification of acid sphingomyelinase by sequential chromatography on octyl-sepharose, concanavalin A-sepharose, blue-sepharose and DEAE-cellulose in the presence of detergent resulted in a 23,000 fold enrichment of enzymatic activity. A single protein of about 70 kDa molecular mass in denaturing SDS-PAGE electrophoresis indicated a homogenous monomeric form. Deglycosylation studies revealed a protein core of about 61 kDa. The native protein has a specific activity of about 2.5 mmol/h/mg at pH 4.5-5.0 and an isoelectric point (pI) near 7.5. Enzyme activity can be stimulated with nonionic detergents, various bile salts or by the addition of certain lipids to a detergent free system. In the presence of taurodeoxycholate acid sphingomyelinase shows also phosphodiesterase activity toward phosphatidylglycerol and phosphatidylcholine in a phospholipase C-like manner (Table 1). This is in agreement with findings suggesting a deficiency of phospholipase C activity acting on PC and 1,2 diacylglycerophosphoglycerol (PG) in cultured skin fibroblasts from Niemann-Pick type A and type B tissue [Wherret and Huterer, 1983].

Table 1: **Substrate specificity of acid sphingomyelinase** [Quintern et al., 1987]

Substrate \ Detergent	Specific activity (*μmol/h/mg*): no additions	Nonidet P-40 0.05%	sodiumtaurodeoxy -cholate 0.075%
sphingomyelin	18	2150	980
phosphatidyl-choline	6	10	105
phosphatidyl-glycerol	205	316	195

(200 mM sodium acetate buffer pH 4.5, 100 μM substrate; 60min, 37°C)

The purified sphingomyelinase is strongly inhibited by phosphatidyl inositol 4',5'-bisphosphate (PIP_2), adenosine 3',5'-bisphosphate and adenine-9-*β*-D-arabino-furanosyl 5'-monophosphate at concentrations in the μM range (Table 2) [Quintern et al., 1987].

Table 2: **Inhibition of lysosomal sphingomyelinase** [Quintern et al., 1987]

Inhibitor	Concentration (*μM*)	Activity (%)
no addition		100
9-*β*-D-arabinofuranosyl-adenine 5'-mono-phosphate	50 500	7 0
3',5'-adenosine-bisphosphate	50 500	8 1
5'-adenosine mono-phosphate	500	10
adenosine 5'-phospho-sulphate	500	6
phosphatidylinositol- 4',5'-bisphosphate	50	1

(in 200 mM sodium acetate buffer pH 4.5, 0.05% Nonidet P-40, 100μM sphingomyelin; 30 min, 37°C)

Recently, we reported upon the production of monospecific anti-sphingomyelinase antibodies raised against recombinant protein [Ferlinz et al., 1994; Hurwitz et al., 1994a]. Studies on biosynthesis and processing of acid sphingomyelinase in human fibroblasts and transfected COS-cells revealed that the enzyme is synthesized as a glycosylated propolypeptide of 75 kDa, yielding a precursor form of 72 kDa after proteolytic cleavage (Figure 2). Within five hours after biosythesis the precursor is processed to the mature lysosomal protein of 70 kDa, which is further converted to a protein of 50 kDa [Hurwitz et al., 1994a]. Since secretion of sphingomyelinase precursor of cultured cells was significantly enhanced in I-cells as well as after treatment of normal cells with lysosomotropic reagents intracellular targeting is apparently mediated by a mannose-6-phosphate-receptor dependent pathway. However, extracellular acid sphingomyelinase is obviously endocytosed via an alternative pathway.

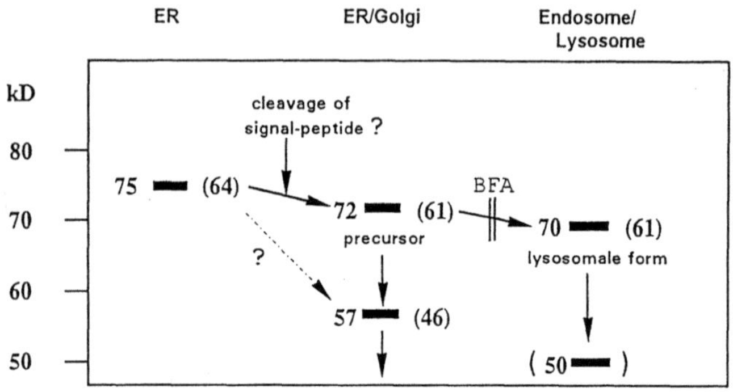

Figure 2: **Processing of human acid sphingomyelinase.** Numbers in brackets refer to molecular masses of N-deglycosylated proteins [Hurwitz et al., 1994a].

Effect of tricyclic antidepressiva on acid sphingomyelinase

Tricyclic antidepressant drugs, phenothiazines and other cationic amphiphilic drugs give rise to a severe and irreversible reduction of the acid sphingomyelinase activity in various cell types [Sakuragawa et al. 1977; Albouz et al., 1981] accompanied by characteristic lysosomal inclusions known from lipid storage cells [Lüllmann-Rauch, 1974]. The effect of the tricyclic antidepressant desipramine on enzyme processing was investigated by pulse-chase studies on labeled cultured human skin fibroblasts [Hurwitz et al. 1994b]. Desipramine induced rapid intracellular degradation of mature acid sphingomyelinase when added to the cells in the micromolar range, concomitantly abolishing the enzyme activity. Pulse chase labeling revealed the disappearance of mature enzyme forms when fibroblasts were treated with desipramine. Incubation of cells with leupeptin, an inhibitor of thiol proteases, prior to desipramine intoxication prevented this drug-induced effect. From these results we concluded that desipramine and possibly also similarly acting tricyclic antidepressants induce proteolytic degradation of acid sphingomyelinase.

Other sphingomyelinases

In addition to acid sphingomyelinase, several other mammalian sphingomyelinases have been identified so far: A Mg^{2+}- dependent, membrane integral, neutral sphingomyelinase with approximate molecular weight ranging from 90 through 600 kD dependent on the method used for mass analysis, and pH optimum around 7.4 [Rao and Spence, 1976; Spence et al., 1981; Levade et al., 1985; Chatterjee and Gosh, 1989; Gosh et al., 1993], a Mg^{2+}-independent neutral sphingomyelinase [Okazaki et al., 1994] and a Zn^{2+}- dependent acidic sphingomyelinase primarly found in fetal serum [Spence et al., 1989]. Furthermore, a placental sphingomyelinase, also acting on sphingomyelin breakdown under acid conditions but lacking any amino acid sequence homology with acid sphingomyelinase was reported [Kurth and Stoffel, 1991].

Defects of lysosomal sphingomyelinase and storage of sphingomyelin

Sphingomyelin originating predominantly from the anti-cytosolic leaflet of cellular membranes is degraded by lysosomal acid sphingomyelinase. Defects of this enzyme result in profound storage of sphingomyelin in patients with acute type A and moderate type B mainly in liver and spleen [Brady et al, 1966]. In contrast to a significant increase of sphingomyelin in visceral tissue, only moderate accumulation was found in brain of patients with type A despite of severe neurological involvement. In addition, storage of large amounts of bis(monoacylglycero)-phosphate in liver, spleen, kidney, lung and heart as well as remarkable accumulation of glucosylceramide and some more complex glycolipids as gangliosides was detected. Significant storage of cholesterol can be explained secondary to deposition of sphingomyelin [see Spence and Callahan, 1989 for review].

3. Molecular aspects of the acid sphingomyelinase gene

Cloning of the sphingomyelinase cDNA

While the clinical course of the classical Niemann-Pick disease type A is quite uniform, patients with type B or intermediate forms show a broad variety concerning onset and clinical phenotype. The residual activity of patients with Niemann-Pick type A was found to be <3% of normal, whereas patients with type B show generally higher but variable residual sphingomyelinase activities (about 1-10% of normal) as assayed under in vitro conditions. To understand the broad spectrum of clinical manifestations properties of normal and mutant acid sphingomyelinases have to be studied in detail.

To investigate the molecular basis of acid sphingomyelinase deficiency, several groups attempted to purify this enzyme and to raise antibodies against it [Jobb and Callahan, 1987; Freeman et al., 1983; Rousson et al., 1986]. However, the final antigen preparations were either not homogenous or their immunogenicity was low so that production of high affinity monospecific polyclonal and monoclonal antibodies failed. Therefore, purification and subsequent sequencing of the protein appeared to be the only way of getting more insight into molecular mechanisms of Niemann-Pick disease. Initially human fibroblast and placental cDNA libraries were screened for authentic cDNA clones using peptide-derived oligonucleotides [Quintern et al., 1989b]. Partial cDNA information, used to isolate full length cDNAs by screening placental, testis, hepatoma, and retinal human cDNA libraries, resulted in at least three different cDNA sequences, obviously derived from a single hnRNA species by alternative splicing. Expression studies of these transcripts indicated only one type of cDNA encoding for catalytically active human acid sphingomyelinase. The other two underrepresented transcripts might be formed due to cryptic intronic splice sites resulting in proteins without any physiological significance [Schuchman et al., 1991]. Similarly, alternative splicing of the transcripts of lysosomal enzymes, β-glucuronidase and β-galactosidase, also did not lead to more than one single catalytically active protein [Oshima et al., 1987; Morreau et al., 1989].

The full-length cDNA of human acid sphingomyelinase spans an open reading frame of 1890 bp which contains two potential initiation codons at positions 1 and 97, respectively. Although both initiation signals would lead to a protein containing the predicted hydrophobic presequence and a cleveage site for ER translocation, the initiation consensus sequence for the second ATG is rather weak [Kozak, 1987]. This is in good agreement with expression studies performed on site specific mutated cDNAs missing either one of both initiation codons or including a frameshift in between both [Ferlinz et al., 1994]. Therefore, protein biosynthesis in vivo is most likely initiated from the first ATG, resulting in a maximum protein length of 629 amino acids. There are six potential N-glycosylation sites, Asn-X-Ser/Thr, in the sphingomyelinase polypeptide. Deglycosylation of purified protein resulted in a reduction of the molecular weight by 11 kDa suggesting a highly glycosylated protein core with as many as five utilized N-glycosylation sites.

The genomic structure of acid sphingomyelinase is organized in six exons and 5 introns in a total length of about 5000 bp. A 1-kb 5'-upstream promotor region, containing several regulatory motives like Sp1 binding sites, GC rich stretches,

TATA, CAAT, AP-1 and NF-1 binding sites suggest that it comprises all or part of the entire sphingomyelinase promotor [Schuchman et al., 1992].

In contrast to previous enzymatic studies on cell hybrids which assigned the gene of acid sphingomyelinase to chromosome 17 [Konrad and Wilson, 1987], a recent report using molecular DNA techniques in somatic human-mouse and human-hamster cell hybrids clearly assigned the gene locus to chromosome 11 [Pereira et al., 1991]. Chromosomal analysis by in situ hybridization could clearly indicate a single locus for human acid sphingomyelinase to 11p15.1-p15.4. Although most of the 14 mapped lysosomal genes are dispersed throughout the human genome, cathepsin D and acid phosphatase have also been assigned to chromosome 11p11 and 11p15, respectively [Henry et al., 1989; Bruns and Gerald, 1974; Jones and Kao, 1978].

Analysis of mutations in acid sphingomyelinase-deficient cells

Analysis of molecular and genetic properties of acid sphingomyelinase and the corresponding gene structure finally provided the tools to identify mutations causing abnormal protein function in tissue of Niemann-Pick patients.

Although Niemann-Pick type A and B disease is a widely spread panethnic inborn error of metabolism, the frequency among individuals of Ashkenazi Jewish descent is significantly higher than in general. In this group, it occurs with an estimated incidence between 1 in 40,000 and 1 in 80,000, respectivly [Levran et al., 1991a]. Recent screening studies based on acid sphingomyelinase cDNA analysis of Jewish type A patients revealed a remarkedly high frequency of a single G →T nucleotide transversion at position 1487. This common point mutation results in a substitution of hydrophobic leucine for normal basic arginine in residue 496 [Arg496Leu]. The allele frequency was estimated to be more than 30% of Niemann-Pick type A alleles in Jewish but only 5% in non-Jewish population. Tissue of homoallelic carriers of this particular defect indicated a residual activity of acid sphingomyelinase < 1% of normal employing a detergent containing assay [Levran et al., 1991a]. A second point mutation observed within the coding region for the enzyme of an obviously homozygous patient with Niemann-Pick type A of unknown origin was assigned to nucleotide position 1729. Transition of G→A resulted in a substitution of serine for normal glycine [Gly577Ser]. The residual enzymatic activity of the patients cultured fibroblasts was less than 1% of to normal. First expression studies of mutagenized cDNA in a COS cell system confirmed the functional lesion of the altered protein [Ferlinz et al., 1991].

Genetic studies on type B Jewish patients so far revealed a deletion of three consequtive bases within position 1820-1825 predicting removal of an arginine residue from residue 608 of acid sphingomyelinase [ΔArg608]. These Jewish probands were exclusively found to be heteroallelic, one compound heterozygote also carried the point mutation Arg496Leu on the other allele [Levran et al., 1991b]. Moreover, a mildly affected patient of Arabic descent was homozygous for the arginine deletion at residue 608 and thus permitted the first genotype/phenotype correlation. Later it turned out that this specific deletion is the predominant mutation of Niemann-Pick type B patients among North African Arabs [Vanier et al., 1993] and presumably the whole Mediterranean area. Clinical investigations about these patients suggest that Arg608 deletion results in a sphingomyelinase exhibiting relatively high residual activity, thus, causing a late onset with mild course of the disease. In contrast, homozygosity for [Arg496Leu] leads to severe course of type A with an almost complete deficiency of acid sphingomyelinase (residual activity 0.7% of normal), and a rather short life expectancy. Compound heterozygots carrying both mutated alleles [Arg496Leu/ Δ Arg608] are expected to show residual enzyme activity in between (measured residual enzyme activity near 5% of normal) [Levran et al., 1991b].

During the last two years various additional mutations within the acid sphingomyelinase gene resulting in different phenotypes of Niemann Pick disease have been identified (Table 4). These are spread over the whole genomic structure and thus do not allow any correlation between specific protein domains and distinct biochemical properties. Future analysis of sphingomyelinase deficient cells using immunochemical methods may hopefully gain more insight into biochemical properties of the enzyme and the impact of mutations described above.

Table 4: **Mutations causing Niemann Pick Disease**

Mutation within ASM-cDNA	Change in aminoacid sequence	Exon	Pheno-type	Origin	Reference
$G^{1487} \rightarrow T$	Arg496Leu	6	NP-A	Askenazi Jewish	Levran et al., 1991, Proc. *Natl. Acad. Sci. USA* **88**, 3748-3762
$T^{905} \rightarrow C$	Leu302Pro	2	NP-A	Askenazi Jewish	Levran et al., 1992, *Blood* **80**, 2081-2087
$G^{1729} \rightarrow A$	Gly577Ser	6	NP-A	unknown	Ferlinz et al., 1991, *Biochem. Biophys. Res. Commun.* **179**, 1187-1191

$A^{1308} \rightarrow C$	Ser436Arg	4	NP-A	Japanese	Takahashi et al., 1992, *J.Biol.Chem.* **267**, 12552-12558
$T^{782} \rightarrow A$	Leu261Stop	2	NP-A	Asian Indian	Takahashi et al., 1992, *J.Biol.Chem.* **267**, 12552-12558
Int2 SA splice-site mutation		intron 2	NP-A	unknown	Levran et al., 1993, *Hum. Mut.* **2**, 205-206
TTdel $G^{1146} \rightarrow A$	fsLeu178 Met382Ile	2	NP-A	European	Takahashi et al., 1992, *J.Biol.Chem.* **267**, 12552-12558
deletion of three consecutive bases within positions 1820-1825	ΔArg608	6	NP-B	Mahgrebenian Arabic	Levran et al., 1991, J. *Clin. Invest.* **88**, 806-810; Vanier et al., 1993, *Hum. Genet.* **92**, 325-330
$G^{724} \rightarrow A$ $A^{1148} \rightarrow G$	Gly242Arg Asn383Ser	2 3	NP-B	European	Takahashi et al., 1992, *J.Biol.Chem.* **267**, 12552-12558

fs= frame shift; Δ, del= deletion

Diagnosis and therapeutic outlook

Although these results cannot explain the enormous variability of late-onset variants, there is some evidence that minor differences in residual enzyme activity may cause a broad spectrum of clinical manifestations. Recent investigations on patients with metachromatic leukodystrophy (MLD) and ß-hexosaminidase A deficiency suggest a correlation of residual enzyme activity with the variability of the phenotype of different clinical forms of a neurolipidosis by a simple kinetic model [Conzelmann and Sandhoff, 1991]. Residual sphingomyelinase activities of Niemann-Pick type A patients often but not systematically tend to be slightly lower than those of type B patients measured in tissue homogenate. Besides the two mutations discussed above, no general genotype-phenotype correlation has been demonstrated due to the few experimental information available so far. The enormous number of possible combinations of different mutated alleles resulting in different genotypes, might explain the marked heterogeneity of the phenotype of late onset forms of Niemann-Pick disease and thus prevent a prediction of clinical manifestations of affected individuals in general. However, identification of prevalent mutations among certain ethnic populations will hopefully provide a reliable basis for prenatal diagnoses [Levran et al., 1991; Vanier et al., 1993].

The availability of cDNA and genomic DNA of human acid sphingomyelinase permits production of large amounts of recombinant protein in eukaryotic cells and may serve as a source of active enzyme for a therapeutic approach. Thus, enzyme replacement therapy of non-neurodegenerative Gaucher patients using *in vitro* modified glucocerebrosidase provided some initially satisfying results [Barton et al., 1991; Murray et al., 1985]. Major disadvantages are expensive production and modification of complex human proteins as well as their potential failure in curing neuronopathic forms of lipid storage diseases.

Knowledge of the structure of the human sphingomyelinase gene provides also information needed to consider somatic cell gene transfer. Insertion of foreign DNA into host cell lines and organisms by retrovirus mediated gene transfer has been investigated in several approches [Cone and Mulligan, 1984; Williams et al., 1984; Soriano et al., 1986]. Preliminary studies on correction of sphingomyelinase deficiency by retroviral mediated gene transfer in cultured cells have more recently been reported [Suchi et al., 1992; Dinur et al., 1992]. However, somatic gene curing by gene transfer and expression of lysosomal enzymes which has been achieved succesfully in cultured cells [Sorge et al., 1987] mostly failed in animal models due to a restricted temporary expression of foreign protein. Genetic complementation of acid sphingomyelinase deficient somatic cells (e.g.bone marrow) and subsequent replantation has been examined in few models [Bembi et al., 1992; Bayever et al., 1992]. Although a variety of problems remains to be overcome, in future, somatic gene therapy using genetically reconstituted cells might be a promising approach to cure even neurodegenerative lysosomal storage disorders like Niemann-Pick disease.

4. References

Albouz, S., Hauw, J.J., Berwald-Netter, Y., Boutry, J.M., Bourdon, R. and Baumann, N. (1981) *Biomed.Express.* **35**, 218-220.

Barton, N.W., Brady, R.O., Dambrosia, J.M., Di Bisceglie, A.M., Doppelt, S.H., Hill, S.C., Mankin, H.J., Murray, G.J., Parker, R.I., Argoff, C.E., Grewal, R.P., Yu, K.-T., and Collaborators (1991). Replacement therapy for inherited enzyme deficiency-macrophage targeted glucocerebrosidase for Gaucher's disease. *N. Engl. J. Med.* **324**: 1464-1470.

Bayever, E., Kamani, N., Ferreira, P., Machin, G.A., Yudkoff, M., Conrad, K., Palmieri, M., Radcliffe, J., Wenger, D.A., and August, C.S. (1992) Bone marrow transplantation for Niemann-Pick type 1A disease. *J. Inher. Met. Dis.* **15**, 919-928.

Bembi, B., Comelli, M., Scaggiante, B., Pineschi, A., Rapelli, S., Gornati, R., Montorfano, G., Berra, B, Agosti, E, and Romeo, D. (1992) Treatment of sphingomyelinase deficiency by repeated implantations of amniotic epethelial cells. *Am. J. Med. Genet.* **44**, 527-533.

Bernert, J. T., and Ullman, M. D. (1981). Biosynthesis of sphingomyelin from erythro- ceramide and phosphatidylcholine by a microsomal cholinephosphotransferase. *Biochim. Biophys. Acta* **666**: 99-109.

Besley, G.T., and Elleder, M. (1986). Enzyme activities and phospholipid storage paterns in brain and spleen samples from Niemann-Pick disease variants: A comparison of neuropathic and non-neuropathic forms. J. Inherited Metab.Dis. 9: 59.

Besley, G.T.N., Hoogeboom, A.J.M., Hoogeveen, A., Kleijer, W.J., and Galjaard, H. (1980). Somatic cell hybridization studies showing different gene mutations in Niemann-Pick variants. *Hum. Genet.* **54**: 409-412.

Blouin, A., Bolender, R.P., and Weibel, E.R. (1977). Distribution of organelles and membranes between hepatocytes and nonhepatocytes in the rat liver parenchyma. *J. Cell.Biol.* **72**: 441-455.

Brady, R.O. (1983). Sphingomyelin lipidosis: Niemann- Pick disease. The Metabolic Basis of Inherited Disease, eds. Stanbury, J.B., Wyngaarden, J.B., and Fredrickson, D.S. (Mc Graw-Hill, New York) 7th ed.: 831-884.

Brady, R.O., Kanfer, J.N., Mock, M.B., and Fredrickson, D.S. (1966). The metabolism of sphingomyelin. Evidence of an encymatic deficiency in Niemann-Pick disease. *Proc. Natl. Acad. Sci. USA* **55**: 367-370.

Bruns, G., and Gerald, P.S. (1974). Human acid phosphatase in somatic cell hybrids. *Science* **184**: 480-481.

Bunza, A., Lowden, J.A., and Charlton, K.M. (1979). Niemann-Pick disease in a poodle dog. *Vet. Pathol.* **16**: 530-538.

Chatterjee, S., and Gosh, N. (1989). Neutral sphingomyelinase from human urine. *J. Biol. Chem.* **264**: 12554-12561.

Cone, R., and Mulligan, R.C. (1984). High efficiency gene transfer into mammalian cells: Generation of helper-free recombinant retrovirus with broad mammalian host range. *Proc. Natl. Acad. Sci. USA* **81**: 6349-6355.

Conzelmann, E., Sandhoff, K. (1991). Biochemical Basis of Late-Onset Neurolipidoses. *Dev. Neurosci.* **13**: 197-204.

Crocker, A.C. (1961). The cerebral defect in Tay-Sachs disease and Niemann-Pick disease. *J Neurochem.* **7**. 69-73.

Dinur, T. Schuchman, E.H., Fibach, E., Dagan, A., Suchi, M., Desnick, R.J., and Gatt, S. (1992) Toward gene therapy for Niemann-Pick diseas (NPD)-Separation of retrovirally corrected and non corrected NPD fibroblasts using a novel fluorescent sphingomyelin, *Hum. Gene Ther.* **3**, 633-639

Diringer, H., and Koch, M.A. (1973). Biosynthesis of sphingomyelin: Transfer of phosphorylcholine from phosphatidylcholine to erythro-ceramide in a cell-free system. *Hoppe-Seyler's Z. Physiol. Chem.* **354**: 1661-1665.

Elleder, M. (1989) Niemann-Pick Disease. *Path. Res. Pract.* **185**: 293-328.

Ferlinz, K., Hurwitz, R., and Sandhoff K. (1991). Molecular Basis of Acid Sphingomyelinase Deficiency in a Patient with Niemann-Pick Disease Type A. *Biochem. Biophys. Res. Commun.* **179**: 1187-1191.

Ferlinz, K., Hurwitz, R., Vielhaber, G., Suzuki, K., and Sandhoff, K. (1994). Occurrence of two molecular forms of human acid sphingomyelinase, *Biochem. J.*, in press.

Fredrickson, D.S., Sloan, H.R. (1972). Sphingomyelin lipidoses: Niemann-Pick disease, in Stanbury, J. B., Wyngaarden, J. B., Fredrickson, D. S. (eds): The Metabolic Bases of Inherited Disease, 3th ed.: 783, New York, Mc Graw-Hill

Freeman, S.J., Davidson, D.J., Shankaran, P., and Callahan, J.W. (1983). Monoclonal antibodies against human placental sphingomyelinase. *Biosci. Rep.* **3**: 545-550.

Fujino, Y., Negishi, T. (1968). Investigations of the enzymatic synthesis of sphingomyelin. *Biochim. Biophys. Acta* **152**: 428-433.

Futerman, A.H., Stieger, B., Hubbard, A.L., and Pagano, R.E. (1990). Sphingomyelin synthesis in rat liver occurs predominantly at the cis and medial cisternae of the Golgi apparatus. *J. Biol. Chem.* **265**: 8650-8657.

Gosh, N., Chatterjee, S., and Sabbadini, R. (1993). Neutral sphingomyelinase from skeletal muskle: It's lokalization, solubilisation and possible involvement in Ca2+ release. *FASEB J.* **7/7**, 1255.

Henry, I., Puech, A., Antignac, C., Couillin, P., Jean-Pierre, M., Ahnine, L., Barichard, F., Boehm, T., Augerau, P., Scrable, H., Rabbitts, T.H., Rochefort, H., Cavenee, W., and Junien, C. (1989). Subregional mapping of BWS, CTSD, MYODI, and T-ALL breakpoint in 11-15. *Cytogenet. Cell. Genet.* **51**: 1013-1016.

Hurwitz, R., Ferlinz, K., and Sandhoff, K. (1994b). The tricyclic antidepressant desipramine causes proteolytic degradation of lysosomal sphingomyelinase in human fibroblasts, *Biol. Chem. Hoppe-Seyler*, in press.

Hurwitz, R., Ferlinz, K., Vielhaber, G., Moczall, H. (1994a), Processing of human acid sphingomyelinase in normal and I-cell fibroblasts. *J. Biol. Chem.* **269**: 5440-5445.

Jeckel, D., Karrenbauer, A., Birk, R., Schmidt, R.R., and Wieland, F. (1990). Sphingomyelin is synthesized in the cis Golgi. *FEBS Lett.* **261**: 155-157.

Jobb, E.A., and Callahan, J.W. (1989) Biosynthesis of sphingomyelinase in normal and Niemann-Pick fibroblasts. *Biochem. Cell Biol.* **67**: 801-807.

Jones, C., and Kao, F.T. (1978). Regional mapping of the gene for lysosomal acid phosphatase (ACP-2) using a hybride clone panel containing segments of human chromosome 11. *Hum. Genet.* **45**: 1-10.

Jones, C.S., Shankaran, P., and Callahan, J.W. (1981). Purification of sphingomyelinase to appearent homogeneity by using hydrophobic chromatography. *Biochem J.* **195**: 373-379.

Klenk, E. (1935). Über die Natur der Phosphatide und anderer Lipide des Gehirns und der Leber bei der Niemann-Pickschen Krankheit. *Z. Physiol. Chem.* **235**: 24-25.

Konrad, R., and Wilson, D. (1987). Assignment of the gene for acid lysosomal sphingomyelinase to human chromosome 17. *Cytogenet. Cell Genet.* **46**: 641-643.

Koval, M., and Pagano, E. (1991). Intracellular transport and metabolism of sphingomyelin. *Biochim. Biophys. Acta* **1082**: 113-125.

Kozak, M. (1987). An analysis of 5'- noncoding sequences from 699 vertebrate messenger RNAs. *Nucleic.Acids Res.* **15**: 8126-8149.

Kurth, J. and Stoffel, W. (1991). Human placental sphingomyelinase. *Biol. Chem Hoppe-Seyler* **372**: 215-223.

Levade, T., Potier, M., Salvayre, R., and Douste-Blazy L. (1985). Molecular weight of human brain neutral sphingomyelinase determined in situ by radiation inactivation method. *J. Neurochem.* **45**: 630-634.

Levade, T., Salvayre, R., and Douste-Blazy, L. (1986). Sphingomyelinases and Niemann-Pick Disease. *J. Clin. Chem. Biochem.* **24**: 205-220.

Levran, O., Desnick, R.J., and Schuchman E.H. (1991a). Niemann-Pick disease: A frequent missense mutation in the acid sphingomyelinase gene of Ashkenazi Jewish type A and B patients. *Proc. Natl. Acad. Sci. USA* **88**: 3748-3752.

Levran, O., Desnick, R.J., and Schuchman, E.H. (1991b). Niemann-Pick Type B Disease. *J. Clin. Invest.* **88**: 806-810.

Levran, O., Desnick, R.J., and Schuchman E.H. (1993) *Hum. Mut.* **2**, 205-206.

Lüllmann-Rauch, R. (1974) Lipidosis-like alterations in spinal cord and cerebrallar cortex of rats treated with chlophentermine or tricyclic anti-depressants. *Acta Neurophathol.* (Berl.) **29**, 237

Malgat, M., Maurice, A., and Baraud, J. (1986). Sphingomyelin and ceramide-phosphoethanolamine synthesis by microsomes and plasma membranes from rat liver and brain. *J. Lipid Res.* **27**: 251-260.

Marggraf, W.D., Zertani, R., Anderer, F.A., and Kanfer, J.N. (1982). The role of endogenous phosphatidylcholine and ceramide in the biosynthesis of sphingomyelin in mouse fibroblasts. *Biochim. Biophys. Acta* **710**: 314.

Maruyama, E.N., and Arima, M. (1989). Purification and characterization of neutral and acid sphingomyelinases from rat brain. *J. Neurochem.* **52**: 611-618.

Morreau, H., Galjart, N.J., Gillemans, N., Willemsen, R., van der Horst, G.T.J., and d'Azzo, A. (1989). Alternative splicing of ß- galactosidase mRNA generates the classic lysosomal enzyme and a ß- galactosidase- related protein. *J. Biol. Chem.* **264**: 20655-20663.

Niemann, A. (1914). Ein unbekanntes Krankheitsbild. *Jahrb. Kinderheilkd.* **79**: 1-3.

Okazaki, T., Bielawska, A., Domae, N., Bell, R.M., and Hannun, Y.A. (1994). Characteristics and partial purification of a novel cytosolic magnesium independent, neutral sphingomyelinase activated the early signal transduction of 1α,25-dihydroxyvitamin D3-induced HL-60 cell differentiation. *J. Biol. Chem.* **269**, 4070-4077.

Oshima, A., Kyle, J. W., Miller, R.D., Hoffmann, J. W., Powell, P.P., Grubb, J.H., Sly, W. S., Tropak, M., Guise, K.S., and Gravel, R.A. (1987). Cloning, sequencing, and expression of cDNA for human ß- glucuronidase. *Proc. Acad. Natl. Sci. USA* **84**: 685-689.

Pereira, L.V., Desnick, R.J., Adler, D.A., Disteche, C.M., and Schuchman, E.H. (1991). Regional Assignment of the Human Acid Sphingomyelinase Gene (SMPD1) by PCR Analysis of Somatic cell Hybrids and in Situ Hybridization to 11p15.1-p15.4. *Genomics* **9**: 229-234.

Pick, L. (1927). Über die kipoidzellige Splenohepatomegalie typus Niemann-Pick als Stoffwechselerkrankung. *Med. Klin.* **23**: 1483-1486.

Quintern,L.E., Zenk,T.S. and Sandhoff,K. (1989a) *Biochim. Biophys Acta*, **1003**, 121-124.

Quintern, L. E., and Sandhoff K. (1991). Human acid sphingomyelinase from human urine. *Methods in Enzymology* **197**: 536-540.

Quintern, L.E., Schuchman, E.H., Levran, O., Suchi, M., Ferlinz, K., Reinke, H., Sandhoff, K., and Desnick, R. J. (1989b). Isolation of cDNA clones encoding human acid sphingomyelinase: occurence of alternatively processed transcripts. *EMBO J.* **8**: 2469-2473

Quintern, L.E., Weitz, G., Nehrkorn, H., Tager, J.M., Schram, A.W., and Sandhoff, K. (1987). Acid sphingomyelinase from human urine: Purification and characterization. *Biochim Biophys Acta* **922**: 323-336.

Rao, B. G., and Spence, M. W. (1976). Sphingomyelinase activity at pH 7.4 in human brain and a comparison to activity at pH 5.0. J. *Lipid Res.* **17**: 506-510.

Rousson, R., Vanier, M.T., and Louisot, P. (1986). Immunologic studies on acidic sphingomyelinases. In Enzymes of Lipid Metabolism II. Freysz, L., Dreyfus, H., Massareli, R., and Gatt, S. (eds). Plenum Publishing Corporation, New York: 273-283.

Sakuragawa, N., Sakuragawa, M., Kuwabara, T., Pentchev, P.G., Barranger, J.A. and Brady, R.O. (1977) *Science* **196,** 317-319.

Schuchman, E.H., Levran, O., Peireira, L.V., and Desnick, R.J. (1992). Structural Organization and Complete Nucleotide Sequence of the Gene Encoding Human Acid Sphingomyelinase (SMPD1). *Genomics* **12**: 197-205.

Schuchman, E.H., Suchi, M., Takahashi, T., Sandhoff, K. and Desnick, R.J. (1991). Human acid sphingomyelinase. *J. Biol. Chem.* **266**: 8531-8539.

Sorge, J., Kuhl, W., West, C., and Beutler, E. (1987). Complete correction of the enzymatic defect of type I Gaucher disease fibroblasts by retroviral-mediated gene transfer. *Proc. Natl. Acad. Sci. USA* **84**: 906-912.

Soriano, P., Cone, R.D., Mulligan, R.C., and Jaenisch, R. (1986). Tissue-specific and ectopic expression of genes introduced into transgenic mice by retroviruses. *Science* **234**: 1409.

Spence, M.W., and Callahan, J.E. (1989). The Niemann-Pick Group of Diseases, in The Metabolic Basis of Inherited Disease, eds. Scriver, C.R., Beaudet, A.L., Sly, W.S., and Valle D. (McGraw-Hill, New York), 8th Ed.: 1655-1676.

Spence, M.W., Burgess, J.K., Sperker, E.R., Hamed, L., and Murphy, M.G. (1981). Neutral sphingomyelinases of brain, in Callahan, J W., and Lowden, J. A. (eds): Lysosomes and Lysosomal Storage Diseases. New York,Raven.

Spence, M.W., Byers, D.M., Palmer, F.B.C., and Cook H.W. (1989). A New Zn2+-stimulated sphingomyelinase in fetal bovine serum. *J. Biol. Chem.* **264**: 5358-5363.

Spence, M.W., Clarke, J.T.R., and Cook, H.W. (1983). Pathways of sphingomyelin metabolism in cultured fibroblasts from normal and sphingomyelin lipidosis subjects. *J. Biol. Chem.* **258**: 8595-8600.

Ullman, M.D., and Radin, N.S. (1974). The enzymatic formation of sphingomyelin from ceramide and lecethin in mouse liver. *J. Biol. Chem.* **249**: 1506-1511.

Vanha-Perttula, T. (1988). Sphingomyelinases in human, bovine and porcine seminal plasma. *FEBS Lett.* **233**: 263-267

Vanier, M.T. (1983). Biochemical studies in Niemann-Pick disease. I. Major sphingolipids in liver and spleen. *Biochim. Biophys. Acta* **750**: 178-184.

Vanier, M.T., Ferlinz, K., Rousson, R., Duthel, S., Louisot, P., Sandhoff, K., and Suzuki, K. (1993). Deletion of Arginine (608) in Acid Sphingomyelinase is the Prevalent Mutation among Arabic Niemann-Pick Disease Type B Patients from Northern Africa. *Hum. Genet.* **92**: 325-330.

Vanier, M.T., Rodriguez-Lafrasse, C., Rousson, R., Duthel, S., Harzer, K., Pentchev, P. G., Revol, A., and Louisot, P. (1991). Type C Niemann-Pick Disease: Biochemical Aspects and Phenotypic Heterogeneiety. *Dev. Neurosci.* **13**: 307-314.

Voelker, D.R., Kennedy, E.P. (1982). Cellular and enzymic synthesis of sphingomyelin. *Biochemistry* **21**: 2753.

Weitz, G., Lindl, T., Hinrichs, U., and Sandhoff, K. (1983). Release of sphingomyelin phosphodiesterase (acid sphingomyelinase) by ammonium chloride from CL 1D mouse L-cells and human fibroblasts in partial purification and characterization of the exported enzymes. *Hoppe-Seyler's Z. Physiol. Chem.* **364**: 863-869.

Wenger, D.A., Sattler, M., Kudoh, T., Snyder, S.P., and Kingston, R.S. (1980). Niemann-Pick disease: A model in Siamese cats. *Science* **208**: 1471-1473.

Wherrett, J.R. and Huterer, S. (1983) Deficiency of taurocholate-dependent phospholipase C acting on phosphatidylcholine in Niemann-Pick Disease. *Neurochem. Res.* **8**, 89-99.

Williams, D.A., Lemischka, I.R., Nathan, D.G., and Mulligan, R.C. (1984). Introduction of new genetic material into pluripotent hematopoietic stem cells of mice. *Nature* **310**: 475-478.

Yamanaka, T., and Suzuki, K. (1982). Acid sphingomyelinase of human brain: Purification to homogeneity. *J. Neurochem.* **38**:1753-1758.

Interaction of liposomes with cells *in vitro*

Kyung-Dall Lee and [1]Demetrios Papahadjopoulos
Department of Neurobiology & Department of Cell Biology
Harvard Medical School
Boston, MA 02115 USA

Introduction

The use of liposomes as model membranes has made significant contributions to our understanding and current knowledge of lipid bilayers and their interaction with various proteins (McConnell, 1983; Papahadjopoulos, et al., 1990). However, the potential use of liposomes as carriers of drugs and macromolecules, has taken much longer than expected to reach a successful conclusion in terms of clinical applications. One of the key problems in using liposomes as carriers is the lack of understanding about the parameters that control the interaction of liposomes with cells.

One important and significant obstacle in using liposomes as drug carriers with targeting capability has been the fast clearance of liposomes from blood circulation by mononuclear phagocytic cells in the reticuloendothelial system (RES) (Gregoriadis & Ryman, 1972; Poste, et al., 1982). Contrary to a naive prediction that the behavior of liposomes in blood would be similar to that of blood cells, the half-lives of various types of liposomes were quite short despite the fact that the liposomes are composed of naturally occurring lipids. Thus the idea of using liposomes as carriers targeted to cells other than macrophages of RES had to wait at least a decade for the advent of liposome formulations with blood circulation time far greater than that of conventional liposomes (Allen and Chonn, 987; Gabizon and

[1] Cancer Research Institute and Department of Pharmacology
University of California
San Francisco, CA 94143 USA

NATO ASI Series, Vol. H 91
Trafficking of Intracellular Membranes
Edited by M.C. Pedroso de Lima N. Düzgüneş and D. Hoekstra

Papahadjopoulos, 1988), the so called sterically stabilized liposomes or stealth liposomes (Papahadjopoulos, et al., 1991) seem to circulate in blood relatively long ($t_{1/2}$ on the order of a day) by avoiding the rapid uptake by macrophages of RES. However, the exact mechanism of how this is achieved is yet to be elucidated.

Thus with the new formulation of liposomes which have greatly reduced uptake by cells of RES, it became quite crucial to revisit and address the question of what determines the uptake of liposomes utilizing *in vitro* model cell systems. Since one of the most important factors that control the pharmacokinetics of liposomes is the uptake by cells in RES as pointed out earlier, it is obvious that understanding of the cellular parameters involved in each step of the liposome uptake by cells is critical for the future design of liposomes which possess specific requirements as a carrier system. In this chapter, we describe our effort to investigate several aspects of the interaction of liposomes with cells using two cultured cell lines. Even though numerous reports studying various parameters controlling the uptake of liposomes have already been published, the literature still lacks in studies which systematically change each parameter and also compare different cell types under the same conditions.

The interaction of liposomes with cells is complex. It involves surface binding, internalization, and possible degradation of liposomes and subsequent release of liposomal contents. Here we focus on the first two processes, and the fate of liposomes and their contents after internalization is beyond the scope of current discussion. It is unclear which cellular components are involved in each step in the processing of liposomes. The primary mode by which liposomes are incorporated into certain types of cells is known to be endocytosis via the coated pit pathway (Straubinger, et al., 1983). The involvement of clathrin-coated vesicles in liposome endocytosis has been further confirmed by microinjection of anti-clathrin antibody into cells (Chin, et al., 1989). Like other ligands endocytosed by cells, liposomes, once endocytosed, come in contact with low pH compartments such as endosomes and lysosomes (Daleke, et al., 1990; Straubinger, et al., 1990). However, the existence or the nature of the membrane component(s) for liposome binding and internalization is unknown at present. In our opinion, the identification of the binding site for liposomes on cell surface may be the most crucial next step toward a better understanding of molecular mechanism of liposome uptake.

In an attempt to identify some of the parameters that control the early events in liposome-cell interaction, we have analyzed the uptake of different types of liposomes with systematically varied surface properties. We also compared the pattern of liposome uptake in two different cell types, CV1 African Green Monkey Kidney cell line and J774 murine macrophage cell line, in order to address the existence of cell type specificity. In addition, we have characterized kinetic and equilibrium parameters of liposome binding and internalization using two liposome compositions which behave completely differently in J774 cells, and compared the parameters to those of well-studied ligands with known receptors. In sum, all the data presented here point to the possibility that cells may have specific receptors for liposomes.

The effect of lipid headgroups on liposome uptake.

In general, liposomes containing negatively charged phospholipids exhibit increased binding and endocytosis over neutral liposomes. Earlier in vitro studies reported that negatively charged liposomes containing phosphatidylserine (PS) or phosphatidylglycerol (PG) are taken up and deliver drugs or DNA more efficiently than neutral liposomes (Fraley, et al., 1981; Heath, et al., 1985; Raz, et al., 1981; Schroit & Fidler, 1982). Also, the *in vivo* clearance rate of negatively charged liposomes from blood circulation is known to be faster than that of neutral liposomes (Senior, et al., 1985). These studies had concluded, on the basis of the behaviors of liposomes containing PS, PG or phosphatidic acid (PA), that simply the negative surface charge of liposomes is the important factor in determining the uptake rate. However, this notion can not be generalized to all the negatively charged lipids, on the basis of newer evidence. Some of our data demonstrate that the negative charge is not a sole determinant for the extent of this increase in binding and endocytosis.

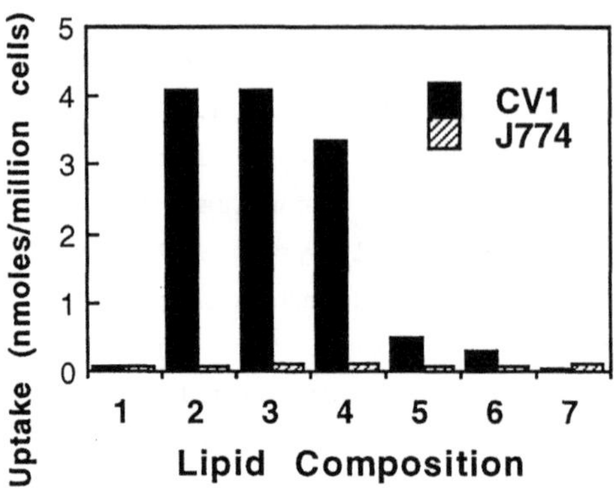

Figure 1. The headgroup of the negatively charged lipids incorporated into PC/Chol (2/1) liposomes (Lipid composition #1) determines the extent of enhancement in their uptake by CV1 and J774. #2 9% of the phospholipid was replaced by PS; #3 PG; #4 PA; #5 PI; #6 G_{M1}; and #7 PEG-PE; and its effect on the uptake by cells was monitored in comparison with neutral PC/Chol liposomes (#1).

The uptake by cells depends on the specific headgroup of the anionic phospholipid. PS, PG, and PA possess distinct properties when compared with monosialoganglioside G_{M1} (G_{M1}), phosphatidylinositol (PI) and PEG-PE (poly(ethylene glycol) conjugated to distearoylphosphatidylethanolamine). Recent *in vivo* studies (Allen & Chonn, 1987; Papahadjopoulos,et al., 1991) have demonstrated that several negatively charged lipids, such as G_{M1}, PI and PEG-PE, when incorporated as 9% of phosphatidylcholine (PC) : cholesterol (Chol) liposomes, produce a drastic decrease in the rate of RES uptake compared to similar liposomes composed of 9% PS or PG. This was also seen in our present study with CV1 cells; none of G_{M1}, PI, and PEG-PE showed any enhancement of the uptake as did PS, PG, or PA, even if introduced at the same mole% (Figure 1). The inclusion of PS, PG, or PA into the neutral PC : Chol liposomes makes a drastic difference in their uptake by CV1 cells. Figure 1 shows the total number of liposomes associated with these cells (i.e., uptake) during 1 hour incubation at 37°C (expressed as nanomoles of lipid per million cells). For liposomes containing 9 mole% PS, PG, or PA, there was more than 20-fold higher uptake by CV1 cells compared with the neutral PC : Chol liposomes (approximately 1,300 liposomes composed of neutral PC : Chol liposomes were taken up by one CV1 cell during 1 hour incubation, while 30,000 liposomes containing 9% PS were taken up per cell for the same period of incubation; based on the average diameter of about 80 nm of liposomes extruded through final pore size 0.05μm). However, the negative charge imparted by the inclusion of either 9% PI, G_{M1}, or PEG-PE did not show any enhancement of the uptake over PC : Chol.

The effect of surface charge density on liposome uptake.

The two cell lines, CV1 and J774, showed differences in the pattern of liposome uptake depending on the type of negatively charged lipids incorporated. Contrary to the intriguing sensitivity of CV1 cells to PS, PG, or PA in the liposome bilayer, J774 cells could not recognize the presence any of the above negatively charged lipids at a concentration of 9 mole% (Figure 1).

Increasing the density of the charged phospholipids, PS or PG (also PA, but not shown) beyond 9% does not further increase uptake by CV1 cells (Figure 2A). In contrast, J774 cells behaved quite differently from CV1 cells with respect to the increase of PS or PG content. PS or PG, if introduced as 9% of phospholipids, had little effect on the uptake by J774 cells, but they induced enhanced uptake as their concentration in the liposome bilayer was increased to more than 30% (Figure 2B).

PI is an extremely interesting case; it showed only a minimal enhancement of uptake in both cell lines when incorporated at 9%, significantly increased uptake in a concentration dependent manner at mole percentages between 9% and 50% (Figure 2A). It should be noted that calculation of zeta potential has shown that liposomes containing 10% PS or PG, are not different from those with 10% PI, although they are significantly different from the liposomes with 10% G_{M1} (Langner, et al., 1990)

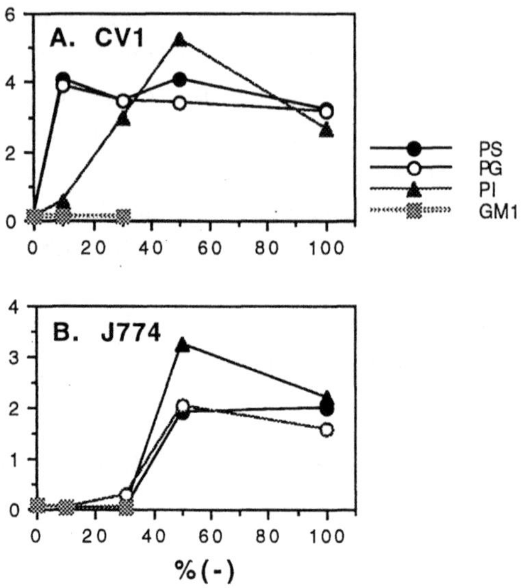

Figure 2. The effect of increased charge density on liposome uptake by cells. Liposomes of lipid composition, PS or PG or PI or G_{M1} : PC : Chol (x : 100 - x : 50) were used with increasing percentage (x) of anionic phospholipids. The amount of liposomal HPTS taken up by CV1 cells (A) and J774 cells (B) after 1 hour incubation at 37°C is monitored as a function of x. The relative increase in the enhancement of uptake with increased negatively charge lipids over neutral liposomes was consistent with standard deviation less than 15% in each experiment.

Although this type of physical measurement can not distinguish 9% PS containing liposomes from 9% PI containing liposomes, CV1 cells have the ability to recognize the different surface property of these two types of liposomes and bind differently. It is also noteworthy that PI behaved more or less similar to PS or PG in J774 cells, but it was distinct from PS and PG in CV1 cells.

G_{M1} showed the same effect as PI at 9% concentration in both cell lines in the sense that it did not induce any enhancement of liposome uptake. Thus we investigated whether G_{M1} enhances the uptake as its concentration in liposome bilayer increases like PI does. Figure 2 shows that G_{M1} did not increase the uptake by either cell type even when 30% of PC was replaced by G_{M1} in the bilayer. Unfortunately, the maximum concentration of G_{M1} one can incorporate in the lipid bilayer is approximately 30%. Beyond that concentration, G_{M1} is known to form micelles in equilibrium with the bilayer form (Felgner, et al., 1983). Therefore we could not test whether it starts to be recognized in a similar fashion like PI if it is incorporated into liposome bilayer at more than 30% of PC.

Binding versus Endocytosis in liposome uptake.

The whole process of liposome association with, and incorporation into, cells (uptake) is a two step process involving binding and endocytosis, and, therefore, it is important to distinguish these two steps and assess to what extent each step controls the overall uptake. When the total amount of liposomes associated with a million cells after 1 hour of incubation at 37°C is compared to that after 1 hour of incubation at 4°C, there is a significant correlation between the two values; the first value is a measure of the total uptake, while the second one is a measure of binding without any endocytosis. For both CV1 and J774 cells, the higher extent of uptake of a certain type of liposomes correlates with a correspondingly higher binding level at 4°C. Neutral PC liposomes show minimal uptake during 1 hour incubation at 37°C and also the binding of PC liposomes at 4°C is minimal. PS promotes increased binding of the liposomes to CV1 cells and concomitantly higher uptake. Higher content of PS (and also PG, and PI) increases the levels of cell associated liposomes by J774 cells both at 4°C and 37°C. Therefore the rate of liposome endocytosis appears to be an intrinsic constant determined by the cell and not a rate determining step in the overall rate of liposome uptake; once a liposome establishes a stable binding and shifts into a state that is ready to be internalized by endocytosis, it goes into an endosome very fast regardless of the liposome formulation. The former step, where the liposome binds and becomes ready to be internalized, seems to depend greatly on the presence of the negatively charged lipid and its specific lipid headgroup. This has been supported by a study quantitatively analyzing the kinetic parameters of binding and endocytosis employing mass action kinetics (Lee, et al., 1993). A detailed description of this method (Nir, et al., 1986) and its applications appears in the another chapter by S. Nir of this book. The overall association of liposomes with cells was viewed as a two step process, a binding step described by a second order reaction followed by an endocytosis step described by a first order reaction.

The parameters of liposome binding and endocytosis by J774 cells were characterized in detail for two types of liposomes: neutral liposomes composed of PC/Chol (2/1), and negatively charged liposomes composed of PS/PC/Chol (1/1/1). In terms of uptake rate by J774 cells, these two lipid compositions are the two extreme cases, and the negatively charged liposomes containing PG, PA or PI instead of PS behave similarly to PS/PC/Chol liposomes as was discussed earlier in this chapter (Lee, et al., 1992). This was the first study to report the equilibrium binding affinity and the kinetic rate constants of liposome-cell interactions as well as the endocytic rate constants (Lee, et al., 1993). The analysis of the binding data has included Scatchard plots as well as determination of the kinetic rates of liposome binding both at 4°C and at 37°C under conditions inhibiting endocytosis. Model calculations employing mass action kinetics were used to simulate the experimental data of binding and endocytosis of liposomes, and determined affinity constants of liposome binding at 37°C and 4°C, on- and off- rate constants of the binding, and endocytic rate constants. It demonstrates that the overall rate of uptake is dictated by the binding step. The order of magnitude excess in total uptake of PS/PC/Chol liposomes over PC/Chol liposomes is due to the difference in binding, since both liposomes exhibited the same rate constant of endocytosis. The conclusion that the amount of PS/PC/Chol liposomes bound to J774 cells exceeds significantly that of PC/Chol liposomes was based on the results of binding experiments at 4°C, and at 37°C in the presence of inhibitors of endocytosis; it was established

Table I. Parameters describing liposome binding to and endocytosis by J774 cells[a].

Experiment Number	T (°C)	ε (s^{-1})	C ($M^{-1}s^{-1}$)	D (s^{-1})	K (M^{-1})	N
<u>PS/PC/Chol</u>						
	4°C					
1		_	$3.7 \cdot 10^9$	0.001	$3.7 \cdot 10^{12}$	3200
2		_	$1 \cdot 10^9$	0.001	$1 \cdot 10^{12}$	1800
3		_	$6 \cdot 10^8$	0.0006	$1 \cdot 10^{12}$	3000
4		_	_	_	$2.2 \cdot 10^{12}$	1000
5		_	_	_	$1.4 \cdot 10^{13}$	11000
	37°C					
2		0.0025	$1.3 \cdot 10^9$	0.0004	$3.2 \cdot 10^{12}$	3000
3		0.001	$2.3 \cdot 10^9$	0.0009	$2.6 \cdot 10^{12}$	3300
6[b]		0.00024	_	_	_	_
7[b,c]		0.0014	_	_	_	_
<u>PC/Chol</u>						
	4°C					
1		_	$1.5 \cdot 10^8$	$2.3 \cdot 10^{-4}$	$6.5 \cdot 10^{11}$	300
2		_	$3.5 \cdot 10^8$	$7 \cdot 10^{-4}$	$5 \cdot 10^{11}$	300
3		_	$2.5 \cdot 10^8$	$2.5 \cdot 10^{-4}$	$1 \cdot 10^{12}$	350
4		_	_	_	$8 \cdot 10^{11}$	500
5		_	_	_	$5.7 \cdot 10^{11}$	2900
	37°C					
2		0.002	$4 \cdot 10^8$	$8 \cdot 10^{-4}$	$5 \cdot 10^{11}$	300
3		0.001	$1.5 \cdot 10^8$	0.0011	$1.4 \cdot 10^{11}$	350

a. The cells used were J774 cells in suspension at the concentration of 10^7/ml unless otherwise stated. The label for the liposomes was Rho-PE, except for the cases specified if HPTS was used. The estimated uncertainties in the parameters N, C, D and ε are 30%, 10%, 50%, and 50%, respectively.
b. The endocytic rate constants for these cases were obtained experimentally using HPTS containing liposomes
c. The endocytic rate constant for this case was for adherent J774 cells using HPTS containing liposomes.

experimentally that no endocytosis occurred at 4°C or at 37°C in the presence of the inhibitors employed.

The strength of the binding between a liposome and a cell is expressed by the equilibrium constant, K. The affinity constant K between a liposome and a cell is equal to the product of k, the affinity of binding between a liposome and a binding site which might involve several receptors, and N, the number of these binding sites in a cell; thus $K=k \cdot N$. Table I shows that at 4°C the values of $K=(0.1 - 1.4) \cdot 10^{13}\ M^{-1}$ for liposomes composed of PS/PC/Chol are between 2- and 14-fold larger than those found for PC/Chol liposomes. At 37°C we found by the use of inhibitors of endocytosis K values around $3 \cdot 10^{12}\ M^{-1}$ for PS/PC/Chol liposomes, and the corresponding values for PC/Chol liposomes are 6- to 18-fold smaller.

These values can be compared to the affinity constants of other well-characterized macromolecules and biological particles: sendai virus binding to erythrocyte ghosts (at 37°C) yields $K=(0.6 - 3) \cdot 10^{12}\ M^{-1}$ (Nir, et al., 1986) whereas its binding to the larger HL-60 cells yields $K= (10^{12} - 10^{13})\ M^{-1}$ (Pedroso de Lima, et al., 1991). For influenza virus binding to the latter cells at neutral pH the value of K is larger than $10^{13}\ M^{-1}$ (Düzgünes, et al., 1992). The binding of large glycophorin-bearing liposomes to fibroblast cells expressing the influenza hemagglutinin gave $k= 7 \cdot 10^{10}\ M^{-1}$ (Ellens, et al., 1990) which yields $K=1.2 \cdot 10^{14}\ M^{-1}$ (N here was approximately 1,700). Hence the strength of binding of PS-containing liposomes to cells is comparable to the binding of certain viruses to cells, but is about an order of magnitude below the high affinity binding of glycophorin to influenza hemagglutinin. The number of binding sites N for virus has been estimated in the range of several thousands (Düzgünes, et al., 1992), hence it is generally comparable to that found here for liposome binding. It is also interesting to note that the affinity constant of liposomes binding to the cells is comparable to, although several fold to an order of magnitude lower than, that of lipoproteins binding to their cell-surface receptors. The affinity constants K for the interaction of low density lipoproteins and apo-E high density lipoproteins with human fibroblasts can be calculated to be $\sim 3.6 \cdot 10^{13}$ and $2.3 \cdot 10^{14}\ M^{-1}$, respectively, from the k and N values in Goldstein and Brown (Goldstein & Brown, 1977) and Pitas *et al.* (Pitas, et al., 1979). The affinity constants K of acetylated low density lipoproteins to scavenger receptors in smooth muscle cells and murine macrophages range from $8 \cdot 10^{12}$ to $2 \cdot 10^{13}$ M^{-1}. The rate constant C of association of PS-containing liposomes to J774 cells is 2- to 10-fold larger than for neutral PC/Chol liposomes. Yet it is one to two orders of magnitude below the values found for virus binding to cells (Düzgünes, et al., 1992; Nir, et al., 1990; Pedroso de Lima, et al., 1991), although it is close to the association rate of lipoprotein particles to their cell surface receptors (Pitas, et al., 1979).

Binding vs total uptake; Calculation of endocytic rate.

The kinetics of total association (uptake) of liposomes with cells was significantly faster at 37°C than at 4°C. This uptake at 37°C was dramatically reduced by metabolic inhibitors strongly suggesting that endocytosis following binding at 37°C accelerates the total uptake under this condition. This is consistent with the notion that the major route of uptake is via endocytosis. Likewise, lowering the incubation temperature to 4°C blocks endocytosis completely as the average pH of the liposomal contents remains neutral. A combination of metabolic inhibitors, Antimycin A plus NaF, with or without sodium azide, inhibited 100% of the endocytosis. The kinetic data of total cell-associated liposomes at 37°C with and without metabolic inhibitors were analyzed by a mass action kinetic model. In this study we first determined at 37°C the parameters C, D (on- and off-rates of a liposome to a cell) and N (number of binding sites per cell) from binding results in the presence of inhibitors of endocytosis. Then, utilizing these values, the rate constant of endocytosis ε was determined from the results at 37°C without inhibitors.

The rate constant of endocytosis varies by an order of magnitude depending on the day of experiment (Table I), although when the rate constant ε was determined by one type of liposomal label, its variation was merely from 0.001 to 0.003 s^{-1}. The remarkable result is that within the same experiment the rate constant of endocytosis of liposomes by suspension J774 cells is independent of the composition of liposomes. Hence, again, the large variation in the total uptake of liposomes by cells is due entirely to the differences in the binding characteristics, whereas the rate constant of endocytosis is a property of the cell under the given conditions. Therefore, the difference in the overall rate of liposome uptake via endocytosis between different liposome compositions appears to be solely due to the differences in binding capacity.

The trend demonstrated by the calculations employing the mass action kinetic model indicates that the amount of cell-surface bound liposomes at 37°C reaches a constant value after less than 30 min. When inhibitors of endocytosis are employed, the time of equilibration of binding is prolonged. Moreover, endocytosis results in a reduction on the amount of liposomes bound to the cell-surface at steady state.

The value ε in the range of ~0.001 sec^{-1} indicates that $t_{1/2}$, the time required for half of pre-bound liposomes to be endocytosed, is ~8 min based on the equation $\varepsilon \cdot t_{1/2} = \ln 2$. This endocytic rate constant of liposomes is at least one order of magnitude smaller than the rate of endocytosis of formadlehyde-treated serum albumin via scavenger receptors (Eskild *et al.*, 1989), several-fold smaller than that of epidermal growth factor receptors (Carpenter and Cohen, 1976), transferrin receptors (Ciechanover *et al.*, 1983), and low density lipoprotein receptors (Goldstein and Brown, 1977), but comparable to that of galactose receptors in hepatocytes (Weigel and Oka, 1982).

The nature of liposome binding to cell surface.

Like other ligands that are taken up by cells, the uptake of liposomes can be assumed to fall into either one of two categories: non-specific (Hsu & Juliano, 1982) or receptor-mediated endocytosis (Pastan & Willingham, 1983; Stahl & Schwarts, 1986). The uptake of liposomes so far has been generally believed to be mediated by non-specific association (or adsorption) of liposomes onto the cell surface and subsequent endocytosis. This has been mostly due to the lack of knowledge about the binding of liposomes to cell surface and the nature of the binding site. The fact that liposome binding is lipid headgroup-specific and cell type dependent, as we have discussed in the above sections, indicates that liposome-cell interactions show such specificity that we should entertain the possibility that it is indeed a receptor-mediated process. Using the term "receptor-mediated" requires the existence of cell surface proteins that bind a specific chemical structure (ligand) on the liposome surface and mediate endosome formation which contains the ligand-receptor complex (receptosome) (Helenius, et al., 1983; Pastan & Willingham, 1981). Although, the binding sites for liposomes have not been identified, there is evidence indicating the involvement of membrane protein(s) which may require Ca^{2+} for the binding (Dijkstra, et al., 1985; Pagano & Takeichi, 1977). The binding of anionic liposomes to the cell surface is trypsin-sensitive to a large extent, suggesting the involvement of cell surface proteins. Neutral PC liposomes have been also shown to bind to trypsin-sensitive membrane proteins in fibroblast cell line (Pagano & Takeichi, 1977). Although it is still not entirely clear what are the determining factors defining the uptake by various cells, the different levels of binding for a given liposome composition by two types of cells suggest that the binding step itself is crucial and is not governed by a simple electrostatic interaction between liposomes and cell membranes.

Competition experiments using high molecular poly-anions gave us additional insights into the nature of the binding of negatively charged liposomes to cells. A range of ligands specific for the scavenger receptor have gotten particular attention because those seemed to compete with negatively charged liposomes in terms of uptake by macrophage cells (Nishikawa, et al., 1990). Further investigation showed that the scavenger receptor is not involved in the liposome uptake (Lee, et al., 1992). However the pattern of PS containing liposome uptake by two cells lines, CV1 and J774, in presence of poly(inosinic acid) (poly-I) and poly(cytidylic acid) (poly-C) was very illuminating (Figure 3). It clearly showed that the uptake of liposomes by CV1 cells does not involve the scavenger receptor and that the mechanism of PS liposome uptake at least by these two cell lines may not be the same (Lee, et al., 1992). The inability of a very high concentration (up to 5mg/ ml; not shown in Fig. 3) of poly-anions (1000-fold excess in terms of number of molecules) to compete with negatively charged liposomes for CV1 cells indicates again that the binding of liposomes is probably not a non-specific adsorption via simple electrostatic interaction; 5 mg/ml of 500,000 molecular weight dextran sulfate corresponds to 10mM particle concentration, while a typical particle concentration of liposomes used in the competition experiment is less than 10nM.

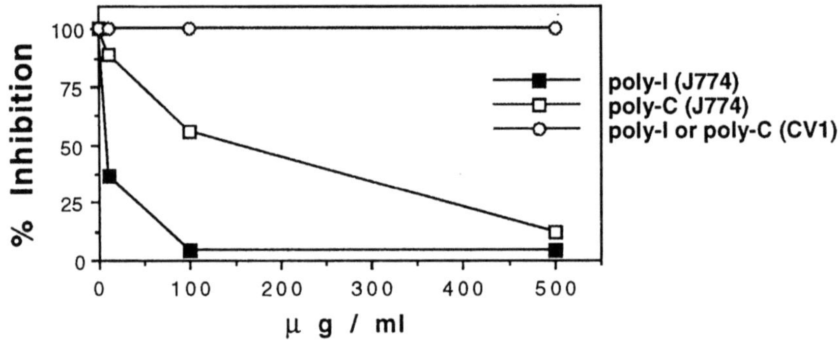

Figure 3. The effect of negatively charged compounds on the uptake of PS containing liposomes. The uptake of PS : PC : Chol (1 : 1 : 1) liposomes by CV1 and J774 cells in the presence of different concentrations of poly-I and poly-C was expressed as percent of the uptake in absence of the compounds. The curve (○) shows that these poly-anions have no effect on the liposome uptake by CV1 cells, contrary to the case of J774 cells (□, ■).

One more supporting evidence for specific binding comes from the binding and uptake of PS containing liposomes relative to neutral liposomes in cultures of monocytes with different history of culture condition and age. The regulation of liposome binding by the culture conditions, as has also been implied from the results using adherent and suspended J774 cells, was assessed with monocytes isolated from human peripheral blood cells. Monocytes cultured for 5 to 11 days in different states, one in suspension and the other in adherent state, showed different levels of binding and uptake of PS/PC/Chol liposomes relative to PC/Chol liposomes. Differences in binding depended on the number of days in culture as well as on the condition of culture (Lee, et al., 1993). Monocytes cultured in adherent state always exhibited a higher binding ratio of PS/PC/Chol to PC/Chol liposomes than the monocytes cultured in suspension. However, this ratio also depended on the possible maturation/differentiation of monocytes in culture since the ratio increases from 5 to ~30-40 as a function of days of culture, irrespectively of whether the cells were cultured in suspension or as adherent. In contrast, the lymphocytes did not show this dependency on the culture duration as this ratio in lymphocytes stayed ~10. Overall, the extent of liposome binding by lymphocytes was 2- to 3-fold less per cell than the extent by monocytes.

These data suggest that the binding and uptake of PS-containing liposomes by monocytes increases ~5-fold as the peripheral monocytes mature and differentiate into macrophages whereas the binding and uptake of PC liposomes stays relatively constant during this process. The rate of this maturation/ differentiation and the increase in the binding and uptake of PS-containing liposomes depend on the culture condition; they are shown to be facilitated by the attachment of monocytes to plastic, which bears similarities to the maturation of circulating blood monocytes into fixed tissue macrophages of RES and also to the extravasation into inflamation sites and activation. Lymphocytes are in general less active in binding and taking up liposomes, and also do not show much change in their ability of liposome uptake over 10 days of culture.

This preferential increase in the binding of PS-containing liposomes over PC liposomes by monocytes bears similarity to the adherence-induced selective expression of various proteins in monocytes, such as several monokines and proto-oncogenes (Haskill *et al.*, 1988; Kaplan and Gaudernack, 1982) as well as the expression of a number of cell-surface receptors including transferrin receptors (Taetle and Honeysett, 1988) upon monocyte maturation and differentiation. The binding and uptake of PS, PG, or PA containing liposomes is probably mediated by cell surface proteins that constitute the binding sites and the expression of these binding sites may be modulated like other cell surface receptors.

The increased capacity of monocytes to bind PS-containing liposomes upon maturation/activation certainly resembles other previously proposed mechanism of specific recognition of PS by monocyte/macrophages. It has been proposed previously that macrophages recognize and remove senescent erythrocytes via the increased expression of PS on the outer leaflet of erythrocyte membranes (Schroit *et al.*, 1985; Allen *et al.*, 1988; Connor *et al.*, 1989). Also, an enhanced recognition of tumor cells by activated human blood monocytes has been shown to have a correlation to an increased expression of PS on the tumor cell surface (Utsugi *et al.*, 1991). In this respect, the significance of the existence of a receptor for liposomes containing certain negatively charged lipids such as PS, PG, and PA may be far greater than just better understanding of liposome uptake and optimal designing of drug carriers. More recently, it has been also brought to attention that the recognition of apoptotic cells and removal by phagocytosis may involve the same receptor or mechanism as the one we have described here (Fadok *et al.*, 1992a and b).

Epilogue

In this chapter, we have described the data which characterize the binding and uptake of liposomes of various lipid compositions and point to the possible presence of cell surface receptor specific for liposomes containing negatively charged lipids such as PS, PG, and PA. All the characteristics of binding and uptake of negatively charged liposomes, the lipid headgroup specificity, cell type dependency, competition with other polyanions, as well as the affinity constant and endocytic rate, are very reminiscent of the uptake of other ligands which have been known to be taken up by receptor-mediated processes. At this point, it is worth noting that even the uptake of neutral liposomes, PC/Chol, is mediated by as yet undefined mechanism which involves binding and endocytosis; the overall uptake rate is approximately 10 times higher than that of random entrapment of liposomes in the pinocytic and endocytic vesicles (Lee *et al.*, 1993). The biological implication of the receptor for the negatively charged lipids seems to be very broad and essential for the removal of unwanted cells in the body. Further characterization of the mechanism of liposome uptake at hemolecular level will give us invaluable insight into the cell biology of how some cells recognize specific features of other cells via a novel mechanism other than through protein-protein interaction not to mention the strategies of designing liposome carriers with better targeting capability.

References

Allen, TM, & Chonn, A (1987). Large unilamellar liposomes with low uptake into the reticuloendothelial system. FEBS Lett., 223, 42-46.
Carpenter, G., & Cohen, S. (1976) 125I-labeled human epidermal growth factor. Binding, internalization, and degradation in human fibroblasts. J. Cell Biol. 71, 59-171.
Chin, D, Straubinger, RM, Acton, S, Nathke, I, & Brodsky, FM (1989). 100-kD polypeptides in peripheral clathrin-coated vesicles are required for receptor-mediated endocytosis. Proc. Natl. Acad. Sci. U.S.A., 86, 9289-9293.
Ciechnover, A., Schwartz, A.L., Dautry-Varsat, A., & Lodish, H.F. (1983) Kinetics of internalization and recycling of transferrin and the transferrin receptor in a human hepatoma cell line. Effect of lysosomotropic agents. J. Biol. Chem. 258, 9681-9689.
Daleke, DL, Hong, K, & Papahadjopoulos, D (1990). Endocytosis of liposomes by macrophages: binding, acidification and leakage of liposomes monitored by a new fluorescence assay. Biochim. Biophys. Acta, 1024, 352-366.
Dijkstra, J, van Galen, M, & Scherphof, G (1985). Influence of liposome charge on the association of liposomes with Kupffer cells *in vitro*. Effects of divalent cations and competition with latex particles. Biochim. Biophys. Acta, 813, 287-297.
Düzgünes, N, Pedroso de Lima, MC, Stamatatos, L, Flasher, D, Alford, D, Friend, DS, & Nir, S (1992). Fusion activity and inactivation of influenza virus: kinetics of low pH-induced fusion with cultured cells. J. Gen. Virology, 73, 27-37.
Ellens, H, Bentz, J, Mason, D, Zhang, F, & White, J (1990).Fusion of influenza hemagglutinin-expressing fibroblasts ith glycophorin-bearing liposomes: role of hemagglutinin suface density. Biochemistry, 29, 9697-9707.
Fadok, VA, Savill, JS, Haslett, C, Bratton, DL, Doherty, DE, Campbell, PA, & Henson, PM (1992) Different populations of macrophages use either the vitronectin receptor or the phosphatidylserine receptor to recognized and remove apoptotic cells. J. Immunnology, 149, 4029-4035.
Fadok, VA, Voelker, DR, Campbell, PA, Cohen, JJ, Bratton, DL, & Henson, PM (1992) Exposure of phosphatidylserine on the surface of apoptotic lymphocytes triggers specific recognition and removal by macrophages. J. Immunology, 148, 2207-2216.
Felgner, PL, Thompson, TE, Barenholz, Y, & Lichtenberg, D (1983). Kinetics of transfer of gangliosides from their micelles to dipalmitoylphosphatidylcholine vesicles. Biochemistry, 22, 1670-1674.
Fidler, IJ, Raz, A, Fogler, WE, R. Kirsh, Bugelski, P, & Poste, G (1980). Design of liposomes to improve delivery of macrophage-augmenting agents to alveolar macrophages. Cancer Res., 40, 4460-4466.
Fraley, RT, Straubinger, RM, Rule, G, Springer, EL, & Papahadjopoulos, D (1981). Liposome-mediated delivery of deoxyribonucleic acid to cells: Enhanced efficiency of delivery related to lipid composition and incubation conditions. Biochemistry, 20, 6978-6987.

Gabizon, A, & Papahadjopoulos, D (1988). Liposome formulations with prolonged circulation time in blood and enhanced uptake by tumors. Proc. Natl. Acad. Sci. USA, 85, 6949-6953.
Goldstein, JL, & Brown, MS (1977). The low-density lipoprotein pathway and its elation to atherosclerosis. Annu. Rev. Biochem., 46, 897-930.
Gregoriadis, G, & Ryman, B (1972). Fate of protein-containing liposomes injected into rats. An approach to the treatment of storage diseases. Eur.J. Biochem., 24, 485-491.
Heath, TD, Lopez, NG, & Papahadjopoulos, D (1985). The effects of liposome size and surface charge on liposome-mediated delivery of methotrexate-g-aspartate to cells in vitro. Biochim. Biophys. Acta, 820, 74-84.
Helenius, A, Mellman, I, Wall, D, & Hubbard, A (1983). Endosomes. Trends Bioche. Sci, 8, 245-250.
Hsu, MJ, & Juliano, RL (1982). Interaction of liposomes with the reticuloendothelial system. II : Nonspecific and receptor-mediated uptake of liposomes by mouse peritoneal macrophages. Biochimica et Biophysica Acta, 720, 411-419.
Lee, K-D, Hong, K, & Papahadjopoulos, D (1992). Recognition of liposome by cells:in vitro binding and endocytosis mediated by specific lipid headgroups and surface charge density. Biochim. Biophys. Acta, 1103, 185-197.
Lee, K-D, Nir, S, & Papahadjopoulos, D (1993). Quantitative analysis of liposome-cell interactions *in vitro*: rate constants of binding and endocytosis with suspension and adherent J774 cells and human monocytes. Biochemistry, 32, 889-899.
Lee, K-D, Pitas, RE, & Papahadjopoulos, D (1992). Evidence that the scavenger receptor is not involved in the uptake of negatively charged liposomes in cells. Biochim. Biophys. Acta, 1111, 1-6.
McConnell, HM (1983). . New York: Academic Press.
Nir, S, Düzgüneş, N, Pedroso de Lima, MC, & Hoekstra, D (1990) Fusion of enveloped viruses ity cells and liposomes. Cell Biophys., 17, 181-201.
Nir, S, Klappe, K, & Hoekstra, D (1986). Kinetics and extent of fusion between Sendai virus and erythrocyte ghosts: application of a mass action kinetic model. Biochemistry, 25, 2155-2161.
Nishikawa, K, Arai, H, & Inoue, K (1990). Scavenger receptor-mediated uptake and metabolism of lipid vesicles containing acidic phospholipids by mouse peritoneal macrophages. J. Biol. Chem., 265(9), 5226-5231.
Pagano, RE, & Takeichi, M (1977). Adhesion of phospholipid vesicles to Chinese hamster fibroblasts. Role of cell suface proteins. J. Cell Biol., 74, 531-546.
Papahadjopoulos, D, Nir, S, & Düzgünes, N (1990).Molecular mechanisms of calcium-induced membrane fusion, J. Bioenergetics and Membranes, 22, 157-179.
Pastan, I, & Willingham, MC (1983). Receptor-mediated endocytosis: coated pits, receptosomes and the Golgi. Trends Biochem. Sci., 8, 250-254.
Pastan, IH, & Willingham, MC (1981). Journey to the center of the cell: role of the receptosome. Science, 214, 504-509.
Pedroso de Lima, MC, Nir, S, Flasher, D, Klappe, K, Hoekstra, D, & Düzgünes, N (1991). Biochim. Biophys. Acta, 1070, 446-454.

Pitas, RE, Innerarity, TL, Arnold, KS, & Mahley, RW (1979). Rate and equilibrium constants for binding of apo-E HDLc (a cholesterol-induced lipoprotein) and low density lipoproteins to human fibroblasts: evidence for multiple receptor binding of apo-E HDLc. Proc. Natl. Acad. Sci. *U.S.A.*, 76, 2311-2315.

Poste, G, Bucana, C, Raz, A, Bugelski, P, Kirsh, R, & Fidler, IJ (1982). Analysis of the fate of systemically administered liposomes and implications for their use in drug delivery. Cancer Res., 42, 1412-1422.

Raz, A, Bucana, C, Fogler, WE, Poste, G, & Fidler, IJ (1981). Biochemical, morphological, and ultrastructural studies on the uptake of liposomes by murine macrophages. Cancer Research, 41, 487-494.

Schroit, AJ, & Fidler, IJ (1982). Effects of liposome structure and lipid composition on the activation of the tumoricidal properties of macrophages by liposomes containing muramyl dipeptide. Cancer Research, 42, 161-167.

Senior, J, Crawley, JCW, & Gregoriadis, G (1985). Tissue distribution of liposomes exhibiting long half-lives in the circulation after intravenous injection. Biochim. Biophys. Acta, 839, 1-8.

Stahl, P, & Schwarts, AL (1986). Receptor-mediated endocytosis. J. Clin. Invest., 77, 657-662.

Straubinger, RM, Hong, K, Friend, DS, & Papahadjopoulos, D (1983). Endocytosis of liposomes and intracellular fate of encapsulated contents: encounter with a low-pH compartment after internalization in coated vesicles. Cell, 32, 1069-1079.

Straubinger, RM, Papahadjopoulos, D, & Hong, K (1990). Endocytosis and intracellular fate of liposomes using pyranine as a probe. Biochemistry, 29, 4929-4939.

Molecular Analysis of Exocytosis in Neurons and Endocrine Cells

Barbara Höhne-Zell and Manfred Gratzl
Department of Anatomy and Cell Biology
University of Ulm
89069 Ulm
Germany.

Abstract

Neurotransmitters and hormones are released in a Ca^{2+} dependent way from neurons and endocrine cells by exocytosis. Key components of the fusion machinery have been recently identified using clostridial neurotoxins as tools. Following uptake of the neurotoxins, the light chains of the neurotoxins inhibit exocytosis by nerve terminals and endocrine cells. The tetanus toxin light chain cleaves synaptobrevin, a membrane protein of secretory vesicles in neurons and endocrine cells. This process is inhibited by removal of zinc from the active site of the tetanus toxin light chain or addition of peptides containing the cleavage site of synaptobrevin. Synaptobrevins bind to SNAP-25 and syntaxins forming, together with synaptotagmin, GTP binding proteins and cytosolic proteins a complex, which plays an crucial role in exocytosis of neurotransmitters and hormones.

Introduction

Exocytosis, discharge of secretory products by nerve and gland cells, involves fusion of a secretory organelle with the cell membrane (Palade, 1975). During regulated exocytosis stored substances such as neurotransmitters, hormones or enzymes are released in response to cell stimulation. In contrast, during constitutive exocytosis, newly translated proteins (e.g. plasma membrane proteins,

NATO ASI Series, Vol. H 91
Trafficking of Intracellular Membranes
Edited by M.C. Pedroso de Lima N. Düzgüneş and D. Hoekstra

extracellular matrix components, immunoglobulines and albumin) are continuously released from the cells without intermediate storage (Kelly, 1985). However, the molecular mechanisms of regulated and constitutive exocytosis are poorly understood.

Release of neurotransmitters is a highly regulated and fast process. Synaptic vesicles, docked at the active zone of the nerve terminal, fuse with the presynaptic plasma membrane upon stimulation. Whereas neurons transmit information through synapses to neighboring cells, endocrine cells release hormones into the circulation. Although endocrine compared to neuronal exocytosis is a slower process the underlying mechanisms are apparently very similar. Both, neurons and endocrine cells possess two different types of secretory organelles characteristic of two different regulated secretory pathways. In neurons the translucent synaptic vesicles containing neurotransmitters (excitatory or inhibitory) can be distinguished from secretory granules which store neuropeptides or amines. These secretory granules have an electron dense core (hence the term dense core vesicles) as do their counterparts in endocrine cells. In the endocrine cells, the secretory granules predominantly contain stored peptide hormones or amines.

Exocytosis of synaptic vesicles and secretory granules is controlled by cytoplasmic Ca^{2+}. The vesicles fuse with the plasma membrane and release their contents in response to an increase of intracellular Ca^{2+}, caused by an influx of Ca^{2+} from the extracellular space and/or release from intracellular stores. Following exocytosis, the vesicle membrane proteins are retrieved from the plasma membrane by endocytosis.

Proteins involved in exocytosis

Different approaches, which include yeast genetics, transport reconstitution, membrane capacitance measurements, introduction of substances into permeabilized neurons and endocrine cells and biochemical characterization of cell components, allowed to identify a number of proteins that are involved in membrane traffic and fusion. The results obtained suggest that the mechanisms involved in synaptic vesicle or secretory granule docking and fusion may be evolutionary conserved (Jahn & Südhof, 1993; Bennett & Scheller, 1993).

Biochemical and genetic studies support the involvement of GTP binding proteins in intracellular membrane trafficking and exocytosis. Using cell free systems, it was demonstrated that the nonhydrolyzable GTP analog GTPγS inhibits vesicle mediated transport between endoplasmatic reticulum, Golgi apparatus and endosomes (Balch, 1989). Some members of the family of small GTP-binding proteins (e. g. Rab3A and B) are specifically expressed in the nervous and endocrine system, where they are found in membranes of synaptic vesicles and secretory granules (Darchen et al 1990; Toutant et al 1987). In nerve terminals synaptic vesicles are used for several cycles of exocytosis and small G proteins control cycling and docking of synaptic vesicles (Fischer von Mollard, Südhof, & Jahn, 1991; Hess, Doroshenko, & Augustine, 1993), possibly together with additional proteins such as rabphilin 3A (Shirataki et al 1993), a member of the synaptotagmin family (see below). Recent data suggest that Rab3A inhibits Ca^{2+} dependent exocytosis in endocrine cells whereas Rab3B facilitates hormone release (Holz et al 1994; Johannes et al 1994; Lledo et al 1993). The inhibition of exocytosis by non-hydrolyzable GTP analogs (Ahnert-Hilger, Bräutigam, & Gratzl, 1987; Bader et al 1989; Burgoyne, Morgan, & Roth, 1994; Sontag, Aunis, & Bader, 1992) and of Rab3A derived peptides (Senyshyn, Balch, & Holz, 1992; Davidson et al 1993) in permeabilized chromaffin cells is in agreement with the proposed inhibitory role of the GTP bound form of Rab3A on exocytosis. In addition to small GTP binding proteins heterotrimeric G proteins have been implicated in the regulation of catecholamine release (Bader et al 1989; Ahnert-Hilger et al 1992). Thus it appears that different types of G proteins are involved in docking and/or fusion of synaptic vesicles and secretory granules with the plasma membrane. The analysis of their precize roles alone and/or in combination is certainly far from being trivial and requires careful analysis of exocytosis of cells endowed both with single and certain sets of G proteins.

Overexpression and antisense strategies as well as investigation of hormone and transmitter release in G protein knockout animals will be instrumental to achieve this goal (Holz et al 1994; Johannes et al 1994; Geppert et al 1994).

Synaptotagmin is an abundant synaptic vesicle protein which has been proposed to play a role in regulating exocytosis of synaptic vesicles and secretory granules (Perin et al 1991; Wendland et al 1991). Several isoforms have been detected in synaptic vesicles of rat brain and in vesicles purified from the electric organ of marine rays (Perin et al 1990; Perin et al 1991; Wendland et al 1991). Synaptotagmin was detected in both synaptic vesicles and in secretory granules, isolated from adrenal chromaffin cells and neurohypophysial terminals (Matthew, Tsavaler, & Reichardt, 1981; Trifaró, Fournier, & Novas, 1989; Walch-Solimena et al 1993). Because of its ability to bind Ca^{2+} in the presence of phospholipids (Brose et al 1992; Chapman & Jahn, 1994) synaptotagmin is considered to act as a Ca^{2+} sensor that triggers neurotransmitter release (Yoshida et al 1992). Synatotagmin binds to syntaxin (Bennett, Calakos, & Scheller, 1992; Bennett & Scheller, 1993; Leveque et al 1994) a synaptic protein participating in docking of synaptic vesicles to the presynaptic plasma membrane. It also interacts with neurexin, the receptor for α-latrotoxin which activates neurotransmitter release (Petrenko et al 1991; Hata et al 1993). Direct evidence for synaptotagmin involvement in exocytosis was reported from nerve terminals and pheochromocytoma cells in which exocytosis is blocked by peptides and larger cytoplasmic domains of synaptotagmin (Bommert et al 1993; Elferink, Peterson, & Scheller, 1993).

Synaptophysin was the first synaptic vesicle protein studied in detail. It is an abundant 38 kDa membrane glycoprotein. Evidence for the involvement of synaptophysin in neurotransmitter release has come from reconstitution experiments in oocytes (Alder et al 1992a) and injection of synaptophysin antibodies into neuromuscular synapses (Alder et al 1992b). Recent observations have shown that synaptophysin exists, besides in synaptic vesicles of mature nerve terminals, in vesicles of developing axons and dendrites (Bergmann et al

1991; Parton, Simons, & Dotti, 1992; Bergmann et al 1993; Mundigl et al 1993; Ovtscharoff et al 1993; Grabs et al 1994). This suggests a role of synaptophysin in constitutive recycling during membrane biogenesis of the axonal arborization and of the dendritic tree in addition to its function during exocytosis/endocytosis of synaptic vesicles.

Besides membrane proteins, cytosolic proteins like NSF (N-ethylmaleimide sensitive factor) and SNAPs (soluble NSF attachment proteins) are involved in exocytosis. NSF, which forms a tetramer composed of four 76 kDa subunits, has been purified from mammalian cells (Block et al 1988). It associates with Golgi membranes but can be released in the presence of ATP (Malhotra et al 1989). NSF is related to the SEC18 gene product which in yeast is essential for the transport between endoplasmatic reticulum and Golgi apparatus (Wilson et al 1989). It is believed that NSF is not a fusogenic protein per se, but is an essential component of the 20 S fusion complex (see below) (Söllner et al 1993b).

NSF requires additional cytoplasmic factors to attach to the Golgi membranes (Weidmann et al 1989). Three species termed α, β and γ-SNAP have been purified from brain (Clary & Rothman, 1990), but only α-SNAP can restore transport in the Golgi apparatus (Clary, Griff, & Rothman, 1990). NSF and α-SNAP associate into a high molecular weight (20 S) "fusion complex" with membrane proteins essential for membrane trafficking (Söllner et al 1993b), including synaptobrevin, syntaxin (also called HPC-1) and SNAP-25, which are associated with synaptic vesicles/secretory granules and the plasma membrane, respectively. It has been postulated that α-SNAP displaces synaptotagmin from a complex formed during docking. In this way synaptotagmin, acting as a clamp, is removed (Söllner et al 1993a). Subsequent ATP hydrolysis by the ATPase activity of complexed NSF could initiate membrane fusion (Söllner et al 1993b; Söllner et al 1993a).

From the vesicle proteins mentioned above synaptobrevin, syntaxin and SNAP-25 are regarded as key components of the fusion machinery involved in Ca^{2+} regulated exocytosis. This conclusion has been drawn from studies with Clostridial neurotoxins (see below) which allowed to investigate the function of these proteins during exocytosis of neurotransmitters and hormones.

Clostridial neurotoxins: mechanisms and targets

Tetanus and botulinum neurotoxins are the most potent toxins known. Tetanus toxin (TeTx) binds to the presynaptic membrane at the neuromuscular junction, it invades the motoneuron and moves retroaxonally. After transynaptic transport it reaches the inhibitory interneurons of the spinal cord. Here the toxin blocks the release of inhibitory neurotransmitters, thus causing spastic paralysis (Niemann, 1991). Botulinum toxins block acetylcholine release at motor endplates resulting in flaccid neuromuscular paralysis.

Clostridial neurotoxins are composed of a heavy chain of about 100 kDa and a light chain of 50 kDa, linked together by a single disulphide bond. As generally accepted, the neurotoxin action involves three steps: a) binding to neuronal surface via the heavy chain; b) internalization and intraneuronal targeting of the toxin; and c) blockade of the transmitter release after translocation of the light chain into the synaptic cytosol (Niemann, 1991).

The sequences of tetanus and botulinum neurotoxins have been elucidated (Niemann, 1991). These neurotoxins show a low degree of homology but a conserved segment is located in the central region of the light chain including the HELIH zinc binding motif of metalloendopeptidases (position 233-237) (Jongeneel, Bouvier, & Bairoch, 1989). In metalloendopeptidases zinc and the HELIH motif are essential for catalytic activity. Zinc is also essential for the inhibition of neurotransmitter release by TeTx in Aplysia neuron. Histidine specific reagents and chelators for divalent cations abolished the TeTx effect (Schiavo et al 1992b). Replacement of amino acid residues within the zinc binding domain completely abolished the effect of tetanus toxin light chain (TeTxL) in permeabilized neurohypophysial nerve terminals and in chromaffin cells (Dayanithi et al 1994; Höhne-Zell, Stecher, & Gratzl, 1993). Selective removal of zinc by dipicolinic acid reversibly abolished the TeTxL activity in both systems. Together these data indicate that the conserved HELIH motif and zinc constitute the active site of the tetanus toxin light chain.

Clostridial neurotoxins have been used as excellent tools for the investigation of the mechanism of exocytosis at the molecular level. Application of neurotoxins to intact synaptosomes and neurohypophysial nerve endings prevented exocytosis of synaptic vesicles and secretory granules (McMahon et al 1992; Halpern et al 1990). Earlier investigations have shown, that the neurotoxin light chains block exocytosis from within adrenal chromaffin cells (Penner, Neher, & Dreyer, 1986; Bittner & Holz, 1988; Bittner, Habig, & Holz, 1989; Bittner, DasGupta, & Holz, 1989; Ahnert-Hilger et al 1989a; Ahnert-Hilger et al 1989b; Stecher, Gratzl, & Ahnert-Hilger, 1989; Stecher et al 1989: Ahnert-Hilger et al 1993). Also neurotransmitter release (Mochida et al 1989; Mochida et al 1990; Kurazono et al 1992; de Paiva & Dolly, 1990; Stecher et al 1992) and vasopressin release from nerve terminals (Dayanithi et al 1990; Dayanithi et al 1992) is inhibited by TeTxL. The data indicate that neurotoxins act intracellularly after separation of their constituent heavy and light chains, which is a prerequisite for the inhibition of exocytosis.

Synaptobrevin (also termed VAMP: vesicle associated membrane protein) is a highly conserved 18 kDa protein anchored to the membrane of synaptic vesicles (Baumert et al 1989; Trimble, Cowan, & Scheller, 1988). Synaptobrevins belong to a family of synaptic vesicle membrane proteins. Cellubrevin is a more widespread member of the synaptobrevin family (McMahon et al 1993). By cloning techniques the isoforms 1 and 2 have been identified, synaptobrevin 1 is found predominantly in motoneurons whereas synaptobrevin 2 is associated with cells of autonomic and neuroendocrine function (Elferink, Trimble, & Scheller, 1989; Trimble et al 1990). Recent studies demonstrated that tetanus and some botulinum toxins block the release of neurotransmitter, as a consequence of the cleavage of the cytoplasmic domain of synaptobrevin 2 (Schiavo et al 1992b; Schiavo et al 1992a; Link et al 1992). In neurohypophysial terminals (Dayanithi et al 1994) synthetic peptides spanning the site of synaptobrevin cleaved by TeTxL in synaptic vesicles inhibits Ca^{2+} dependent exocytosis. Also captopril, a zinc peptidase inhibitor, blocks the vasopressin release (Dayanithi et al 1994). Highly purified adrenal chromaffin vesicle membranes contain synaptobrevin which is cleaved by TeTxL (Höhne-Zell et al. unpublished). This proteolysis was inhibited by chelating agents and synaptobrevin derived peptides (Höhne-Zell et al. unpublished). These recent data indicate that TeTxL acts as a zinc dependent protease that cleaves synaptobrevin of secretory granules and synaptic vesicles, identifying synaptobrevin as an essential component of the exocytosis machinery in chromaffin cells and neurons.

Also syntaxin is a target of clostridial neurotoxins which is proteolyzed by botulinum C1 toxin (Blasi et al 1993b; Schiavo et al 1993c). Syntaxins belong to a family of related proteins that are anchored to the cell membrane (Bennett et al 1993). Syntaxins interact with synaptobrevin and synaptotagmin (Bennett, Calakos, & Scheller, 1992; Yoshida et al 1992; Inoue, Obata, & Akagawa, 1992) and bind to a complex formed by the cytosolic ATP-binding protein NSF, a protein required in fusion reactions between intracellular membranes (see above), and soluble NSF attachment proteins (SNAPs) (see above) (Söllner et al 1993b; Südhof et al 1993). Six different syntaxin-like proteins have been characterized in rat. While syntaxins 1A and 1B are only expressed in brain, other forms are expressed in chromaffin cells (Hodel et al 1994) and in various peripheral, nonneuronal tissues (Bennett et al 1993), implicating a general role in the regulation of membrane traffic and in regulated exocytosis.

SNAP-25, a synaptosomal-associated protein of 25 kDa, and syntaxin are mainly localized in the plasma membrane. Together these proteins with synaptobrevin, syntaxin and cytosolic proteins, form a 20 S complex. It was postulated, that SNAP-25 within this complex serves as a receptor, probably conserved from yeast to neurons (Söllner et al 1993b). Recently it was reported that other proteins like unc-18 (Hata, Slaughter, & Südhof, 1993) and n-sec1 (Pevsner, Hsu, & Scheller, 1994; Garcia et al 1994) bind to syntaxin. Sec1 is a yeast homologue of syntaxin

1A and sec4, a homologue of rab3A (Aalto, Ronne, & Keränen, 1993). Interestingly the syntaxin-binding proteins are also found in chromaffin cells (Hodel et al 1994).

The mechanism of botulinum A toxin action closely resembles that of tetanus toxin. Following binding, endocytic uptake and separation of the constituent chains, the light chain blocks exocytosis (Stecher, Gratzl, & Ahnert-Hilger, 1989; Stecher et al 1989; Bittner, DasGupta, & Holz, 1989). However, the target of botulinum A toxin is different to that of tetanus toxin. The light chain of botulinum A toxin cleaves SNAP-25 (Schiavo et al 1993a; Blasi et al 1993a; Schiavo et al 1993b), which results in about 50 % inhibition of exocytosis (de Paiva & Dolly, 1990; Dayanithi et al 1990; de Paiva et al 1993; Stecher, Gratzl, & Ahnert-Hilger, 1989; Stecher et al 1989; Bittner, DasGupta, & Holz, 1989). This observation suggests that other yet unknown isoforms of this protein not cleavable by botulinum A toxin or different proteins, can fulfill the function of SNAP-25 in neurons or endocrine cells.

The targets of tetanus and botulinum toxins are summarized in Table 1. It is remarkable that only few of the proteins implicated in the regulated aspect of neurotransmitter release have been found to be targets of the clostridial neurotoxins. The observations suggest that these proteins are directly involved in vesicle docking and fusion. However, the exact mechanisms and the functions of these components during exocytosis remain to be elucidated.

Acknowledgement

The studies of the authors of this review have been supported by Deutsche Forschungsgemeinschaft (Gr 681).

Table 1

Neurotoxin	Target	References
BoTx A	SNAP-25	(Blasi et al 1993a; Schiavo et al 1993a; Schiavo et al 1993b; Binz et al 1994)
BoTx B	Synaptobrevin	(Schiavo et al 1992a)
BoTx C1	Syntaxin	(Blasi et al 1993b; Schiavo et al 1993c)
BoTx D	Synaptobrevin	(Schiavo et al 1993a; Yamasaki et al 1994a)
	Cellubrevin	(Yamasaki et al 1994a)
BoTx E	SNAP-25	(Schiavo et al 1993a; Schiavo et al 1993b; Binz et al 1994)
BoTx F	Synaptobrevin	(Schiavo et al 1993c)
	Cellubrevin	(McMahon et al 1993)
BoTxG	Synaptobrevin2	(Yamasaki et al 1994b)
TeTx	Synaptobrevin	(Schiavo et al 1992a; Schiavo et al 1992b; Link et al 1992; Dayanithi et al 1994; Höhne-Zell et al unpublished)
	Cellubrevin	(McMahon et al 1993; Link et al 1993)

References

Aalto MK, Ronne H, & Keränen S (1993) Yeast syntaxins Sso1p and Sso2p belong to a family of related membrane proteins that function in vesicular transport. EMBO J 12: 4095-4104

Ahnert-Hilger G, Bader MF, Bhakdi S, & Gratzl M (1989a) Introduction of macromolecules into bovine adrenal medullary chromaffin cells and rat pheochromocytoma cells (PC12) by permeabilization with streptolysin O: Inhibitory effect of tetanus toxin on catecholamine secretion. J Neurochem 52: 1751-1758

Ahnert-Hilger G, Weller U, Dauzenroth M-E, Habermann E, Gratzl M (1989b) The tetanus toxin light chain inhibits exocytosis. FEBS Lett 242: 245-248

Ahnert-Hilger G, Stecher B, Beyer C, & Gratzl M (1993) Exocytotic membrane fusion as studied in toxin-permeabilized cells. Methods Enzymol 221: 139-149

Ahnert-Hilger G, Wegenhorst U, Stecher B, Spicher K, Rosenthal W, & Gratzl M (1992) Exocytosis from permeabilized bovine adrenal chromaffin cells is differently modulated by guanosine 5'-[τ-thio]triphosphate and guanosine 5'-[βy-imido]triphosphate. Biochem J 284: 321-326

Ahnert-Hilger G, Bräutigam M, & Gratzl M (1987) Ca2+-stimulated catecholamine release from alpha-toxin permeabilized PC12 cells: Biochemical evidence for exocytosis and its modulation by protein kinase C and G-proteins. Biochemistry 26: 7842-7848

Alder J, Lu B, Valtorta F, Greengard P, & Poo M-m (1992a) Calcium-dependent transmitter secretion reconstituted in Xenopus oocytes: Requirement for synaptophysin. Science 257: 657-661

Alder J, Xie Z-P, Valtorta F, Greengard P, & Poo M-m (1992b) Antibodies to synaptophysin interfere with transmitter secretion at neuromuscular synapses. Neuron 9: 759-768

Bader MF, Sontag J M, Thiersé D, & Aunis D (1989) A reassessment of guanine nucleotide effects on catecholamine secretion from permeabilized adrenal chromaffin cells. J Biol Chem 264: 16426-16434

Balch WE (1989) Biochemistry of interorganelle transport. J Biol Chem 264: 16965-16968

Baumert M, Maycox PR, Navone F, De Camilli P, & Jahn R (1989) Synaptobrevin: an integral membrane protein of 18 000 daltons present in small synaptic vesicle of rat brain. EMBO J 8: 379-384

Bennett MK, Garcia-Arrarás JE, Elferink LA, Peterson K, Fleming AM, Hazuka CD, & Scheller RH (1993) The syntaxin family of vesicular transport receptors. Cell 74: 863-873

Bennett MK, Calakos N, & Scheller RH (1992) Syntaxin: A synaptic protein implicated in docking of synaptic vesicles at presynaptic active zones. Science 257: 255-259

Bennett MK & Scheller RH (1993) The molecular machinery for secretion is conserved from yeast to neurons. Proc Natl Acad Sci USA 90: 2559-2563

Bergmann M, Lahr G, Mayerhofer A, & Gratzl M (1991) Expression of synaptophysin during the prenatal development of the rat spinal cord: Correlation with basic differentiation processes of neurons. Neuroscience 42: 569-582

Bergmann M, Schuster T, Grabs D, Marquèze-Pouey B, Betz H, Traurig H, Mayerhofer A, & Gratzl M (1993) Synaptophysin and synaptoporin expression in the developing rat olfactory system. Dev Brain Res 74: 235-244

Binz T, Blasi J, Yamasaki S, Baumeister A, Link E, Südhof TC, Jahn R, & Niemann H (1994) Proteolysis of SNAP-25 by type-E and type-A botulinal neurotoxins. J Biol Chem 269: 1617-1620

Bittner MA, DasGupta BR, & Holz RW (1989) Isolated light chains of botulinum neurotoxins inhibit exocytosis. J Biol Chem 264: 10354-10360

Bittner MA, Habig WH, & Holz RW (1989) Isolated light chain of tetanus toxin inhibits exocytosis: Studies in digitonin-permeabilized cells. J Neurochem 53: 966-968

Bittner MA & Holz RW (1988) Effects of tetanus toxin on catecholamine release from intact and digitonin-permeabilized chromaffin cells. J Neurochem 51: 451-456

Blasi J, Chapman ER, Link E, Binz T, Yamasaki S, De Camilli P, Südhof TC, Niemann H, & Jahn R (1993a) Botulinum neurotoxin A selectively cleaves the synaptic protein SNAP-25. Nature 365: 160-163

Blasi J, Chapman ER, Yamasaki S, Binz T, Niemann H, & Jahn R (1993b) Botulinum neurotoxin C1 blocks neurotransmitter release by means of cleaving HPC-1/syntaxin. EMBO J 12: 4821-4828

Block MR, Glick BS, Wilcox CA, Wieland FT, & Rothman JE (1988) Purification of an N-ethylmaleimide-sensitive protein catalyzing vesicular transport. Proc Natl Acad Sci USA 85: 7852-7856

Bommert K, Charlton MP, DeBello W, Chin GJ, Betz H, & Augustine GJ (1993) Inhibition of neurotransmitter release by C2 domain peptides implicates synaptotagmin in exocytosis. Nature 363: 163-165

Brose N, Petrenko AG, Südhof TC, & Jahn R (1992) Synaptotagmin: A calcium sensor on the synaptic vesicle surface. Science 256: 1021-1025

Burgoyne RD, Morgan A, & Roth D (1994) Characterization of proteins that regulate calcium-dependent exocytosis in adrenal chromaffin cells. Ann N Y Acad Sci 710: 333-346

Chapman ER & Jahn R (1994) Calcium-dependent interaction of the cytoplasmic region of synaptotagmin with membranes. Autonomous function of a single C_2-homologous domain. J Biol Chem 269: 5735-5741

Clary DO, Griff IC, & Rothman JE (1990) SNAPS, a family of NSF attachment proteins involved in intracellular membrane fusion in animals and yeast. Cell 61: 709-721

Clary DO & Rothman JE (1990) Purification of three related peripheral membrane proteins needed for vesicular transport. J Biol Chem 265: 10109-10117

Darchen F, Zahraoui A, Hammel F, Monteils M-P, Tavitian A, & Scherman D (1990) Association of the GTP-binding protein Rab3A with bovine adrenal chromaffin granules. Proc Natl Acad Sci USA 87: 5692-5696

Davidson JS, Eales A, Roeske RW, & Millar RP (1993) Inhibition of pituitary hormone exocytosis by a synthetic peptide related to the rab effector domain. FEBS Lett 326: 219-221

Dayanithi G, Ahnert-Hilger G, Weller U, Nordmann JJ, & Gratzl M (1990) Release of vasopressin from isolated permeabilized neurosecretory nerve terminals is blocked by the light chain of botulinum A toxin. Neuroscience 39: 711-715

Dayanithi G, Weller U, Ahnert-Hilger G, Link H, Nordmann JJ, & Gratzl M (1992) The light chain of tetanus toxin inhibits calcium dependent vasopressin release from permeabilized nerve endings. Neuroscience 46: 489-493

Dayanithi G, Stecher B, Höhne-Zell B, Yamasaki S, Binz T, Weller U, Niemann H, & Gratzl M (1994) Exploring the functional domain and the target of the tetanus toxin light chain in neurohypophysial terminals. Neuroscience 58: 423-431

de Paiva A, Ashton AC, Foran P, Schiavo G, Montecucco C, & Dolly JO (1993) Botulinum A like type B and tetanus toxins fulfills criteria for being a zinc-dependent protease. J Neurochem 61: 2338-2341

de Paiva A & Dolly JO (1990) Light chain of botulinum neurotoxin is active in mammalian motor nerve terminals when delivered via liposomes. FEBS Lett 277: 171-174

Elferink LA, Peterson MR, & Scheller RH (1993) A role for synaptotagmin (p65) in regulated exocytosis. Cell 72: 153-159

Elferink LA, Trimble WS, & Scheller RH (1989) Two vesicle-associated membrane protein genes are differentially expressed in the rat central nervous system. J Biol Chem 264: 11061-11064

Fischer von Mollard G, Südhof TC, & Jahn R (1991) A small GTP-binding protein dissociates from synaptic vesicles during exocytosis. Nature 349: 79-81

Garcia EP, Gatti E, Butler M, Burton J, & De Camilli P (1994) A rat brain Sec1 homologue related to Rop and UNC18 interacts with syntaxin. Proc Natl Acad Sci USA 91: 2003-2007

Geppert M, Bolshakov VY, Siegelbaum SA, Takei K, De Camilli P, Hammer RE, & Südhof TC (1994) The role of Rab3A in neurotransmitter release. Nature 369: 493-497

Grabs D, Bergmann M, Schuster T, Fox PA, Brich M, & Gratzl M (1994) . Eur J Neurosci (in press)

Halpern JL, Habig WH, Trenchard H, & Russell JT (1990) Effect of tetanus toxin on oxytocin and vasopressin release from nerve endings of the neurohypophysis. J Neurochem 55: 2072-2078

Hata Y, Davletov B, Petrenko AG, Jahn R, & Südhof TC (1993) Interaction of synaptotagmin with the cytoplasmic domains of neurexins. Neuron 10: 307-315

Hata Y, Slaughter CA, & Südhof TC (1993) Synaptic vesicle fusion complex contains unc-18 homologue bound to syntaxin. Nature 366: 347-351

Hess SD, Doroshenko PA, & Augustine GJ (1993) A functional role for GTP-binding proteins in synaptic vesicle cycling. Science 259: 1169-1172

Hodel A, Schäfer T, Gerosa D, & Burger MM (1994) In chromaffin cells, the mammalian Sec1p homologue is a syntaxin 1A-binding protein associated with chromaffin granules. J Biol Chem 269: 8623-8626

Holz RW, Brondyk WH, Senter RA, Kuizon L, & Macara IG (1994) Evidence for the involvement of Rab3A in Ca^{2+}-dependent exocytosis from adrenal chromaffin cells. J Biol Chem 269: 10229-10234

Höhne-Zell B, Stecher B, & Gratzl M (1993) Functional characterization of the catalytic site of the tetanus toxin light chain using permeabilized adrenal chromaffin cells. FEBS Lett 336: 175-180

Inoue A, Obata K, & Akagawa K (1992) Cloning and sequence analysis of cDNA for a neuronal cell membrane antigen, HPC-1. J Biol Chem 267: 10613-10619

Jahn R & Südhof TC (1993) Synaptic vesicles and exocytosis. Annu Rev Neurosci 17: 216-246

Johannes L, Lledo P-M, Roa M, Vincent J-D, Henry J-P, & Darchen F (1994) The GTPase Rab3a negatively controls calcium-dependent exocytosis in neuroendocrine cells. EMBO J 13: 2029-2037

Jongeneel CV, Bouvier J, & Bairoch A (1989) A unique signature identifies a family of zinc-dependent metallopeptidases. FEBS Lett 242: 211-214

Kelly RB (1985) Pathways of protein secretion in eukaryotes. Science 230: 25-32

Kurazono H, Mochida S, Binz T, Eisel U, Quanz M, Grebenstein O, Wernars K, Poulain B, Tauc L, & Niemann H (1992) Minimal essential domains specifying toxicity of the light chains of tetanus toxin and botulinum neurotoxin type A. J Biol Chem 267: 14721-14729

Leveque C, El Far O, Martin-Moutot N, Sato K, Kato R, Takahashi M, & Seagar MJ (1994) Purification of the N-type calcium channel associated with syntaxin and synaptotagmin. A complex implicated in synaptic vesicle exocytosis. J Biol Chem 269: 6306-6312

Link E, Edelmann L, Chou JH, Binz T, Yamasaki S, Eisel U, Baumert M, Südhof TC, Niemann H, & Jahn R (1992) Tetanus toxin action: Inhibition of neurotransmitter release linked to synaptobrevin proteolysis. Biochem Biophys Res Commun 189: 1017-1023

Link E, McMahon H, Fischer von Mollard G, Yamasaki S, Niemann H, Südhof TC, & Jahn R (1993) Cleavage of cellubrevin by tetanus toxin does not affect fusion of early endosomes. J Biol Chem 268: 18423-18426

Lledo P-M, Vernier P, Vincent J-D, Mason WT, & Zorec R (1993) Inhibition of Rab3B expression attenuates Ca2+ -dependent exocytosis in rat anterior pituitary cells. Nature 364: 540-544

Malhotra V, Serafini T, Orci L, Shepherd JC, & Rothman JE (1989) Purification of a novel class of coated vesicles mediating biosynthetic protein transport through the golgi stack. Cell 58: 329-336

Matthew WD, Tsavaler L, & Reichardt LF (1981) Identification of a synaptic vesicle-specific membrane protein with a wide distribution in neuronal and neurosecretory tissue. J Cell Biol 91: 257-269

McMahon H, Ushkaryov YA, Edelmann L, Link E, Binz T, Niemann H, Jahn R, & Südhof TC (1993) Cellubrevin: A ubiquitous tetanus-toxin substrate homologous to a putative synaptic vesicle fusion protein. Nature 364: 346-349

McMahon HT, Foran P, Dolly JO, Verhage M, Wiegant VM, & Nicholls DG (1992) Tetanus toxin and botulinum toxins type A and B inhibit glutamate, y-aminobutyric acid, aspartate, and met-enkephalin release from synaptosomes. J Biol Chem 267: 21338-21343

Mochida S, Poulain B, Weller U, Habermann E, & Tauc L (1989) Light chain of tetanus toxin intracellularly inhibits acetylcholine release at neuro-neuronal synapses, and its internalization is mediated by heavy chain. FEBS Lett 253: 47-51

Mochida S, Poulain B, Eisel U, Binz T, Kurazono H, Niemann H, & Tauc L (1990) Exogenous mRNA encoding tetanus or botulinum neurotoxins expressed in Aplysia neurons. Proc Natl Acad Sci USA 87: 7844-7848

Mundigl O, Matteoli M, Daniell L, Thomas-Reetz A, Metcalf A, Jahn R, & De Camilli P (1993) Synaptic vesicle proteins and early endosomes in cultured hippocampal neurons: Differential effects of Brefeldin A in axon and dendrites. J Cell Biol 122: 1207-1221

Niemann H (1991) in Sourcebook of bacterial protein toxins, Molecular biology of clostridial neurotoxins (Alouf, J.E.& Freer, J.H.,eds.),pp. 299-344, Academic Press, N.Y.

Ovtscharoff W, Bergmann M, Marquèze-Pouey B, Knaus P, Betz H, Grabs D, Reisert I, & Gratzl M (1993) Ontogeny of synaptophysin and synaptoporin expression in the central nervous system: Differential expression in striatal neurons and their afferents during development. Dev Brain Res 72: 219-225

Palade GE (1975) Intracellular aspects of the process of protein synthesis. Science 189: 347-358

Parton RG, Simons K, & Dotti CG (1992) Axonal and dendritic endocytic pathways in cultured neurons. J Cell Biol 119: 123-137

Penner R, Neher E, & Dreyer F (1986) Intracellularly injected tetanus toxin inhibits exocytosis in bovine adrenal chromaffin cells. Nature 324: 76-78

Perin MS, Fried VA, Mignery GA, Jahn R, & Südhof TC (1990) Phospholipid binding by a synaptic vesicle protein homologous to the regulatory region of protein kinase C. Nature 345: 260-263

Perin MS, Brose N, Jahn R, & Südhof TC (1991) Domain structure of synaptotagmin (p65). J Biol Chem 266: 623-629

Petrenko AG, Perin MS, Davletov BA, Ushkaryov YA, Geppert M, & Südhof TC (1991) Binding of synaptotagmin to the α-latrotoxin receptor implicates both in vesicle exocytosis. Nature 353: 65-68

Pevsner J, Hsu S-C, & Scheller RH (1994) n-Sec1: A neural-specific syntaxin-binding protein. Proc Natl Acad Sci USA 91: 1445-1449

Schiavo G, Benfenati F, Poulain B, Rossetto O, de Laureto PP, DasGupta BR, & Montecucco C (1992a) Tetanus and botulinum-B neurotoxins block neurotransmitter release by proteolytic cleavage of synaptobrevin. Nature 359: 832-835

Schiavo G, Poulain B, Rossetto O, Benfenati F, Tauc L, & Montecucco C (1992b) Tetanus toxin is a zinc protein and its inhibition of neurotransmitter release and protease activity depends on zinc. EMBO J 11: 3577-3583

Schiavo G, Rossetto O, Catsicas S, Polverino de Laureto P, DasGupta BR, Benfenati F, & Montecucco C (1993a) Identification of the nerve terminal targets of botulinum neurotoxin serotypes A,D, and E. J Biol Chem 268: 23784-23787

Schiavo G, Santucci A, DasGupta BR, Mehta PP, Jontes J, Benfenati F, Wilson MC, & Montecucco C (1993b) Botulinum neurotoxins serotypes A and E cleave SNAP-25 at distinct COOH-terminal peptide bonds. FEBS Lett 335: 99-103

Schiavo G, Shone CC, Rossetto O, Alexander FCG, & Montecucco C (1993c) Botulinum neurotoxin serotype F is a zinc endopeptidase specific for VAMP/synaptobrevin. J Biol Chem 268: 11516-11519

Senyshyn J, Balch WE, & Holz RW (1992) Synthetic peptides of the effector-binding domain of rab enhance secretion from digitonin-permeabilized chromaffin cells. FEBS 1: 41-46

Shirataki H, Kaibuchi K, Sakoda T, Kishida S, Yamaguchi T, Wada K, Miyazaki M, & Takai Y (1993) Rabphilin-3A, a putative target protein for smg p25A/rab3A p25 small GTP-binding protein related to synaptotagmin. Mol Cell Biol 13: 2061-2068

Sontag J-M, Aunis D, & Bader M-F (1992) Two GTP-binding proteins control calcium-dependent exocytosis in chromaffin cells. Eur J Neurosci 4: 98-101

Söllner T, Bennett MK, Whiteheart SW, Scheller RH, & Rothman JE (1993a) A protein assembly-disassembly pathway in vitro that may correspond to sequential steps of synaptic vesicle docking, activation, and fusion. Cell 75: 409-418

Söllner T, Whiteheart SW, Brunner M, Erdjument-Bromage H, Geromanos S, Tempst P, & Rothman JE (1993b) SNAP receptors implicated in vesicle targeting and fusion. Nature 362: 318-324

Stecher B, Weller U, Habermann E, Gratzl M, & Ahnert-Hilger G (1989) The light chain but not the heavy chain of botulinum A toxin inhibits exocytosis from permeabilized adrenal chromaffin cells. FEBS Lett 255: 391-394

Stecher B, Hens JJH, Weller U, Gratzl M, Gispen WH, & De Graan PNE (1992) Noradrenaline release from permeabilized synaptosomes is inhibited by the light chain of tetanus toxin. FEBS Lett 312: 192-194

Stecher B, Gratzl M, & Ahnert-Hilger G (1989) Reductive chain separation of botulinum A toxin - a prerequisite to its inhibitory action on exocytosis in chromaffin cells. FEBS Lett 248: 23-27

Südhof TC, De Camilli P, Niemann H, & Jahn R (1993) Membrane fusion machinery: Insights from synaptic proteins. Cell 75: 1-4

Toutant M, Aunis D, Bockaert J, Homburger V, & Rouot B (1987) Presence of three pertussis toxin substrates and Goα immunoreactivity in both plasma and granule membranes of chromaffin cells. FEBS Lett 215: 339-344

Trifaró J-M, Fournier S, & Novas ML (1989) The p65 protein is a calmodulin-binding protein present in several types of secretory vesicles. Neuroscience 29: 1-8

Trimble WS, Gray TS, Elferink LA, Wilson MC, & Scheller RH (1990) Distinct patterns of expression of two VAMP genes within the rat brain. J Neurosci 10: 1380-1387

Trimble WS, Cowan DM, & Scheller RH (1988) VAMP-1: A synaptic vesicle-associated integral membrane protein. Proc Natl Acad Sci USA 85: 4538-4542

Walch-Solimena C, Takei K, Marek KL, Midyett K, Südhof TC, De Camilli P, & Jahn R (1993) Synaptotagmin: A membrane constituent of neuropeptide-containing large dense-core vesicles. J Neurosci 13: 3895-3903

Weidmann PJ, Melancon P, Block MR, & Rothman JE (1989) Binding of an n-ethylmaleimide-sensitive fusion protein to golgi membranes requires both a soluble protein(s) and an integral membrane receptor. J Cell Biol 108: 1589-1596

Wendland B, Miller KG, Schilling J, & Scheller RH (1991) Differential expression of the p65 gene family. Neuron 6: 993-1007

Wilson DW, Wilcox CA, Flynn GC, Chen E, Kuang W-J, Henzel WJ, Block MR, Ullrich A, & Rothman JE (1989) A fusion protein required for vesicle-mediated transport in both mammmalian cells and yeast. Nature 339: 355-359

Yamasaki S, Baumeister A, Binz T, Blasi J, Link E, Cornille F, Roques B, Fykse EM, Südhof TC, Jahn R, & Niemann H (1994a) Cleavage of members of the synaptobrevin/VAMP family by types D and F botulinal neurotoxins and tetanus toxin. J Biol Chem 269: 12764-12772

Yamasaki S, Binz T, Hayashi T, Szabo E, Yamasaki N, Eklund MW, Jahn R, & Niemann H (1994b) Botulinum neurotoxin type G proteolyses the Ala^{81}-Ala^{82} bond of rat synaptobrevin 2. Biochem Biophys Res Commun 200: 829-835

Yoshida A, Oho C, Omori A, Kuwahara R, Ito T, & Takahashi M (1992) HPC-1 is associated with synaptotagmin and omega-conotoxin receptor. J Biol Chem 267: 24925-24928

Control of Intracellular Free Calcium in Neurons and Endocrine Cells

Karl Josef Föhr, Artur Mayerhofer[1] and Manfred Gratzl[2]
Natural and Medical Science Institute
at the University of Tübingen in Reutlingen
72762 Reutlingen
Germany.

Abstract

Intracellular free calcium concentrations in both neurons and endocrine cells are regulated by transporters and ion channels present either in the cell membrane or in the membrane of intracellular organelles. Low intracellular free Ca^{2+} levels in resting cells are maintained by active Ca^{2+} extrusion from the cytoplasm into the extracellular space via the Na^{+}/Ca^{2+}-exchanger and the plasma membrane Ca^{2+}-ATPase (PMCA), as well as Ca^{2+} uptake from the cytoplasm into various intracellular compartments. Sequestration by intracellular compartments is accomplished by uni- and antiporter systems present in mitochondria or secretory granules and a family of endoplasmic and sarcoplasmic reticulum Ca^{2+}-ATPases termed SERCA. Upon cell stimulation extracellular Ca^{2+} may enter the cell via voltage operated (VOCS), receptor operated (ROCS) or second messenger operated (SMOCS) calcium channels. An increase of cytosolic free Ca^{2+} may also be caused by Ca^{2+} release from intracellular compartments by diffusible second messengers activating IP_3 or ryanodine receptor calcium channels.

---

[1] Division of Neuroscience, Oregon Regional Primate Research Center, Beaverton, USA; [2] Department of Anatomy and Cell Biology, University of Ulm, 89069 Ulm, Germany

NATO ASI Series, Vol. H 91
Trafficking of Intracellular Membranes
Edited by M.C. Pedroso de Lima N. Düzgüneş and D. Hoekstra

Introduction

In the present review we attempt to summarize basic mechanisms that control cellular Ca^{2+} which plays a pivotal role for a variety of physiological and pathological processes of all cells including endocrine and neuronal cells. Calcium ions are involved in a plethora of diverse functions such as cell metabolism, cell division, gene expression, cytoskeleton rearrangements, secretion/transmitter release, excitation, synaptic plasticity, control of ion channels, cell survival and programmed cell death. Calcium ions are neither generated nor metabolized, as are other second messenger. They are however compartimentalized, which requires ion gradients generated and maintained by closely related ATPases, the Na^+/K^+-ATPase and the plasmalemma type Ca^{2+}-ATPase. In addition the electrogenic Na^+/Ca^{2+}-exchange contributes to assure the huge calcium gradient of about 1:10000 across the plasma membrane and the intracellular free calcium of approximately 0,1 μmolar. Calcium gradients also exist between the cytosol and intracellular organelles which are established by similar mechanisms. During cell stimulation extra- and/or intracellular calcium sources may be used separately or in combination.

Control at the plasma membrane

Removal of Ca^{2+} out of a cell is an energy consuming process, since Ca^{2+} ions have to be transported against a large electrochemical gradient. This is achieved in most cells via the Ca^{2+}-ATPase of the plasma membrane and the Na^+/Ca^{2+}-exchanger. The energy for the latter is supplied by the Na^+ gradient generated by the plasma membrane Na^+/K^+-ATPase. The absence of a Na^+/Ca^{2+}- exchange system in the erythrocyte membrane and of intracellular compartments greatly facilitated the study of the plasma membrane Ca^{2+}-ATPase. These enzymes, also namend PMCAs, are obligatory components of all eukaryotic plasma membranes (for references see Carafoli, 1992)). The PMCAs exist in four isoforms PMCA 1-4 with molecular weights of about 130 kD encoded by a multigene familiy sharing up to 85-90% similarity in amino acid sequences. Among the isoforms splicing variants are described (Strehler, 1991). They belong to the P-type ATPases, which become temporarily phosphorylated at an aspartate residue during the transport cycle. One Ca^{2+} ion is transported for one ATP hydrolyzed (Schatzmann, 1973). The Ca^{2+} pump is activated by calmodulin (Kd about 1 nM) which increases the Ca^{2+} affinity from about 20 μM in the absence of calmodulin to 0.5 μM in its presence (Niggli, Penniston, & Carafoli, 1979). As a P-type ATPase the Ca^{2+} pump is inhibited by orthovanadate.

In most cell membranes the Na^+/Ca^{2+}-exchanger and the ATP driven Ca^{2+} pump coexist. The Ca^{2+} transport capacity of the Na^+/Ca^{2+}-exchanger is about 10 times higher than that of the Ca^{2+}-ATPase and thus can deal with heavy loads of Ca^{2+} during cell stimulation. The Na^+/Ca^{2+}-exchange, first observed in the frog heart (Lüttgau & Niedergerke, 1958), was studied in a variety of cells (for references see Lagnado & McNaughton, 1990). The protein from cardiac muscle (970 amino acids) has been cloned and expressed in oocytes (Nicoll, Longoni, & Philipson, 1990). Due to a stoichiometry of 1 Ca^{2+}-ion exchanged for 3 Na^+-ions, the transporter operates electrogenically. Thus the activity of the exchanger is determined by the transmembrane Ca^{2+}- and Na^+-gradients, as well as the membrane potential. The Ca^{2+}-affinitiy of the exchanger also depends on ATP and ranges from 0.8 to 8 μM (DiPolo & Beaugé, 1987). Since nonhydrolysable ATP derivatives are not active, ATP may act via phosphorylation of the exchanger. The Na^+/Ca^{2+}-exchange is blocked by mM concentrations of La^{3+} (DiPolo & Beaugé, 1987) and amiloride derivatives (Taglialatela et al 1990).

Table I: Properties of plasmalemmal Ca^{2+}-transporters

	Ca^{2+}-ATPase	Na^+/Ca^{2+}-exchanger
Activation	Calmodulin, PKA, PKC	Ca_i; Na_o; ATP_i
Inhibition	Vanadate (μM), La (μM)	La^{3+} (mM); amiloride
Charge balance	H^+	no, electrogenic
Stoichiometry	ATP/Ca^{2+} 1:1	Na^+/Ca^{2+} 3:1
Ca^{2+}-affinity	0.5 - 10 μM	0.8 - 8 μM
Transport capacity	0.1 (relative to exchange)	1

The observed rapid increase in cytosolic free Ca^{2+} during stimulation of neuronal and endocrine cells may arise from Ca^{2+} influx from the extracellular space which is controlled by the opening of ion channels and the driving force of the electrochemical Ca^{2+} gradient across the plasma membrane. The main types of channels involved can be classified as voltage operated (VOCS), receptor operated (ROCS) or second messenger operated (SMOCS) calcium channels (Bertolino & Llinás, 1992).

The entry of Ca^{2+} through voltage operated calcium channels is well characterized. A channel typically consists of four subunits: an α_1-subunit forming the ion-conducting pore, which also contains the binding site for calcium channel blockers, an intracellularly located ß-subunit, a transmembrane γ-subunit and an α_2-subunit (c.f. Hofmann, Biel, & Flockerzi, 1994). Depending on the extent of membrane depolarization by which they become triggered, high voltage activated (HVA) channels can be distinguished from low voltage activated (LVA) ones. They differ in their conductance and responsiveness to organic and inorganic channel blockers and agonists (Spedding & Paoletti, 1992). L-type Ca^{2+} channel stands for long lasting, T for transient, N for neither L nor T, and P for Purkinje-type calcium channel. L-type channels are blocked by the classical calcium channel blockers such as dihydropyridine. N and P type channels are blocked by ω-conotoxin MVIIC and P type channels by ω-Agatoxin IVA. For T type channels no selective blocker is known.

Voltage dependent Ca^{2+} channels are found in all excitable cells. Their subclasses are believed to be involved in different processes. While T-type channels may participate in the generation of pacemaker activity, L- and N-type channels are thought to be involved in excitation-response coupling. Furthermore the different channel types are modulated in their function selectively by agonists, neurotransmitters and second messengers (c.f. Tsien et al 1988; Anwyl, 1991)).

Table II: Properties of voltage operated calcium channels (VOCS)

Channel type	L	N	P	T
Activation type	HVA	HVA	LVA-HVA	LVA
Inactivation rate	very slow	moderate	fast	fast
Conductance (pS)	25	10-20	10-12	8
Dihydropyridine	+	-	-	-
ω-Conotoxin MVIIC	-	+	±	-
ω-Agatoxin IVA	-	-	+	-

The second major class of Ca^{2+}-channels are receptor operated calcium channels (ROCS) which are activated by the extracellular binding of agonists including neurotransmitters like acetylcholine, glutamate or ATP, thereby opening the receptor integrated ion channel. Most structural and functional information is available for the nicotinic acetylcholine (Ach) receptor of skeletal muscles (Changeux et al 1992; Strehler, 1991). The Ach receptor is composed of five subunits (α_2,ß,γ,δ) with one binding site for acetylcholine on each of the two α-subunits. All subunits contain four membrane spanning domains (M1-4) with the M2 domain lining the pore. Neuronal Ach receptors can be distinguished from the muscle Ach receptor in respect to pharmacological sensitivity and channel properties including higher Ca^{2+} permeability of the neuronal type (Barnard, 1992). The difference are likely to be explained by variants of the α-subunit of the neuronal ACh receptor that have been cloned from brain. Neuronal Ach receptors can be further subdivided into α-bungarotoxin sensitive and -insensitive forms (Sargent, 1993). Nicotinic Ach receptors can be inhibited by d-tubocurarine.

Ionotropic glutamate receptors are subdivided into NMDA (N-methyl-D-aspartate), kainate and AMPA (O-amino-3-hydroxy-5methylisoxazole-4-propionic acid) receptors. NMDA receptors are composed of two different types of subunits NR1, which occur in different splicing forms and NR2A-D. NR1 mRNA is found in almost all neurons, while NR2 mRNA shows distinct spatial and temporal expression profiles in the developing and adult brain. The unique properties of the NMDA receptors which are mainly determined by the NR2 subunit, include requirement for glycine as cofactor (nM - μM), voltage dependent block by Mg^{2+} (μM - mM), large single channel conductance (50 pS), slow kinetics and high Ca^{2+} permeability. It should be mentioned that the Ca^{2+}-permeability of NMDA receptors is low compared to voltage operated Ca^{2+} channels. Due to the extreme Ca^{2+}-selectivity of voltage operated Ca^{2+} channels (permeability coefficient (P): Ca^{2+}/Na^{+}: approx. 1000) the ion influx through these channels is carried almost exclusively by Ca^{2+} ions. For the NMDA receptor permeability coefficients (P: Ca^{2+}/Na^{+}) of about 10 have been reported. Considering the distribution of Ca^{2+} and Na^{+} across the plasma membrane only about 7% of the current flowing through the NMDA receptor complex is carried by Ca^{2+} ions (Schneggenburger et al 1993). The NMDA receptor is modulated/inhibited by H^{+}, polyamine, arachidonic acid, protein kinase C, nitric oxide (NO), dithiothreitol, Zn^{2+} and Ca^{2+} (Ben-Ari, Aniksztejn, & Bregestovski, 1992). D-AP5 (2-amino-5-phosphono-pentanoic acid) is commonly used as an antagonist and (+)-MK-801 (dizocilpine maleate) as an open channel blocker.

AMPA-receptors participate in fast excitatory neurotransmission and are composed of up to four different subunits GluR1-4. Due to various combinations of subunits a large heterogenity in channel properties is possible, with the Ca^{2+} permeability mainly determined by GluR2 (Nakanishi, 1992). In contrast to the other subunits, GluR2 is modified by RNA-editing by which a single codon is exchanged posttrans-criptionally (Higuchi et al 1993). AMPA receptors of hippocampal and retinal bipolar cells, which apparently lack the GluR2 subunit or receptors with an unedited GluR2 subunit, show a significant Ca^{2+} permeability (Burnashev et al 1992). In kainate receptors the subunits GluR5 and GluR6 exist in the edited and unedited form. Again RNA editing reduces the Ca^{2+} permeability of this channel type (Egebjerg & Heinemann, 1993). Ca^{2+} permeability coefficients (P: Ca^{2+}/Na^{+}) of AMPA and kainate receptors range between 0.05 and 1.3. Both receptors are competitively antagonized by CNQX (6-cyano-7-nitroquinoxaline-2,3dione) and DNQX (6,7-dinitroquinoxaline-2,3-dione) (Egebjerg et al 1991).

ATP dependent cation and Ca^{2+}-selective ion channels have been described for smooth muscle cells, secretory cells and neuronal cells. In the latter the purinergic receptors participate in the process of fast synaptic transmission (Evans, Derkach, & Surprenant, 1992). The permeability coefficient Ca^{2+}/Na^{+} ranges from 0.3 to 3. Main inhibitors of this channel type are suramin and reactive blue (Bean, 1992).

Table III: Properties of receptor operated calcium channels (ROCS)

Transmitter	Ach	————————	Glutamate	————————	ATP
Agonist	nicotine	AMPA	Kainate	NMDA	ATP
Antagonist	d-tuboc.	CNQX/DNQX	CNQX/DNQX	D-AP5	suramin
Conductance	40 pS	8/35 pS	~ 1 pS	50 pS	5-17 pS
Kinetics	fast	fast	slow	slow	fast

The term second messenger operated calcium channels (SMOCS) has been introduced to describe channels acting in a voltage independent manner to increase plasma membrane Ca^{2+} permeability (Meldolesi & Pozzan, 1987). SMOCS also differ from ROCS by the receptor binding site localized intracellularly. The existence of SMOCS was experimentally confirmed in lymphoctes by the discovery of inositol 1,4,5-trisphospahte (IP_3) receptors located in the plasma membrane (Khan et al 1992). Also IP_4, which like IP_3 may be generated during cell stimulation as described below, may activate SMOCS (Lückhoff & Clapham, 1992).

Control by intracellular Ca^{2+} stores

The intracellular compartments considered as major Ca^{2+} stores are mitochondria, secretory vesicles and the endoplasmic reticulum (ER). They markedly differ from each other in their Ca^{2+} storage capacity, Ca^{2+} concentration at which their uptake mechanisms become activated and their mechanisms of Ca^{2+} uptake and Ca^{2+} release.

Mitochondria have a high Ca^{2+} capacitiy and are activated at about 1 μM free Ca^{2+} or higher concentrations. The movement of Ca^{2+} across the inner mitochondrial membrane is driven by the electrochemical H^+ gradient built up during cellular respiration. Ca^{2+}-uptake occurs by an electrophoretic uniporter and a Na^+/Ca^{2+}- or a H^+/Ca^{2+}-exchange system (c.f. (Miller, 1991)). Secretory vesicles are another intracellular compartment with a high capacity for Ca^{2+}-storage (Bulenda & Gratzl, 1985; Reiffen & Gratzl, 1986a; Reiffen & Gratzl, 1986b). Calcium uptake into these organelles occurs via Na^+/Ca^{2+}-exchange which may be coupled to the vacuolar proton pump (Saermark, Krieger-Brauer, & Gratzl, 1983; Krieger-Brauer & Gratzl, 1982; Krieger-Brauer & Gratzl, 1983; Haigh & Phillips, 1993). Ca^{2+}-transport into the endoplasmic reticulum is directly driven by Ca^{2+}-ATPases termed SERCA. In contrast to plasma membrane Ca^{2+}-ATPases (PMCA), which are regulated by calmodulin, the Ca^{2+}-affinity of SERCAs in cardiac and smooth muscle is regulated by phospholamban, a protein which increases the Ca^{2+}-affinitiy of the SERCAs in its phosphorylated form. An additional difference between PMCAs and SERCAs is their stoichiometry of ATP hydrolyzed per Ca^{2+} ions transported (1 for PMCA and 0.5 for SERCA). Both, PMCAs and SERCAs are inhibited by vanadate. SERCAs can be specifically inhibited by thapsigargin (Jackson et al 1988) and cyclopiazonic acid which results in the discharge of Ca^{2+} stores. In contrast to mitochondria the Ca^{2+} uptake by the ER occurs already at resting free Ca^{2+} concentrations (around 0,1 μM). Mitochondria and ER also differ in their kinetics of Ca^{2+} uptake with the ER being about 10 times faster than mitochondria.

Ca^{2+} release from the ER is triggered by the intracellular messenger IP_3 which is formed as a consequence of binding of a large variety of different extracellular agents to their receptors at the plasma membrane. This results in the activation of phospholipase C (PLC) at the plasma membrane, which is either mediated by G-proteins or tyrosine kinases. PLC hydrolyses phosphatidylinositol 4,5-bisphosphate to give diacylglycerol and IP_3. The water soluble IP_3 diffuses to the intracellular stores, binds to IP_3-receptors thereby opening the receptor integrated Ca^{2+}-channel (Berridge, 1993). The IP_3 receptor, as well as the related Ca^{2+} activated ryanodine receptor (RYR) of the sarcoplasmic reticulum are homotetrameric Ca^{2+}- channels with a molecular mass of the monomer of 313 and 565 kDa, respectively. For both types of channels at least three different isoforms have been described. The molecular diversity of the IP_3R-familiy is further enlarged due to alternatively spliced forms and the existence of fetal and adult forms (Ferris & Snyder, 1992). Ca^{2+}-release by IP_3 has first been found in exocrine pancreatic cells (Streb et al 1983) and is now characterized in detail in a large variety of different cell types. It turned out that the Ca^{2+} release depends on the amount of stored Ca^{2+} and cytosolic free Ca^{2+} (Föhr et al 1989; Föhr et al 1991). The amount of Ca^{2+} released by IP_3 increases with the charge of the store. The use of permeabilized cell preparations showed that 1 molecule of IP_3 released up to 60 Ca^{2+} ions. This capacity, disregarding cellular Ca^{2+} buffering, could give rises to cytosolic free Ca^{2+} levels up to 500 μM and higher. However, such high cytosolic free Ca^{2+} concentrations are unlikely to occur since IP_3-induced Ca^{2+} release is strongly inhibited by Ca^{2+}. A complete block was observed at about 5-10 μM free Ca^{2+} (Föhr et al 1991; Engling et al 1991). Other important regulators of the IP_3-induced Ca^{2+} release are GTP and ATP. Stimulation of IP_3-induced Ca^{2+} release by GTP may be caused by transfer of Ca^{2+} from IP_3-insensitive to IP_3-sensitive Ca^{2+} stores or GTP dependent phosphorylation. ATP may act in different ways. At low concentrations (μM) it potentiates Ca^{2+} mobilization by binding to a high affinity ATP binding site thereby increasing channel open probability while at high concentrations (mM) it is inhibitory by competing with IP_3. Furthermore ATP is necessary for phosphorylation of the IP_3-receptor which modifies Ca^{2+} mobilization. Phosphorylation by protein kinase A provides a means for cross-talk between the second messengers IP_3 and cAMP and may explain synergistic interactions between adrenergic receptors and receptors coupled to cAMP (Föhr et al 1993). Most potent inhibitors of IP_3-induced Ca^{2+} relase are polyanions like heparin or decavanadate (Taylor & Marshall, 1992). IP_3 receptors have also be found in the membrane of secretory granules (Nathanson et al 1994; Yoo, 1994), which may also release stored calcium during the process of exocytosis. Thus, IP_3 produced upon receptor activation at the plasma membrane could also diffuse to secretory granules near the site of exocytosis and release calcium from the organelles, which in turn directly could trigger exocytosis.

Ca^{2+} activated ryanodine receptors are named according to a membrane permeant plant alcaloid which opens this channel type at low concentrations (nM) and inhibits it at high concentrations (μM). This receptor was extensively studied in muscle cells and more recently was also found in neurons and endocrine cells. Calcium ions entering the cell via voltage operated ion channels bind to the receptor and cause the release of Ca^{2+} from the sarcoplasmic reticulum contacting the T-tubule. Cyclic ADP-ribose (cADPR), synthesized from NAD^{+}, also activates release of Ca^{2+} via the ryanodine receptor (Galione, 1993). In addition the ryanodine receptor can be activated by millimolar concentrations of caffeine, which does not activate the IP_3 receptor.

Acknowledgement

The studies of the authors of this review have been supported by Deutsche Forschungsgemeinschaft and Fonds der Chemischen Industrie.

References

Anwyl R (1991) Modulation of vertebrate neuronal calcium channels by transmitters. Brain Res Rev 16: 265-281

Barnard EA (1992) Receptor classes and the transmitter-gated ion channels. TIBS 17: 368-374

Bean BP (1992) Pharmacology and electrophysiology of ATP-activated ion channels. TiPS 13: 87-90

Ben-Ari Y, Aniksztejn L, & Bregestovski P (1992) Protein kinase C modulation of NMDA currents: An important link for LTP induction. TINS 15: 333-339

Berridge MJ (1993) Inositol trisphosphate and calcium signalling. Nature 361: 315-325

Bertolino M & Llinás R (1992) The central role of voltage-activated and receptor-operated calcium channels in neuronal cells. Annu Rev Neurosci 32: 399-421

Bulenda D & Gratzl M (1985) Matrix free calcium in isolated chromaffin vesicles. Biochemistry 24: 7760-7765

Burnashev N, Monyer H, Seeburg PH, Sakmann B (1992) Divalent ion permeability of AMPA receptor channels is dominated by the edited form of a single subunit. Neuron 8: 189-198

Carafoli E (1992) The Ca^{2+} pump of the plasma membrane. J Biol Chem 267: 2115-2118

Changeux J-P, Galzi J-L, Devillers-Thiéry A, & Betrand D (1992) The functional architecture of the acetylcholine nicotinic receptor explored by affinity labelling and site-directed mutagenesis. Quart Rev Biophys 25: 395-432

DiPolo R & Beaugé L (1987) Characterization of the reverse Na^{+}/Ca^{2+} exchange in squid axons and its modulation by Ca^{2+} and ATP. J Gen Physiol 90: 505-525

Egebjerg J, Bettler B, Hermans-Borgmeyer I, Heinemann S (1991) Cloning of a cDNA for a glutamate receptor subunit activated by kainate but not by AMPA. Nature 351: 745-748

Egebjerg J, Heinemann SF (1993) Ca2+ permeability of unedited and edited versions of the kainate selective glutamate receptor GluR6. Proc.Natl.Acad.Sci.USA 90: 755-759

Engling R, Föhr KJ, Kemmer TP, & Gratzl M (1991) Effect of GTP and calcium on inositol 1,4,5,-trisphosphate induced calcium release from permeabilized rat exocrine pancreatic acinar cells. Cell Calcium 12: 1-9

Evans RJ, Derkach V, & Surprenant A (1992) ATP mediates fast synaptic transmission in mammalian neurons. Nature 357: 503-507

Ferris CD & Snyder SH (1992) Inositol 1,4,5-trisphosphate-activated calcium channels. Annu Rev Physiol 54: 469-488

Föhr KJ, Scott J, Ahnert-Hilger G, & Gratzl M (1989) Characterization of the inositol 1,4,5-trisphosphate-induced calcium release from permeabilized endocrine cells and its inhibition by decavanadate and p-hydroxymercuribenzoate. Biochem J 262: 83-89

Föhr KJ, Ahnert-Hilger G, Stecher B, Scott J, & Gratzl M (1991) GTP and calcium modulate the inositol 1,4,5-trisphosphate dependent calcium release in streptolysin O permeabilized bovine adrenal chromaffin cells. J Neurochem 56: 665-670

Föhr KJ, Mayerhofer A, Sterzik K, Rudolf M, Rosenbusch B, & Gratzl M (1993) Concerted action of human chorionic gonadotrophin and norepinephrine on intracellular-free calcium in human granulosa-lutein cells: evidence for the presence of a functional α-adrenergic receptor. J Clin Endocrinology and Metabolism 76: 367-373

Galione A (1993) Cyclic ADP-ribose: a new way to control calcium. Science 259: 325-326

Haigh J & Phillips JH (1993) Indirect coupling of calcium transport in chromaffin granule ghosts to the proton pump. NeuroReport 4: 571-574

Higuchi M, Single FN, Köhler M, Sommer B, Sprengel R, Seeburg PH (1993) RNA editing of AMPA receptor subunit GlurR-B: a base-paired intron-exon strucutre determines postion and efficiency. Cell 75: 1361-1370

Hofmann F, Biel M, & Flockerzi V (1994) Molecular basis for Ca^2+ channel diversity. Annu Rev Neurosci 17: 399-418

Jackson TR, Patterson SI, Thastrup O, & Hanley MR (1988) A novel tumour promoter, thapsigargin, transiently increases cytoplasmic free Ca^{2+} without generation of inositol phosphates in NG115-401L neuronal cells. Biochem J 253: 81-86

Khan AA, Steiner JP, Klein MG, Schneider MF, & Snyder SH (1992) IP_3 receptor: localization to plasma membrane of T cells and cocapping with the T cell receptor. Science 257: 815-818

Krieger-Brauer HI & Gratzl M (1982) Uptake of calcium by isolated secretory vesicles from adrenal medulla. Biochim Biophys Acta 691: 61-70

Krieger-Brauer HI & Gratzl M (1983) Effects of monovalent and divalent cations on calcium fluxes accross chromaffin secretory membrane vesicles. J Neurochem 41: 1269-1283

Lagnado L & McNaughton PA (1990) Electrogenic properties of the Na:Ca exchange. J Membr Biol 113: 177-191

Lückhoff A & Clapham (1992) Inositol 1,3,4,5-tetrakishposhphate activates an endothelial Ca^{2+}-permeable channel. Nature 355: 356-358

Lüttgau HC & Niedergerke R (1958) The antagonism between Ca^{2+} and Na^+ ions on the frog heart. J Physiol 143: 486-505

Meldolesi J & Pozzan T (1987) Pathways of Ca^{2+} influx at the plasma membrane: Voltage-, receptor-, and second messenger-operated channels. Exp Cell Res 171: 271-283

Miller RJ (1991) The control of neuronal Ca^{2+} homeostasis. Prog Neurobiol 37: 255-285

Nakanishi S (1992) Molecular diversity of glutamate receptors and implications for brain function. Science 258: 597-603

Nathanson MH, Fallon MB, Padfield PJ, & Maranto AR (1994) Localization of the type 3 inositol 1,4,5-triphosphate receptor in the Ca^{2+} wave trigger zone of pancreatic acinar cells. J Biol Chem 269: 4693-4696

Nicoll DA, Longoni S, & Philipson KD (1990) Molecular cloning and functional expression of the cardiac sarcolemmal Na^+-Ca^{2+} exchanger. Science 250: 562-565

Niggli V, Penniston JT, & Carafoli E (1979) Purification of the $(Ca^{2+}+Mg^{2+})$-ATPase from human erythrocyte membranes using a calmodulin affinity column. J Biol Chem 254: 9955-9958
Reiffen FU & Gratzl M (1986a) Chromogranins, widespread in endocrine and nervous tissue, bind calcium. FEBS Lett 195: 327-330
Reiffen FU & Gratzl M (1986b) Calcium binding to chromaffin vesicle matrix proteins: Effect of pH, magnesium, and ionic strength. Biochemistry 25: 4402-4406
Saermark T, Krieger-Brauer HI, & Gratzl M (1983) Calcium uptake to purified secretory vesicles from bovine neurohypophyses. Biochim Biophys Acta 727: 239-245
Sargent PB (1993) The diversity of neuronal nicotinic acetylcholine receptors. Annu Rev Neurosci 16: 403-443
Schatzmann HJ (1973) Dependence on calcium concentration and stoichiometry of the calcium pump in human red cells. J Physiol 235: 551-569
Schneggenburger R, Zhou Z, Konnerth A, & Neher E (1993) Fractional contribution of calcium to the cation current through glutamate receptor channels. Neuron 11: 133-143
Spedding M & Paoletti R (1992) Classification of calcium channels and the sites of action of drugs modifying channel function. Pharmacological Reviews 44: 363-376
Streb H, Irvine RF, Berridge MJ, & Schulz I (1983) Release of calcium from a nonmitochondrial intracellular store in pancreatic acinar cells by inositol-1,4,5-trisphosphate. Nature 306: 67-69
Strehler EE (1991) Recent advances in the molecular characterization of plasma membrane Ca^{2+} pumps. J Membr Biol 120: 1-15
Taglialatela M, Canzoniero LMT, Cragoe EJ, Di Renzo G, Annunziato L (1990) Na^{+}/Ca^{2+} exchange activity in central nerve endings II. Relationship between pharmacological blockade by amiloride analogues and dopamine release from tubero-infundibular hypothalamic neurons. Molec Pharmac 38: 393-400
Taylor CW, & Marshall CB (1992) Calcium and inositol 1,4,5-trisphosphate receptors: a complex relationship. TiBS 17: 403- 407.
Tsien RW, Lipscombe D, Madison DV, Bley KR, & Fox AP (1988) Multiple types of neuronal calcium channels and their selective modulation. TINS 11: 431-438
Yoo SH (1994) pH-dependent interaction of chromogranin A with integral membrane proteins of secretory vesicle including 260-kDa protein reactive to inositol 1,4,5-triphosphate receptor antibody. J Biol Chem 269: 12001-12006

MEMBRANE FUSION IN THE EXOCYTOTIC RELEASE OF NEUROTRANSMITTERS

Catarina R. Oliveira, M. Teresa Almeida[1,2] and Maria C. Pedroso de Lima[1,2]

Center for Neurosciences
University of Coimbra
3049 Coimbra Codex
Portugal

INTRODUCTION

The functioning of the brain depends on the flow of information through elaborated circuits consisting of networks of neurons. Information is transferred from one cell to another at specialized points of contact, the synapses, where one nerve cell comes into appropriate apposition to its target cell.

Information flows from the presynaptic to the postsynaptic cell, according to the principle of dynamic polarization (Ramon e Cajal, 1911). However, it is now clear that essential information is also transmitted retrogradelly, from the target cell to the nerve terminal. Candidate molecules that mediate retrograde information transfer have been identified, such as nitric oxide (Bredt and Snyder, 1992) and polypeptide growth proteins (Ip et al., 1992). This concept of a bidirectional and self-modifiable form of cell-to-cell communication is the basis to understand the plasticity of synapses, not only during development but also in the mature brain.

---

1. Department of Biochemistry, University of Coimbra, 3049 Coimbra Codex, Portugal.
2. Center for Neurosciences of Coimbra, University of Coimbra, 3049 Coimbra Codex, Portugal.

NATO ASI Series, Vol. H 91
Trafficking of Intracellular Membranes
Edited by M.C. Pedroso de Lima N. Düzgüneş and D. Hoekstra

SYNAPTIC TRANSMISSION

Synaptic transmission across the majority of synapses is mediated by the interaction of chemical signals, the neurotransmitter molecules released from the presynaptic terminal, with postsynaptic receptors.

The "vesicle hypothesis" for neurotransmitter release (del Castillo and Katz, 1956), proposes that the transmitter is stored in synaptic vesicles which fuse with the presynaptic membrane, releasing their contents by exocytosis upon depolarization of the presynaptic plasma membrane. The "Ca^{2+} hypothesis" (Katz and Miledi, 1965) states that depolarization of the presynaptic nerve terminal opens Ca^{2+} channels, resulting in an influx of Ca^{2+} and a transient rise in intracellular Ca^{2+} concentration (200-300 μM) in microdomains called the active zones. The active zones are located in the presynaptic terminal, adjacent to the plasma membrane, near the synaptic vesicle release sites (Llinás et al., 1992). Figure 1 illustrates that the influx of Ca^{2+} into the presynaptic terminal induces an increase in the transmitter release, probably by enhancing the fusion of the vesicle with the plasma membrane at the release site, an essential step for exocytosis (Kelly, 1993) However, only a fraction of synaptic vesicles are docked at the release sites, the remaining vesicles constituting the reserve pool. The release of transmitters is shown to involve four steps (Kelly, 1993): the transport of vesicles from the reserve to the releasable pool, which requires Ca^{2+} (Koenig et al., 1993); the docking of vesicles to the release sites at the actives zones; the fusion of the vesicle membrane with the plasma membrane in response to an increase in intracellular Ca^{2+} and the recycling of vesicle membrane following exocytosis.

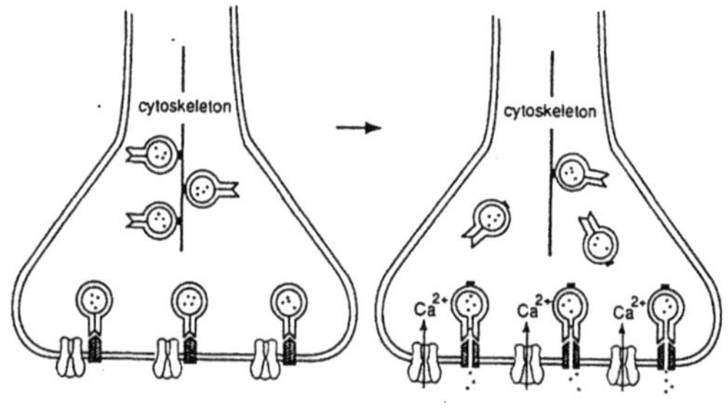

Figure 1. Ca^{2+} control of synaptic vesicle fusion and pore opening. Synaptic vesicles are tethered to the cytoskeleton (storage pool) and docked to the fusion pore complex (releasable pool). Adapted from Jessel and Kandel (1993).

The cytoskeleton of the neuron plays an interesting role in the availability of releasable synaptic vesicles . Under resting conditions, a cluster of synaptic vesicles close to the presynaptic membrane is available for secretion. At a distance, another set of synaptic vesicles are linked to each other, to actin and other anchoring proteins, such as synapsin I and fodrin, constituting the reserve pool. Nerve depolarization causes the opening of L and N-type voltage-gated Ca^{2+}-channels, an increase in intracellular Ca^{2+} concentration and the exocytotic release of neurotransmitters. The activation of protein kinases induces the phosphorilation of synapsin I and the activation of proteins that sever actin filaments, resulting in a partial disassembly of the actin filament network (Trifaró and Vitale, 1993). Synaptic vesicles are now free to move towards the active zone.

In conclusion, an increase in Ca^{2+} and phosphorilation are two important mechanisms in determining docking availability of synaptic vesicles to be released, that is, they determine the efficiency of synaptic transmission.

PLASMA MEMBRANE PROTEINS IN EXOCYTOSIS

Exocytosis can be defined as the process by which the content of specific membrane-bound secretory organelles (granules or vesicles) is released in bulk to the extracellular medium. It requires the fusion fission of the limiting membrane of the organelle with the plasma membrane (Palade, 1975; Meldolesi and Ceccarelli, 1988).

Exocytosis in synapses is a very fast process with a delay time of about 200 μs between Ca^{2+} entry and the postsynaptic response, suggesting that the molecular components that underlie this specialized stimulation/secretion coupling are highly organized (Almers and Tse, 1990) and located close to the voltage-dependent Ca^{2+} channels in the presynaptic membrane (Smith and Augustinc, 1988, Robitaille et al., 1990). Capacitance patch clamp recordings have shown the transient and reversible formation of a pore at the onset of exocytosis. This pore has the conductance of a large ion channel (Monck and Fernandez, 1992).

The role of plasma membrane proteins in exocytosis is far from being elucidated. However, interactions between presynaptic and synaptic vesicle membrane proteins have been identified which might be important in organizing the fast neurotransmitter release mechanism. The first evidence for protein/protein interactions was given for the interaction between synaptophysin, a synaptic vesicle protein, and physophilin, a presynaptic plasma membrane protein (Thomas and Betz, 1990). More recently, a receptor for α-latrotoxin was identified at the plasma membrane, which interacts with synaptotagmin, another synaptic vesicle protein (Petrenko et al., 1991). This latter protein has been shown to exist as a multimeric complex which includes syntaxin, a presynaptic membrane protein (Bennet et al., 1992). Syntaxins and SNAP-25 (a synaptosomal associated protein) are localized in the plasma membrane, and together with synaptobrevin and cytosolic proteins form a 20 S complex (Sollner et al., 1993b). This complex has been shown to be involved in vesicle targeting and docking on the presynaptic membrane.

In order to gain insight into the involvement of presynaptic plasma membrane proteins in neurotransmitter release, we have studied the fusion of synaptosomes with phosphatidylserine liposomes. Fusion was monitored using a fluorescence resonance energy transfer assay (Struck et al., 1981). A significant inhibition of the interaction between synaptosomes and liposomes by trypsin pretreatment of synaptosomes was found (Fig. 2). The inhibitory effect of trypsin at mild acidic conditions was not so strong as that at physiological pH, suggesting that, in addition to the involvement of proteins, nonspecific interactions between membranes may also play a role in the fusion process. Proteins and electrostatic forces were also shown to be involved in membrane aggregation (Almeida et al., 1994).

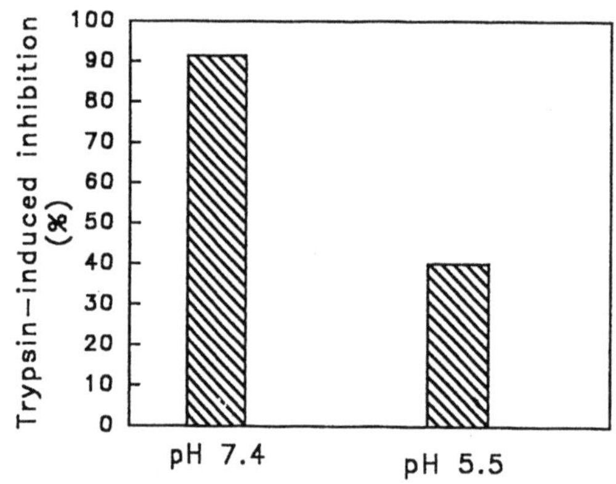

Figure 2. Trypsin-induced inhibition of the fusion reaction. Percentages of inhibition were calculated from the extents of fusion, between synaptosomes and liposomes, after 14 min incubation. Fluorescence was measured at 37° C in a final volume of 2ml. Final liposome and synaptosome concentrations were 16.5 μM and 275 μg/ml, respectively.

It can be concluded that presynaptic plasma membrane proteins play an important role in both stages of the fusion process. The synaptic plasma membrane proteins, syntaxins and neurexins, are potential acceptor sites for synaptic vesicles. However, the interactions between plasma membrane proteins and synaptic vesicle proteins and the way how they control and /or modulate neurotransmitter release need further clarification.

VESICLE PROTEINS IN EXOCYTOSIS

Membrane proteins of exocytotic vesicles are believed to play a key role in the molecular events underlying secretory fusion (Südhof and Jahn, 1991; Jahn and Südhof, 1993).

As previously referred to in the text, several lines of evidence suggest that only a small pool of synaptic vesicles participates in exocytosis and recycling, whereas a large pool filled with transmitter is kept in reserve within the nerve terminal. Electron microscopic studies have shown that synaptic vesicles are suspended in a filamentous network composed mainly of spectrin and actin (Landis et al., 1988; Hirokawa et al., 1989).

Synapsin, a membrane vesicle protein, has been demonstrated to be involved in the release of synaptic vesicles from this reticular network. Synapsins are a family of four homologous proteins (Synapsin I_a, I_b, II_a and II_b) (Südhof et al., 1989). Synapsin I was shown to be bound to vesicle associated Ca^{2+}/calmodulin-dependent kinase II (Benfenati et al., 1992) and was also demonstrated to bind to various cytoskeletal proteins, including spectrin and actin (Valtorta et al., 1992; Sikorski et al., 1991). Phosphorilation of synapsin I by Ca^{2+}/calmodulin-dependent kinase II weaks the binding of synapsin I to synaptic vesicles and actin filaments, resulting in the release of the vesicles from the cytoskeletal network and their transfer to an active releasable pool in response to an increase in intracellular Ca^{2+} concentration (Sikorski et al., 1991).

Docking and fusion of synaptic vesicles with the plasma membrane involve recognition and establishment of tight binding to the membrane. It is not clear whether docking proteins are identical to those catalyzing membrane fusion or whether the two processes are performed by different molecules. Recent evidence suggests that docking is mediated by synaptotagmin, a synaptic vesicle protein which is a putative Ca^{2+} sensor in exocytosis (Perin et al., 1990; Brose et al., 1992) that associates with N type calcium channels (Leveque et al., 1992) in a complex with syntaxin (Yoshida et al., 1992; Bennet et al., 1992) and neurexin (O'Connor et al., 1993). This interaction may locate synaptic vesicles in a zone accessible to rapid calcium transients. Synaptotagmin binds Ca^{2+} with an affinity (Kd) of 10^{-6}-10^{-7}M and binding is strictly dependent on the presence of negatively charged phospholipids (Brose et al., 1992). The homology found between protein kinase C and synaptotagmin has led to the assumption that the Ca^{2+}-binding capacity of the protein can be regulated by phosphorilation (Mochly-Rosen et al., 1992). However, the role of synaptotagmin in exocytosis is not yet clear. Recent studies using PC12 cells have shown that Ca^{2+}-dependent exocytosis can be observed in synaptotagmin-deficient cells (Shoji-Kasay et al., 1992) whereas Elferink et al. (1993) reported that exocytosis from PC12 cells that have been treated with nerve growth factors is partially inhibited when cells are injected either with anti-synaptotagmin antibodies or with recombinant fragments of synaptotagmin. Other synaptic vesicle proteins, namely synaptophysin and synaptobrevin (VAMP), have been shown to play a role in exocytosis. Synaptophysin is a transmembrane glycoprotein of synaptic vesicles whose function is unclear. Betz (1990) has proposed that synaptophysin may form the fusion pore in a complex with a docking protein in the plasma membrane (Fig. 3).

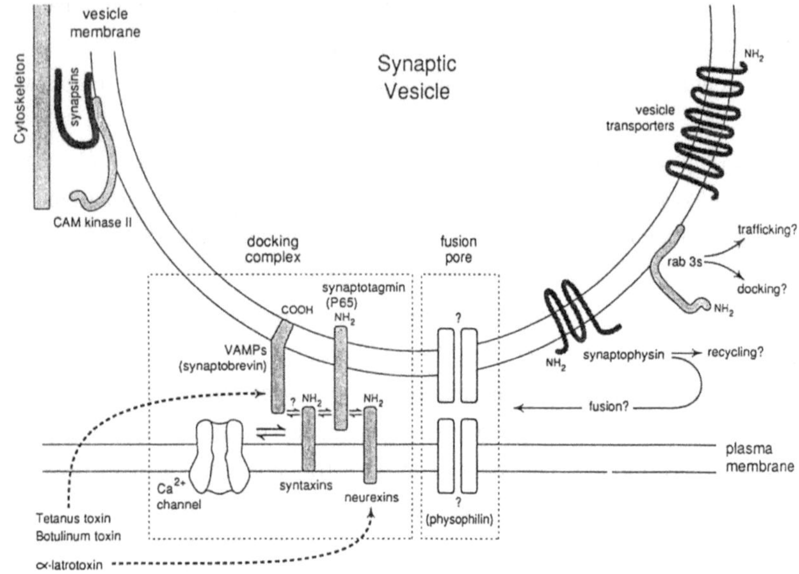

Figure 3. Potential functions of synaptic vesicle proteins and plasma membrane proteins in fusion. Reproduced from Jessell and Kandel (1993), with permission.

Söllner et al. and Bennet and Scheller (1993), drew our attention to synaptobrevin, another synaptic vesicle protein as a candidate for a docking protein. It has been shown that tetanus toxin and botulinum B neurotoxin selectively cleave synaptobrevin (Schiavo et al., 1992). As these toxins are known to block exocytosis, it is suggested that synaptobrevin is part of the exocytotic fusion apparatus. Recently, it was reported that syntaxin plus the synaptic vesicle membrane-associated proteins, synaptobrevin and SNAP 25, and the soluble protein factors, NSF and SNAPS, form the core of a fusion complex operating in synaptic vesicle exocytosis (Söllner et al., 1993).

GTP-binding proteins, besides regulating vesicular transport, are also involved in vesicle docking, and GTP hydrolysis has been shown to occur before Ca^{2+} can trigger exocytosis (Hess et al., 1993). Intracellular docking and fusion of membranes is regulated by members of the rab family of GTPases. The form associated with synaptic vesicles is rab 3A, which remains bound to the synaptic vesicle membrane while it fuses with the presynaptic terminal, but dissociates thereafter (Fisher von Mollard et al., 1991). Recently, rab 3A has been shown to be modulated by Ca^{2+}, and GTP provides a sustained activation of this protein (Robinson et al., 1994).

In conclusion, the assembly of a protein complex is required for fusion to occur. Interactions exist between synaptic vesicle and presynaptic membrane proteins which are determinant for vesicle-plasma membrane association. Exocytosis occurs when the phospholipid bilayer of the docked secretory vesicle becomes continuous with that of the plasma membrane. Peripheral membrane proteins may prime exocytosis by holding the vesicle membrane in close proximity to the presynaptic plasma membrane. The next step is the formation of a fusion pore between the two membranes.

THE FUSION PORE

The fusion pore is the structure that transiently joins the membrane of two fusing compartments, immediatly after membrane fusion and before complete membrane merging.

Three models have been suggested for the structure of the fusion pore:

1. Apposition of the two bilayers and perturbation of the lipid organization lead to fusion and opening of a pore that spans the whole thickness of both bilayers.

2. Fusion of the two external leaflets of the bilayers (hemifusion) generates a region where a single bilayer separates the two compartments; this is followed by the opening of a pore across the single bilayer.

3. Interaction of specific proteins on both membranes creates a proteinaceous pore that spans both bilayers; this is followed by expansion and progressive disassembly of the pore through the inclusion of mobile lipid leading to full fusion.

The idea of protein-mediated fusion has gained popularity following the development of the patch-clamp technique (Monck and Fernandez, 1992). Properties of the initial fusion pore, particularly its flickering behaviour (Breckenridge and Almers, 1987), suggest the involvement of an ion-channel-like structure, possibly composed of synaptic vesicle-specific proteins such as synaptophysin (Thomas et al., 1988).

Recently, biochemical (Söllner et al., 1993) and genetic (Bennet and Scheller, 1993) studies provided strong evidence that synaptic vesicle exocytosis requires the assembly of a supramolecular complex working as a fusion machinery (synaptobevrin, syntaxin and SNAP25 are acceptors of NSF/SNAPs and promote fusion). Therefore, soluble and membrane-associated proteins play an important role in synaptic vesicle fusion. It cannot be excluded that the supramolecular protein complex, rather than mediating fusion *per se*, is involved in recognition and docking steps that precede the actual formation of a fusion pore.

For a lipid pore fusion mechanism, protein-protein interactions might simply be involved in bringing the lipid bilayers close enough and in bending or perturbing the bilayers so as to induce hemifusion and the opening of the lipid pore. Proteins might equally regulate the probability of closure or expansion of the pore (Monck and Fernandez, 1992).

The model of a protein scaffold directing lipid fusion has been recently proposed by Monck and Fernandez (1994). Rab3, synaptotagmin and synaptobrevin, forming complexes with syntaxins, Ca^{2+}-channels and proteins of the presynaptic membranes (Bennet et al., 1992; David et al., 1933; O'Connor et al., 1993; Horikawa et al., 1993), are suggested to play a role in fusion pore scaffold, which is activated by GTP- and Ca^{2+}-binding proteins (Oberhauser et al., 1992; Lledo et al., 1993). It is commonly assumed that following fusion and pore expansion, the contents of secretory granules are released by simple diffusion. However, evidence suggests that a proteoglycan gel, responding to pH, ionic composition of the medium and electric fields, regulates neurotransmitter release after fusion (Carlson and Kelly, 1983).

Clear vesicles, containing only ATP, H^+, Ca^{2+} and low molecular weight neurotransmitters, can discharge almost all their content during the opening of the fusion pore, raising the possibility that vesicles may release their quantum of neurotransmitters without undergoing full fusion (Neher, 1993). Electrochemical measurements of neurotransmitter release from chromaffin cells (Chow et al., 1992) and capacitance and pore-conductance studies in mast cells (Alvarez De Toledo et al., 1993), indicate that an early phase of release precedes the bulk of secretion. This small and fast release corresponds to a period of fluctuating openings of the exocytotic pore prior to full fusion, during which the secretory vesicle fuses with and then detaches several times from the plasma membrane preceding exocytosis (Oberhauser et al., 1992). This capacitance flicker suggests an ion-channel like-structure and is attributed to opening and closure of the fusion pore.

In conclusion, although the molecular structure and function of the fusion pore are still controversial, different experimental approaches support the view that quantal neurotransmitter release may occur by a similar flickering mechanism, which is more efficient than the classically described exocytotic and endocytotic cycle.

REGULATION OF NEUROTRANSMITTER RELEASE

A remarkable feature of synaptic transmission in the brain is that this is not fixed but can be regulated. As discussed before, the exocytotic release of neurotransmitters requires the mobilization of vesicles from a reserve pool. Exocytosis is regulated by Ca^{2+}, whereas vesicle mobilization is controlled by protein phosphorilation and by dissociation of an actin-based cytoskeleton.

Another mechanism which has been shown to regulate the efficiency of synaptic transmission is the frequency of synapse stimulation. The persisting use-dependent synaptic enhancement, that follows either brief high frequency stimulation of certain synapses or the pairing of postsynaptic depolarization with lower frequency synaptic activation, is referred as long-term potentiation (LTP) and is thought to underlie certain forms of learning and memory (Bliss and Collingridge, 1993).

Although the trigger for LTP is generally accepted to occur in the postsynaptic neuron (Manabe et al., 1992), an enhancement in neurotransmitter release is also shown to occur (Larkman et al., 1992). Using quantal analysis techniques, these authors showed that enhanced synaptic responses result from changes in both the release of glutamate from presynaptic terminals and the postsynaptic response to glutamate. The axon terminal with glutamate-filled synaptic vesicles ends on dendritic spines containing NMDA (N-methyl-D-aspartate) and non-NMDA receptors for glutamate. The triggering mechanism for LTP is the influx of Ca^{2+} through NMDA-receptor channels when an adequate number of glutamatergic synapses are simultaneously active.The increase in postsynaptic Ca^{2+} concentration induces the activation of protein kinases, namely Ca^{2+}/calmodulin-dependent kinase II, protein kinase C and one or more tyrosine kinases (Silva et al., 1992; Grant et al., 1992). Protein kinase C was shown to be a positive modulator of NMDA currents, determining the threshold of LTP induction (Ben-Ari et al., 1992) and the production of a retrograde message that causes an increase in the presynaptic release of the neurotransmitter for each nerve impulse arrival (Bliss et al., 1990).

Two freely diffusible molecules are candidates to retrograde messengers: arachidonic acid and nitric oxide (NO). Arachidonic acid is produced in neurons by activation of NMDA receptors through the action of calcium-dependent phospholipase A_2

(Dumuis et al., 1988). Miller et al., (1992), found that arachidonic acid increases the responses mediated by NMDA receptors in cultured cerebellar granule cells. Furthermore, glutamate that is released from the presynaptic terminal activates autoreceptors coupled to inositol 1, 4, 5-triphosphate (which promotes the release of Ca^{2+} from intracellular stores) and to diacylglycerol (which activates protein kinase C). The increase in intracellular Ca^{2+} concentration and the activation of protein kinase C further enhance the release of glutamate from the presynaptic terminal. This effect is observed only when glutamate binds to the receptor in the presence of arachidonic acid (Herrero et al., 1992).The second candidate for a retrograde messenger in NMDA-receptor dependent-LTP is NO. NO is synthesized from L-arginine by NO synthase.

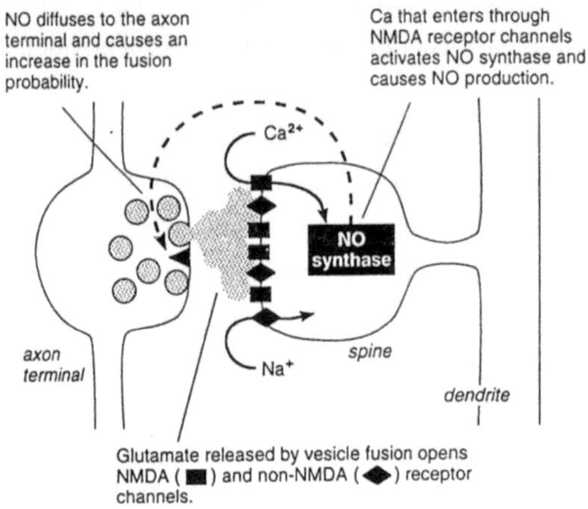

Figure 4. Involvement of nitric oxide (NO)in the control of neurotransmitter release. Reproduced from Stevens (1993), with permission.

The activation of NMDA receptors causes the release of NO from cerebellar granule cells (Furchgott and Zawalzki, 1980). Specific inhibitors of NO synthase block the induction of LTP (O'Dell et al., 1991; Böhme et al., 1991). Recently, NO was shown to induce a marked transient stimulation of Ca^{2+}-independent synaptic vesicle release from hippocampal synaptosomes (Meffert et al., 1994). Thus, as proposed by Gally et al. (1990) the degree of NO release may be a biochemical index of the correlation between presynaptic and postsynaptic activities.

LTP in the hippocampus has become the dominant model of activity-dependent synaptic plasticity in the mammalian brain. LTP has been found in all excitatory pathways in the hippocampus, as well as in other regions of the brain and it underlies at least certain forms of memory.

Finally, it can be concluded that, function in the brain depends less on the properties of single molecules and more on the coordination of a series of molecular events: the targeting and the docking of vesicles at release sites and the assembly of molecules for fusion and exocytosis. The understanding of exocytosis and its control mechanisms is interesting not only in itself, but also to analyse learning and memory mechanisms.

REFERENCES

AlmeidaT, Santos JR, Oliveira CR, Lima MCP (1994) Parameters affecting fusion between liposomes and synaptosomes. Role of proteins, lipid peroxidation, pH and temperature. J. Memb. Biol. in press

Almers W, Tse F W (1990)Transmitter release from synapses: does a preassembly fusion pore initiate exocytosis? Neuron, 4: 813-818

Alvarez de Toledo G, Fernandez-Chacan R, Fernandez JM (1993) Release of secretory products during transient vesicle fusion. Nature 363: 554-557

Ben-Ari Y, Aniksztejn L, Bregestovski P (1992) Protein kinase C modulation of NMDA currents: an important link for LTP induction . Trends Neurosc. 15(9): 333-339.

Benfenati F, Valtorta F, Rubenstein J L, Gorelick FS, Greengard P, Czernik AJ (1992) Synaptic vesicle-associated Ca^{2+}/calmodulin-dependent protein kinase II is a binding protein for synapsin I. Nature 359: 417-420

Bennet M K, Calakos N, Scheller R H (1992) Syntaxin a synaptic protein implicated in docking of synaptic vesicles at presynaptic active zones. Science, 257: 255-259

Bennet MK, Scheller RH (1993) The molecular machinery for secretion is conserved from yeast to neurons. Proc. Natl. Acad. Sci. USA, 90: 2559-2563

Betz H (1990) Homology and analogy in transmembrane channel design: lessons from synaptic membrane-proteins. Biochemistry 29: 3591-3599

Bliss TVP, Clements MP, Errington ML, Lynch MA, Williams JH (1990) Presynaptic changes associated with long-term potentiation in the dentate gyrus. Semin. Neurosci., 2: 345-354

Bliss TVP, Collingride GL (1993) A synaptic model of memory: long-term potentiation in the hippocampus. Nature, 361: 31-39

Böhme GA, Bon C, Slutzmann J-M, Doble A, Blanchard Y-C (1991) Possible involvement of nitric oxide in long-term potentiation. Eur. J. Pharmacol, 199: 379-381

Bourle HR, Sanders DA, McCormick F (1991) The GTPase superfamily-conserved structure and molecular mechanism. Nature, 349: 117-127

Breckenridge LJ, Almers W (1987) Currents through the fusion pore that forms during exocytosis of a secretory vesicle. Nature, 328: 814-817

Bredt DS, Snyder SH (1992) Nitric Oxide, a Novel Neuronal Messenger. Neuron 8, 3-11

Brose N, Petrenko AG, Südhof TC, Jahn R (1992) Synaptotagmin: a Ca^{2+} sensor on the synaptic vesicle surface. Science, 256: 1021-1025

Carlson SS, Kelly RB (1983) A highly antigenic proteoglycan-like component of cholinergic synaptic vesicles. J. Biol. Chem. 258: 11082-11091

Ceccarelli B, Hurlbut WP (1980) Vesicle hypothesis of the release of quanta acetylcholine. Physiol. Rev. 60: 396-441

Chow RH, von Ruden L, Neher E (1992) Delay in vesicle fusion revealed by electrochemical monitoring of single secretory events in adrenal chromaffin cells. Nature, 356: 60-63

David P, El Far O, Martin-Mouto N, Poupon MF, Takahashi M, Seagar MJ (1993) Expression of synaptotagmin and syntaxin associated with N-type Ca^{2+} channels in small cell lung cancer. FEBS Letters, 326: 135-139

del Castillo J, Katz B (1956) Biochemical aspects of neuro-muscular transmission. Prog. Biophys. Chem. 6: 122-170

Dumuis A., Sebben M, Haynes L, Pin JP, Bockaert J (1988) NMDA receptors activate the arachidonic acid cascade system in striatal neurons. Nature, 336: 68-70.

Elferink LA., Petersen MR, Sheller RH (1993) A role for synaptotagmin (p65) in regulated exocytosis Cell, 72: 153-159

Fischer von Molhard G, Südhof T C, Jahn R (1991) A small GTP-binding protein (rab 3A) dissociates from synaptic vesicles during exocytosis. Nature, 349: 79-81

Furchgott R F, Zawalzki, JV (1980) The obligatory role of endothelial cells in the relaxation of arterial smooth muscle by acetylcholine. Nature, 288: 373-376

Gally J A., Montague PR, Reeke GM Jr, Edelman GM (1990) The NO hypothesis: possible effects of a short-lived rapidly diffusible signal in the development and function of the nervous system. Proc. Natl Acad. Sci. USA, 87: 3547-3551

Grant SGM, O'Dell TJ, Karl KA, Stein PA, Soriano P, Kandel ER (1992) Impaired long-term potentiation, spatial learning and hippocampal development in *fyn* mutant mice. Science, 258: 1903-1910

Herrero I, Miras-Portugal MT, Sanchez-Prieto J (1992) Positive feedback of glutamate exocytosis by metabotropic presynaptic receptor stimulation. Nature, 360: 113-166

Hess SD, Doroshenko PA, Augustine GJ (1993) A functional role for GTP-binding proteins in synaptic vesicle cycling. Science, 259: 1169-1172

Hirokawa N, Sobue K, Kanda A, Harada A, Yorifuji H (1989) The cytoskeletal architecture of the presynaptic terminal and molecular structure of synapsin I. J. Cell Biol. 108: 111-126

Horikawa HPM, Saisu H, Ishizuka T, Sekine Y, Tsugita A, Ondani S, Abe T (1993) A complex of rab 3A, SNAP-25, VAMP/synaptobrevin-2 and syntaxins in brain presynaptic terminals. FEBS Lett. 330: 236-240

Ip NY, Nye SH, Boulton TG, Davis S, Taga T, Li Y, Birren SJ, Yasukawa K, Kishimoto T, Anderson DJ, Stahl N, Yancopoulos GD (1992) CNTF and LIF act on neuronal cells via shared signaling pathways that involve the IL-6 signal transducing receptor component gp 130. Cell 69, 1121-1132

Jahn R, Südhof TC (1993) Synaptic vesicle traffic: rush hour in the nerve terminal. J. Neurochem. 61(1): 12-21

Jessell TM, Kandel ER (1993) Synaptic transmission: a bidirectional and self-modifiable form of cell-cell communication Cell/Neuron 10 (Suppl): 1-30

Katz B, Miled R (1965) The effect of calcium on acetylcholine release from motor never terminals. Proc. R. Soc. Lond (Biol) 161: 496-503

Kelly RB (1993) Storage and release of neurotransmitters Cell 72/Neuron 10 (Suppl,) 43-53

Koenig JH, Yayaoka K, Ikeda K (1993) Calcium-induced translocation of synaptic vesicles to the active site. J. Neurosc. 13(6): 2313-2322

Landis DMD, Hall AK, Weinstein LA, Rease TS (1988) The organization of cytoplasm at the presynaptic active zone of a control nervous system synapse. Neuron, 1: 201-209

Larkman A, Hannay T, Stratford K, Jack J (1992) Presynaptic release probability influences the locus of long-term potentiation. Nature, 360: 70-73

Leveque C, Hoshino T, David P, Shoji-Hasay Y, Leys K, Omori A, Lang B, El Far O, Sato K, Martin-Moutot N (1992) The synaptic vesicle protein synaptotagmin associates with calcium channels and is a putative Lambert-Eaton myasthenic syndrome antigen. Proc. Natl. Acad. Sci. USA, 89: 3625-3629

Lledo PM, Vernier P, Vincent JD, Mason WT, Zorec R (1993) Inhibition of rab 3B expression attenuates Ca^{2+}-dependent exocytosis in rat pituitary cells. Nature, 364: 540-544

Llinás R, Sugimori M, Silver RB (1992) Microdomains of high calcium concentration in a presynaptic terminal. Science 256: 677-679

Manabe T, Renner P, Nicoll RA (1992) Post synaptic contribution to long-term potentiation revealed by the analysis of miniature synaptic currents. Nature, 355: 50-55

Meldolesi J, Ceccarelli B (1988) Exocytosis and Membrane recycling. In: Current Topics in Membrane and Transport, vol. 32, Chap 5, 139-163, Acad. Press Inc.

Miller B, Sarantis M, Traynelis SF, Attwell D (1992) Potentiation of NMDA receptor currents by arachidonic acid. Nature, 355: 722-725

Mochly-Rosen D, Miller KG, Scheller RH, Kahner H, Lopez J, Smith BL (1992) p65 fragments, homologous to the C2 region of protein kinase C, bind to the intracellular receptors for protein kinase C. Biochem., 31: 8120-8124

Monck JR, Fernandez JM (1992) The exocytotic fusion pore. J. Cell Biol. 119: 1395-1404

Monck JR, Fernandez JM (1994) The exocytotic fusion pore and neurotransmitter release. Neuron, 12: 707-716

Neher E (1993) Secretion without full fusion. Nature 363: 497-498

O'Connor VM, Shamotienko O, Grishin E, Betz H (1993) On the structure of the "synaptosecretosome". Evidence for a neurexin/synaptotagmin/syntaxin/Ca^{2+} channel complex. FEBS Lett., 326: 255-260

O'Dell TJ, Hawkins RD, Kandel ER, Arancia O (1991a) Tests of the roles of two diffusible substances in long-term potentiation: evidence for nitric oxide as a possible early retrograde messenger. Proc. Natl Acad. Sci. USA, 88: 11285-11289

Oberhauser AF, Monck JR, Balch WE, Fernandez JM (1992) Exocytotic fusion pore is activated by rab 3a peptides. Nature, 360: 270-273

Palade GE (1975) Intracellular aspects of the process of protein synthesis. Science, 189: 347-358

Perin MS, Fried VA, Mignery GA, Jahn R, Südhof TC (1990) Phospholipid binding by a synaptic vesicle protein homologous to the regulatory region of protein kinase C. Nature, 345: 260-263

Petrenko AG, Perin MS, Davletov BA, Vshkaryov YA, Geppert M, Südhof TC (1991) Binding of synaptotagmin to the α-latrotoxin receptor implicates both in vesicle exocitosis. Nature, 353: 65-68

Ramon y Cajal SR (1911) Histologie du Système Nerveux de l'Homme et des Vertébrés. Paris

Robinson IM, Oberhauser AF Fernandez JM (1994) Is the activity of the fusion of the fusion pore scaffold regulated by a coincidence detector? Ann. NY Acad. Sci., in press

Robitaille R, Adler EM, Charlton MP (1990) Location of calcium channels at transmitter release sites of frog neuromuscular synapses. Neuron, 5: 773-779

Schiavo G, Benfenati F, Poulain B, Rossetto O, De Laureto PP, Das Gupta BR, Montecucco C (1992) Tetanus and Botulinum-B neurotoxins block neurotransmitter release by proteolytic cleavage of synaptobrevin. Nature, 359: 832-835

Shoji-Kasai J, Yoshida A, Sato K, Hoshino T, Ogura A, Kondo S, Fujimato Y, Kuwahara R, Kato R, Takahashi M (1992) Neurotransmitter release from synaptotagmin-deficient clonal variants of PC12 cells. Science, 256: 1820-1823

Sikorski A. F, Terlecki G, Zagon IS, Goodman SR (1991) Synapsin I-mediated interaction of brain synaptic vesicles. J. Cell Biol. 114: 313-318

Silva AJ, Stevens CF, Tonegawa S, Wang Y (1992b) Deficient hippocampal long-term potentiation in α-calcium-calmodulin kinase II mutant mice. Science, 257: 201-206

Smith¡SJ, Augustine GJ (1988) Calcium ions, active zones and synaptic transmitter release. Trends Neurosci., 11: 458-464

Söllner T, Whiteheart SW, Brunner M, Erdjument-Bromage H, Geromanos S, Tempst P, Rothman JE (1993) SNAP receptors implicated in vesicle targeting and fusion. Nature, 362: 318-324

Stevens CF (1993) Quantal release of neurotransmitter and long-term potentiation Cell/Neuron 10 (Suppl): 55-63

Struck DK, Hoekstra D, Pagano RE (1981) Use of resonance energy transfer to monitor membrane fusion. Biochemistry, 20: 4093-4099

Südhof TC, Czernik AJ, Hung-Teh K, Takei K, Johnston PA, Horiuchi A, KanazirS. D,Wagner MA, Perin MS, De Camilli P, Greengard P (1989) Synapsins: Mosaics of shared and individual domains in a family of synaptic vesicle phosphoproteins. Science, 245: 1474-1480

Südhof TC, Jahn R (1991) Proteins of synaptic vesicles involved in exocytosis and membrane recycling. Neuron, 6: 665-677

Thomas L, Betz H (1990) Synaptophysin binds to physophilin, a putative synaptic plasma membrane protein. J. Cell Biol., 111; 2041-2052

Thomas L, Hartung K, Langosh D, Rahm H, Bamberg E, Franke WW, Betz (1988) Identification of synaptophysin as a hexameric channel protein of the synaptic vesicle membrane. Science, 242: 1050-1053

Trifaró J-M, Vitale ML (1993) Cytoskeleton dynamics during neurotransmitter release. TINS, 16(11) 466:472

Valtorta F, Benfenati F, Greengard P (1992) Structure and Function of the synapsins. J. Biol. Chem. 267: 7195-7198

Yoshida A, Oho C, Akire O, Kuwahara R, Ito T, Takahashi M (1992) HPC-1 is associated with synaptotagmin and ω-conotoxin receptor. J. Biol. Chem. 267: 24925-24928

Recent Advances in Defining Mannose Receptor Structure and Function

Suzanne E. Pontow and Philip D. Stahl
Department of Cell Biology and Physiology
Washington University School of Medicine
660 S. Euclid Avenue
St. Louis, Missouri, 63110

KEY WORDS/ ABSTRACT: MANNOSE RECEPTOR/ C-TYPE LECTIN/ MRC1/ MACROPHAGE/ PHAGOCYTOSIS/ ENDOCYTOSIS/ HEPATIC SINUSOIDAL ENDOTHELIUM/ TRACHEAL SMOOTH MUSCLE/ YEAST MANNAN/ HOST DEFENSE/ RICIN/ INTERFERON-GAMMA/ TUMOR NECROSIS FACTOR-ALPHA/ INTERLEUKIN 4/ PLA_2 RECEPTOR

The macrophage mannose receptor (MR) mediates the binding and internalization of glycoproteins, viruses, bacteria and fungi through recognition of terminal mannose residues on exposed oligosaccharides (Pontow *et al.*, 1992). A member of the calcium-dependent (C-type) animal lectin family, the MR has been assigned to a distinct subgroup in view of an atypical protein and gene structure (Drickamer, 1993). Because the MR functions in both endocytosis and phagocytosis, it is also unique among the group of pH-dependent recycling endocytic receptors to which it is related. The first decade of research characterizing the MR has been reviewed in detail (Pontow *et al.*, 1992; Stahl, 1992; Stahl, 1990, Ezekowitz and Stahl, 1988), and will be summarized next. The remainder of this review will focus on the recent and exciting observations that have underlined both exceptional and clinically significant qualities of MR structure and function.

MR activity was first observed during studies on the clearance rates for different forms of the lysosomal hydrolase β-glucuronidase (β; Stahl *et al.*, 1976a). Enzymes bearing high mannose chains were rapidly eliminated from the bloodstream by specific uptake (Stahl *et al.*, 1976b), mainly in the liver and spleen, by macrophages and hepatic nonparenchymal cells (Achord *et al.*, 1977; Schlesinger *et al.*, 1978). Subsequent studies determined the binding and internalization characteristics of the MR through assays utilizing primary macrophage cultures and both natural and synthetic glycoprotein ligands. In the presence of calcium and a neutral pH, the MR binds glycoproteins with affinities in the nanomolar range (Stahl *et al.*, 1980). The MR exhibits a range

NATO ASI Series, Vol. H 91
Trafficking of Intracellular Membranes
Edited by M.C. Pedroso de Lima N. Düzgüneş and D. Hoekstra

of specificity, binding fucose (Shepherd *et al.*, 1981) and mannose, and with lower affinity, N-acetylglucosamine and glucose (Stahl *et al.*, 1978). At the cell surface MR ligands are bound, internalized through coated pits and vesicles (Tietze *et al.*, 1982), and released upon exposure to the acidic millieu of endosomes (Tietze *et al.*, 1980; Wileman *et al.*, 1984). Here the ligands are either proteolyzed (Diment and Stahl, 1985) or targetted to lysosomes, while the MR returns to the cell surface to participate in another round of endocytosis (Tietze *et al.*, 1982; Wileman *et al.*, 1984). The MR cycles constitutively through the cell, with 10-20% a cell's complement exposed at the cell surface. The itinerary of the MR during phagocytosis is less well understood, although it has been shown to recycle out of phagosomes and presumably back to the plasma membrane (Pitt *et al.*, 1992b), possibly via endosomes (Pitt *et al.*, 1992a).

Isolation of the MR by affinity chromatography has allowed a detailed charaterization of the receptor protein and the message encoding it (Lennartz *et al.*, 1987). The initial translation product contains a signal sequence which is cleaved following translation on the rough endoplasmic reticulum (Taylor *et al.*, 1990). The mature MR has an apparent molecular weight of 180 kilodaltons (kDa) and bears N- and O-linked sugars (Lennartz, *et al.*, 1989). Greater than 95% of the protein is extracytoplasmic; this portion of the MR contains the area mediating ligand recognition and two domains of unknown function (Taylor *et al.*, 1992; Harris *et al.*, 1992). The cytoplasmic tail and membrane anchor are crucial to the endocytic and phagocytic function of the MR (Kruskal *et al.*, 1991), targetting it to a recycling pathway and likely interacting with the cell's signalling machinery.

Initially thought to be specific to cells of the reticuloendothelial system (Achord *et al.*, 1977), the MR is expressed at varying levels by several other cell types. Differentiated macrophages (Stahl *et al.*, 1978; Shepherd *et al.*, 1982) and liver sinusoidal endothelial cells (Magnusson and Berg, 1993) express an abundance of MR, while tracheal smooth muscle cells (Lew *et al.*, 1994), retinal pigment epithelium (Shepherd *et al.*, 1991), cells of the giant cell tumor of bone (Clohisy *et al.*, 1993) and possibly others, including dendritic cells (Reis e Sousa *et al.*, 1993), express varying amounts. The reasons for differential expression patterns and the function of the MR in nonmacrophage cells are unclear. MR function in macrophages has long been hypothesized to involve scavenging deleterious molecules and organisms, protecting the host from tissue damage and infection (Ezekowitz and Stahl, 1988). However, recent evidence suggests a broader role for this receptor in macrophages and nonphagocytic cells, which includes modulating the activity of both MR-positive (Lew *et al.*, 1994; Marodi *et al.*, 1993) and negative cells (Garner *et al.*, 1994; Kimura *et al.*, 1992).

Structure of the Mannose Receptor and MRC1

The amino acid and nucleotide sequences of both the human (Taylor *et al.*, 1990) and mouse (Harris *et al.*, 1992) MRs have been determined, revealing much about the structure, function and evolutionary history of this protein. MRs from human and murine sources are closely related, exhibiting 75-80% homology at the nucleotide level and 82% identity in their amino acid sequences (Harris *et al.*, 1994). Both the human (Taylor *et al.*, 1990) and mouse (Harris *et al.*, 1992) proteins contain 1438 amino acids translated from an approximately 5 kilobase message. As assessed by SDS-PAGE, each receptor exhibits an apparent molecular weight of about 175 kDa. The similarity between these two proteins extends to their gene organization. Both the human (Kim *et al.*, 1992) and mouse (Harris *et al.*, 1994) MR genes contain 30 exons and 29 introns. Comparison of the intron positions in the two genes suggests that the events generating the MR occured in an ancestor common to both primates and rodents (Harris *et al.*, 1994). Recent work has localized the MR gene (MRC1) to the proximal end of murine Chromosome 2 and to the corresponding loci in the human genome on Chromosome 10 (Harris *et al.*, 1994). As the human and mouse MR are nearly identical, further discussion will refer to the receptor of both, and perhaps other species.

Structurally, mature MR can be divided into 4 domains: the amino-terminal region following the cleaved signal sequence that shares homology with ricin B chain (Harris *et al.*, 1994), a domain containing fibronectin type II repeats, the area of ligand binding consisting of eight tandom CRDs, and the transmembrane and cytoplasmic carboxy-terminal domain (Taylor *et al.*, 1990). The gene encoding the MR reflects these structural divisions. The CRD-encoding exons are separated from those encoding other portions of the receptor, which is common to all C-type animal lectins whose genes have been analyzed (Kim *et al.*, 1992). The three regions not involved in ligand recognition are each encoded by a single exon (Harris *et al.*, 1994; Kim *et al.*, 1992). A detailed description of each MR domain follows.

Amino Terminus

The amino terminal portion of the mature mannose receptor (139 amino acids) contains six cysteine residues, and has been termed the Cys-rich domain (Taylor *et al.*, 1990). Truncated receptors lacking this domain have been expressed in fibroblasts with no loss of binding or endocytic capacity (Taylor *et al.*, 1992). Therefore, this domain does not function in internalization of glycoprotein ligands and has not been assigned any other function. Initially thought to resemble no other protein of known sequence, it recently has been shown to be positively aligned with a segment of the galactose-binding B chain of the plant toxin, ricin (Harris *et al.*, 1994).

Ricin B chain can be divided into N-terminal and C-terminal domains, each of which is further divisible into 3 subdomains, termed α, β and γ (Harris *et al.*, 1994). These subdomains are homologous to one another as they were originally derived from the same forty amino acid peptide. When amino acids 151-226 of B chain are aligned with residues 17-87 of the murine MR, several points of similarity are observed (Harris *et al.*, 1994). Four of the six cysteine residues of the MR are conserved in ricin B chain, and sixteen other amino acids are identical. This corresponds to 28% identity, and when conservative substitutions are considered, there is a 36% similarity between these sequences. In addition, the sequence Gln-Lys-Trp which is found in both the mouse and human MR is conserved as Gln-X-Trp in five of the six subdomains of B chain. However, neither the MR nor this area of B chain participates in galactose binding, rendering the function of this domain in the MR an unanswered question.

Ironically, A chain, the toxic subunit of ricin bears high-mannose chains and is a ligand for the MR (Simmons *et al.*, 1986) As such, ricin has been a useful tool in MR research. An early study showed that injected ricin depletes liver of Kupffler cells, which it intoxicates following uptake by the MR (Simmons *et al.*, 1987). More recently, ricin has been utilized to demonstrate the existence of the MR on hepatic sinusoidal endothelial cells (Magnusson *et al.*, 1991; Magnusson and Berg, 1993). Mannose-dependent internalization of ricin exceeds galactose-dependent uptake by these cells *in vivo* (Magnusson and Berg, 1993). The MR has been employed in the study of how ricin gains access to the cytoplasm of cells. Following MR-mediated internalization into endosomes of macrophages, A chain undergoes proteolytic cleavage that facilitates its transport across the endosomal membrane (Fiani *et al.*, 1993). This observation further delineates the functions of endosomal proteases (Blum *et al.*, 1991).

Fibronectin Type II Repeats

The second domain in the MR sequence, encoded by exon 3, shares a motif common to several proteins, fibronectin Type II repeats (Taylor *et al.*, 1990; Harris *et al.*, 1992). Other proteins exhibiting this motif include the cation-independent mannose-6-phosphate receptor and clotting factor XII (Petersen *et al.*, 1983). In fibronectin, these repeats are localized to the portion of the protein that interacts with collagen Petersen *et al.*, 1983). However, this sequence has not been assigned a function in fibronectin or in any other protein in which it is present (Harris *et al.*, 1992). It has been suggested that this domain may play a role in macrophage spreading through interaction with the extracellular matrix (Harris *et al.*, 1992). Of all the domains in the MR, the fibronectin repeat shows the greatest homology (93% identity) conserved between the human and mouse sequences (Harris *et al.*, 1992). While this observation alludes to an important role for this domain, the fibronectin type II repeats must participate in some function other than ligand recognition and internalization. Like the ricin-like domain described above, MR lacking this domain functions normally in the uptake of mannose-terminal proteins when expressed in cells (Taylor *et al.*, 1992).

Carbohydrate Recognition

The best characterized region of the MR follows the fibronectin type II domain and is responsible for interaction with carbohydrate moieties (Taylor, 1993). The binding domain contains 8 homologous CRDs connected in series by intervening stretches of amino acids (Taylor *et al.*, 1990; Harris *et al.*, 1992). CRDs are common to all C-type (calcium dependent) animal lectins, and are characterized by certain invariant residues and structural features (Drickamer, 1993). Of the 37 conserved amino acids that define an approximately 130 residue CRD, 27-34 are also present in MR CRDs (Taylor *et al.*, 1990; Harris *et al.*, 1992). The degree of identity between mouse and human MR CRDs ranges from 76-92%, with CRD 4 exhibiting the highest homology Harris *et al.*, 1992).

The invariant and conserved amino acids of a C-type CRD form important structural domains that allow for recognition of sugar and calcium binding (reviewed in Drickamer, 1993). The crystal structure of the rat serum mannose binding protein CRD has been analyzed recently (Weis *et al.*, 1991a; Weis *et al.*, 1991b), revealing many features that are likely shared by all C-type CRDs (Drickamer, 1993). Two binding sites for calcium and possibly a third are evident, along with a site for carbohydrate binding. When the CRD crystals were analyzed in the presence of ligand, a weak interaction of the monosaccharide with the binding site was observed (Weis *et al.*, 1992). This limited contact is thought to reflect the ability of CRDs to bind multiple sugars and the inability of single CRD's to bind oligosaccharide ligands with high affinity (Drickamer, 1993). CRDs from different dimers of the serum mannose binding protein were also observed to bind to separate branches of a single oligosaccharide, creating a network among the proteins (Weis, *et al.*, 1992). This type of interaction between ligand and receptors may explain how the MR is able to bind large particulate ligands and could form the basis for a signalling mechanism.

The carbohydrate binding properties of the MR and various MR CRDs have been studied in detail through the use of *in vitro*, bacterial and insect translation systems (reviewed in Taylor, 1993). Of the eight MR CRDs, only CRD 4 exhibits sugar binding activity when expressed as a single CRD (Taylor *et al.*, 1992). In experiments designed to test the ability of various monosaccharides and naturally occuring glycoproteins to inhibit the binding of ^{125}I-Man-BSA to CRDs, the activity of CRD4 was found to be limited to monosaccharide ligands (Taylor *et al.*, 1992). With a K_I in the millimolar range, CRD4 binds D-mannose and L-fucose, and with decreasing affinities, N-acetyl- glucosamine and glucose (Taylor *et al.*, 1992). This range of affinity for monosaccharides correlates well with the values determined for the intact receptor (Stahl *et al.*, 1978). Ligands containing high mannose chains were poor inhibitors for CRD 4, which apparently cannot bind physiological ligands on its own (Taylor,*et al*, 1992). It has been shown previously that branched chain mannose oligosaccharides are more potent inhibitors than linear oligosaccharides or monosaccharides of MR binding to ligand coated plates (Kery *et al.*, 1992).

Excepting the MR, all C-type animal lectins possess one CRD per polypeptide and these proteins require oligomerization prior to high affinity binding of ligand (Drickamer and Taylor, 1993). Since each CRD of these polymeric lectins is capable of binding only monosaccharide, clustering must occur in order to accomodate with high affinity the multivalent, branched ligands found *in vivo*. Not all CRDs of the MR are capable of binding sugar when expressed alone (Taylor *et al.*, 1992). However, the intact receptor displays a Kd for oligosaccharide ligands in the nanomolar range (Stahl *et al.*, 1980) suggesting that either the receptor exists as an oligomer or the clustering of several low affinity CRDs confers upon the MR the ability to bind its ligands with high affinity (Taylor *et al.*, 1992) Recent evidence confirms the latter hypothesis. First, the MR appears to exist as a monomer in both detergent solution and in the membrane (Taylor and Drickamer, 1993). Second, certain groups of MR CRDs can be expressed together that bind to natural glycoproteins with affinities comparable to that of the intact receptor (Taylor *et al.*, 1992; Taylor and Drickamer, 1993).

CRDs 1-3 exhibit no binding activity when translated together and exposed to Man-sepharose or other sugars normally recognized by the MR (Taylor *et al.*, 1992). The function of these CRDs, if any, has yet to be determined. CRDs 4-5 through 4-8 and 5-7 and 5-8 do bind mannose-sepharose and other ligands (Taylor *et al.*, 1992). CRDs 4-5 are the smallest unit to bind invertase, a glycoprotein bearing high mannose chains, with high affinity. These two CRDs have been suggested to act as a ligand binding core, although additional CRDs (6-8) are required for tight interaction with extended polymeric ligands such as mannan (Taylor and Drickamer, 1993). Results from proteolysis experiments are consistent with these findings. One characteristic of C-type CRDs is resistance to digestion in the presence of calcium (Weis *et al.*, 1991a). When CRDs 4-7 are digested with increasing concentrations of the enzyme subtilisin in the presence of calcium, CRDs 7 and 6 are sequentially removed, while CRDs 4 and 5 remain intact (Taylor and Drickamer, 1993).

Efficient endocytosis of Man-BSA can occur when truncated forms of the MR are expressed in fibroblasts (Taylor *et al.*, 1992). CRDs 4-8 plus the tramsmembrane and cytoplasmic domains endocytose MR ligands with kinetics indistinguishable from that of the whole receptor. CRDs 5-8 also bind and internalize ligand, although at a much slower rate, while CRDs 6-8 are incapable of binding. The authors of this study conclude that while CRD 5 is important to the recognition of multivalent ligands, CRD 4 is critical to the normal endocytic function of the MR (Taylor *et al.*, 1992).

The complexity of the MR carbohydrate recognition system is reflected in the CRD-encoding region of MRC1 (Harris *et al*, 1994; Kim *et al.*, 1992). All of the CRDs are encoded by multiple exons ranging from 2 to 4 per domain. In some areas of the gene the exon structure represents the CRDs encoded, however certain CRDs share an exon with a neighboring CRD. Again the MR is unique, as the CRDs of other C-type animal lectins are encoded by either one or three exons (Harris *et al.*, 1994). By examining the exon-intron structure of the MR gene and comparing the pattern to those of other lectin genes, the following conclusions about MR evolution have been drawn. An ancient CRD-encoding sequence was duplicated to generate the multiple CRDs found in the MR primary structure (Kim *et al.*, 1992). This region of the receptor was formed prior to exon shuffling events that added accessory domains to the MR, the hepatic lectins and proteoglycan core protein (Kim *et al.*, 1992). As these events presumably occurred early in our history, the MR may have functioned as part of a primordial immune system (Stahl, 1990).

Transmembrane and Cytoplasmic Domain

The sugar binding region of the MR is proximal to the extracellular face of cells and is adjacent to the membrane spanning hydrophobic sequence that anchors the receptor in the membrane (Taylor *et al.*, 1990; Harris *et al.*, 1992). The remaining carboxy-terminal amino acids make up the intracellular 40 amino acid tail of the receptor. This region of the receptor contains the sequence FENTLY, which is similar to the motif FXNPXY that is thought to direct certain endocytic receptors into clathrin coated pits (Ezekowitz *et al.*, 1990). The importance of the cytoplasmic tail in MR function is clear. Truncated MRs lacking the cytoplasmic tail are present at the cell surface of transfected cells, where they can bind, but not internalize soluble and particulate ligands (Ezekowitz *et al.*, 1990). Site-directed mutagenesis of the cytoplasmic tyrosine reduces uptake of ligand by 50% (Kruskal *et al.*, 1992). However, a recent study has shown that the cytoplasmic tail of the MR works in concert with the transmembrane domain to allow efficient uptake of soluble and particulate ligands (Kruskal *et al.*, 1992).

Chimeric receptors have been generated which combine the transmembrane and cytoplasmic domains of the MR with the binding, transmembrane and intracellular domains of the high-affinity immunoglubulin receptor, $F_C\gamma RI$ (CD64). Unlike the MR, $F_C\gamma RI$ loses the capacity to internalize bound ligand upon transfection into heterologous cells (Ezekowitz *et al.*, 1990). When the F_CRI ectodomain is expressed as a chimera in conjunction with the transmembrane and cytoplasmic domains of the MR, the ability of the transfected receptor to internalize opsonized erythrocytes and IgG is restored (Kruskal *et al.*, 1992). The presence of either MR domain alone in a chimera was insufficient in restoring function. Thus, the transmembrane and cytoplasmic regions of the MR can be considered a functional unit and not surprisingly, they are encoded together by a single exon within the MR gene. This is in contrast with other transmembrane receptors with short cytoplasmic tails which have these domains encoded by multiple exons (Kim *et al.*, 1992).

Regulation

The MR is expressed on resident and elicited macrophages (Stahl *et al.*, 1980), and is progressively expressed by cultured macrophage precursors (Shepherd *et al.*, 1982; Clohisy *et al.*, 1987). Because the MR is not expressed by circulating monocytes (Shepherd *et al.*, 1982), it has been accepted as a marker for terminally differentiated macrophages. MR expression is highly responsive to the environment of the macrophage, and can be modulated *in vitro* by exposure to bacterial pathogens (Ezekowitz *et al.*, 1981; Shepherd *et al.*, 1990) and certain secretory products of immune cells, including cytokines (Moekoena and Gordon, 1985; Schreiber *et al.*, 1993; Stein *et al.*, 1992) and IgG2A (Schreiber *et al.*, 1991). These observations suggest that expression of the MR can also be correlated with the functional state of the macrophage. By determining which factors affect MR activity, researchers have begun to compile compelling evidence that the MR plays significant roles in immune and inflammatory responses.

Exposure of macrophages to IFN-γ during an immune response stimulates proinflammatory cytokine and superoxide release and increases expression of MHC class II antigen (Nathan, 1992). MR expression is down-regulated by IFN-γ *in vitro* (Ezekowitz and Gordon 1982; Moekoena and Gordon, 1985; Schreiber *et al.*, 1993), although the affinity of the receptor for ligand is not affected (Marodi *et al.*, 1993). This effect of IFN-γ is negated by prostaglandin E (Schreiber *et al.*, 1993), which has been shown to hasten expression of the MR on differentiating bone marrow macrophages (Schreiber *et al.*, 1990). IFNγ modulates MR expression by significantly reducing transcription of the MR gene (Harris *et al.*, 1992), which results in a decreased capacity for treated macrophages to endocytose Man-BSA. Such a response seems to contradict the hypothesis that the MR plays an important role in host defense. However, recent evidence indicates that the quantitative reduction in receptor protein may not be accompanied by a qualitative reduction in phagocytic function; IFN-γ-treated monocyte-derived macrophages exhibit an increased capacity to kill the yeast *Candida albicans* in a MR-dependent manner (Marodi *et al.*, 1993). IFN-γ treated macrophages also experience a rise in cytosolic calcium and superoxide anion release upon stimulation with *Candida* that is not observed in untreated cells. This suggests that the MR in treated macrophages is coupled to the cell's signalling machinery in a manner different from that of untreated macrophages (Marodi *et al.*, 1993). In addition, it is possible that IFN-γ treatment selects for a distinct subpopulation of MRs responsible for the phagocytic activity of this receptor. While the existence of two MR populations would explain this receptor's ability to mediate uptake of both soluble and particulate ligands, there is no evidence to date to support this notion.

In contrast to the effects of IFN-γ, interleukin-4 (IL-4) inhibits production of proinflammatory cytokines by macrophages, and can inhibit release of superoxide anions (Paul, 1991). Treatment of macrophages with IL-4 enhances MR expression significantly, increasing surface levels 7-12 fold as assessed by the binding and endocytosis of man-BSA (Stein *et al.*, 1992). Again, this effect on MR expression is mediated through effects on transcription. Whether or not IL-4 treatment also increases MR phagocytic activity is unknown. It has been suggested that the upregulation of MR serves to increase the clearance capacity of newly recruited inflammatory macrophages and to maintain high levels of MR expression in certain subpopulations of tissue macrophages (Stein *et al.*, 1992). Therefore, IL-4 promotes MR anti-inflammatory function, increasing its expression and consequently the scavenginging capacity of macrophages, while IFN-γ enhances MR immune function, decreasing MR levels while upregulating its phagocytic and signalling capacity.

Phagocytosis and Signalling

The MR was definitively shown to mediate the uptake of certain bacteria and fungi through functional expression of the receptor in non-phagocytic Cos-1 cells (Ezekowitz *et al.*, 1990; Ezekowitz *et al.*, 1991). Currently, research involving MR-mediated phagocytosis is focused on the ability of particulate ligands to trigger events such as calcium spikes, respiratory burst and secretion, and on the clinical relevance of MR function. Signalling by the MR in response to particulate ligands has been difficult to assess due to the presence of other lectin-like receptors on the macrophage surface, which include receptors for a cleavage product of the third component of complement (CR3; Wright, 1992) and β-glucan receptors (Stahl, 1992). However, careful production of ligands (Garner *et al.*, 1994; Kimura *et al.*, 1992) and use of CR3-blocking antibodies (Marodi *et al.*, 1993; Wilson and Pearson, 1988) have facilitated the interpretation of results which clearly link the MR to events occuring after exposure of cells to particles bearing mannose terminated oligosaccharides.

The ability of glucan-free mannan isolated from the cell walls of *Candida albicans* to induce secretion of the inflammatory cytokine tumor necrosis factor-α (TNF-α) from alveolar macrophages in a time- and concentration-dependent manner was demonstrated recently (Garner *et al.*, 1994). Significant stimulation of TNF-α secretion was observed with mannan from this fungal source only; mannan from *C. tropicalis* induced only slight secretion of TNF-α, while *S. cerevisiae* mannan did not have any measurable effect. While the mechanism behind this observation is unclear, it may form the basis for the differential pathogenicities of various fungi (Garner *et al.*, 1994). By secreting TNF-α in response to various foriegn agents, macrophages and other cells are thought to promote the clearance activity of neutrophils, thereby enhancing the host's ability to fight infection. In the case of pathogens recognized by the MR, those triggering TNF-α release would be eliminated efficiently while others might not. Consistent with this view is the finding that *Candida albicans* causes far fewer cases of fungal pneumonia in AIDS patients than other fungi (Garner *et al.*, 1994), including *Pneumocystis carinii* (Ezekowitz, 1992). The latter has been shown to be poorly recognized by the MR of macrophages from AIDS patients, although it is readily internalized via the MR in macrophages from uninfected donors (Koziel *et al.*, 1993). However, this observation has not been linked to production of TNF-α.

The mechanism of MR signalling is unknown; MR tail(s) may interact directly with a regulatory protein or may be coupled indirectly to signalling machinery through an accessory molecule. It is clear that a multivalent ligand and participation of multiple CRDs, perhaps from different MRs, are required for coupling of these receptors to downstream effectors. The soluble fraction of boiled zymosan stimulates secretion of peptide leukotrienes via the mannose receptor in perfused liver (Kimura *et al.*, 1992). The release of leukotrienes induces glycogenolysis in parenchymal cells (Kimura *et al.*, 1992), providing another instance in which the MR participates in regulating the function of a different cell type. Man-BSA and lysosomal enzymes trigger secretion of TNF-α from macrophages (Lefkowitz *et al.*, 1991), and stimulate DNA synthesis in bovine tracheal smooth muscle cells (Lew and Rattazzi, 1991). Release of lysosomal enzymes occurs following exposure of macrophages to MR-targetted neoglycoproteins (Oshumi and Lee, 1987). However, monosaccharides are not able to generate these signals, which is consistent with the binding properties of the component CRDs (Taylor, 1993), and with the idea that clustering of CRDs and perhaps cytoplasmic tails, is required for transmission of a signal through the MR.

Phospholipase A_2 Receptor

Recently, the MR was shown to exhibit remarkable structural similarity to the high affinity receptor for the secretory phospholipases A_2 (PLA_2R; Ishizaki *et al.*, 1994; Lambeau *et al.*, 1994). Secretory PLA_2 group I is found in a variety of animal cells and secretory fluids, and displays a wide range of effects, including toxicity, stimulation of DNA synthesis, contraction and chemotaxis, depending on the cell type studied (Ishizaki *et al.*, 1994). While these events are mediated by a specific and possibly ubiquitous receptor, the physiologic role for this type of PLA_2 remains unclear.

The PLA_2R has been identified in several cell types, and cDNAs encoding the receptor have been cloned and sequenced from both rabbit skeletal muscle (Lambeau *et al.*, 1994) and bovine corpus lutea (Ishizaki *et al.*, 1994). The amino acid sequences of rabbit and bovine PLA_2Rs are highly homologous to each other (personal observation), and share approximately 30% homology overall with the human MR (Ishizaki *et al.*, 1994; Lambeau *et al.*, 1994). The PLA_2R and the MR are almost identical in predicted and apparent molecular weight. A striking resemblance exists between the predicted protein structures of the PLA_2R and the MR (below), suggesting that these two proteins are related.

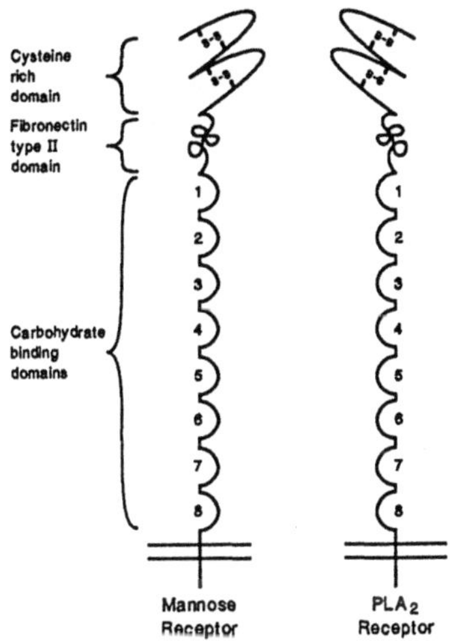

Predicted Topological Structures of the MR and PLA_2R

Like the MR, the PLA_2R consists of an N-terminal Cys-rich domain, a region cantaining fibronectin type II repeats, eight potential CRDs, a single transmembrane-spanning region and a short cytoplasmic sequence. Although secretory PLA_2 bears no saccharide moieties, its binding site lies within the region containing the CRDs; cells transfected with only the cytoplasmic, transmembrane and CRD regions of the PLA_2R bind and internalize ligand (Ishizaki *et al*, 1994). Man-BSA and galactose-BSA, which is not a MR ligand, both inhibit binding of a snake venom PLA_2 to its receptor, although other MR ligands did not compete for binding (Lambeau *et al.*, 1994). One explanation for this observation is that the PLA_2R contains at least one competent

CRD with broad sugar specificity that binds monosaccharide-conjugated protein which sterically inhibits PLA_2 binding.

The apparent relation of the PLA_2R to the MR raises a myriad of questions. Do the MR and the PLA_2R belong to a family of receptors and are there any other members? Are these receptors functionally or physically related? Are they antigenically similar? Does the PLA_2R bind to sugar? Does the MR have a binding site for a noncarbohydrate ligand? Does the distribution of the MR and the PLA_2R overlap? Is the expression of PLA_2Rs regulated similarly to the MR? What is the evolutionary history of these receptors in relation to one another, and what does this tell us about their function? This unexpected relationship between functionally diverse receptors has generated new avenues for investigation which will hopefully yield information about their physiological significance. In addition, the study of the relatioship between these receptors may lead to the production of new recognition moieties for drug targetting.

REFERENCES

Achord D, Brot F, Gonzalez-Noriega A, Sly W, Stahl PD (1977) Human β-glucuronidase. II. Fate of infused human placental β-glucuronidase in the rat. *Pediat Res* 11:816-822

Blum JS, Fiani ML, Stahl PD (1991) Localization of cathepsin D in endosomes: characterization and biological importance. *Adv Exp Med Biol* 306:281-287

Brech A, Magnussone S, Stang E, Berg T, Roos N (1993) Receptor-mediated endocytosis of ricin in rat liver endothelial cells. An immunocytochemical study. *Eur J Cell Biol* 60:154-162

Clohisy DR, Bar-Shavit Z, Chappel JC, Teitelbaum SL (1987) 1,25-dihydroxyvitamin D3 modulates bone marrow macrophage precursor proliferation and differentiation: upregulation of the mannose receptor *J Biol Chem* 262:15922-15929

Clohisy DR, Vorlicky L, Oegma TR Jr., Snover V, Thomson RC Jr. (1993) Histochemical and immunohistochemical characterization of cells constituting the giant cell tumor of bone. *Clin Orthopaed Rel Res* 287:259-265

Diment S, Stahl PD (1985) Macrophage endosomes contain proteases which degrade endocytosed protein ligands. *J Biol Chem* 260(28):15311-15317

Drickamer K (1993) Recognition of complex carbohydrates by calcium-dependent animal lectins. *Biochem Soc Trans* 21(2):468-473

Drickamer K, Taylor ME (1993) Biology of animal lectins. *Annu Rev Cell Biol* 9:237-264

Ezekowitz RAB (1992) The mannose receptor and phagocytosis. *In* "Mononuclear phagocytes" (R. van Furth ed.) p. 208-213 Kluwer Academic Publishers, Netherlands.

Ezekowitz RAB, Austyn J, Stahl PD, Gordon S (1981) Surface properties of bacillus Calmette-Gucrin-activated mouse macrophage. Reduced expression of mannospecific endocytosis, Fc receptors, and antigen F4/80 accompanies induction of Ia. *J Exp Med* 154:60-76

Ezekowitz RAB, Gordon S (1982) Down-regulation of mannosyl receptor-mediated endocytosis and antigen F4/80 in bacillus Calmette-Guerin-activated mouse macrophages. Role of T lymphocytes and lymphokines. *J Exp Med* 155(6):1623-1637

Ezekowitz RAB, Sastry K, Bailly P, Warren A (1990) Moleculor characterization of multiple carbohydrate recognition domains and phagocytosis of yeasts in Cos-1 cells. *J Exp Med* 172:1785-1795

Ezekowitz RAB, Stahl PD (1988) The structure and function of vertebrate mannose lectin-like proteins. *J Cell Sci* 9(Suppl.):121-133

Ezekowitz RAB, Williams D, Koziel H, Armstrong MY, Warner A., Richards FF, Rose RM (1991) *Pneumocystis carinii* uptake by alveolar macrophages is mediated via the mannose receptor. *Nature* 351:155-158

Fiani ML, Blum JS, Stahl PD (1993) Endosomal proteolysis precedes ricin A-chain toxicity in macrophages. *Arch Biochem Biophys* 307(2):225-230

Garner RE, Rubanowice K, Sawer RT, Hudson JA (1994) Secretion of TNF-α by alveolar macrophages in response to *Candida albicans* mannan. *J Leuk Biol* 55:161-168

Harris N, Peters LL, Eicher EM, Rits M, Raspberry D, Eichbaum QG, Super M, Ezekowitz RAB (1994) The exon-intron structure and chromosomal localization of the mouse macrophage mannose receptor gene Mrc1: identification of a ricin-like domain at the N-terminus of the receptor. *Biochem Biophys Res Comm* 198(2):682-692

Harris N, Super M, Rits M, Chang G, Ezekowitz RAB (1992) Characterization of the murine macrophage mannose receptor: Demonstration that the downregulation of receptor expression mediated by interferon-γ occurs at the level of transcription. *Blood* 80(9):2363-2373

Ishizaki J, Hanasaki K, Higashino K, Kishino J, Kikuchi N, Ohara O, Arita H (1994) Molecular cloning of pancreatic group I phospholipase A_2 receptor (1994) *J Bio Chem* 268(8):5897-5904

Kery V, Krepinsky JJF, Warren CD, Capek P, Stahl PD (1992) Ligand recognition by purified human mannose receptor. *Arch Biochem Biophys* 298(1):49-55

Kim SJ, Ruiz N, Bezouska K, Drickmaer K (1992) Organization of the gene encoding the human macrophage mannose receptor (MRC1). *Genomics* 14:721-727

Kimura K, Schota M, Mochizuki K, Ohta M, Sugano T (1992) Different preparations of zymosan induced glycogenolysis independently in the perfused rat liver. Involvement of mannose receptors, peptide-leukotrienes and prostaglandins. *Biochem J* 283(3):773-779

Koziel H, Kruskal BA, Ezekowitz RA, Rose RM (1993) HIV impairs alveolar macrophoage mannose receptor function against *Pneumocystis carinii*. *Chest* 103(2-Suppl):111s-112s

Kruskal BA, Sastry K, Warner AB, Mathiew CE, Ezekowitz RAB (1992) Phagocytic chimeric receptors require both transmembrane and cytoplasmic domains from the mannose receptor. *J Exp Med* 176:1673-1680

Lambeau G, Ancian P, Barhanin J, Lazdunski M (1994) Cloning and expression of a membrane receptor for secretory phospholipases A_2. *J Biol Chem* 269(3):1575-1578

Lefkowitz DL, Mills K, Castro A, Lefkowitz SS (1991) Induction of tumor necrosis factor and macrophage-mediated cytotoxicity by horseradish peroxidase and other glycosylated proteins: the role of enzymatic activity and LPS. *J Leuk Biol* 50:615-623

Lennartz MR, Cole FS, Shepherd VL, Wileman TE, Stahl PD (1987) Isolation and characterization of a mannose-specific endocytosis receptor from human placenta. *J Biol Chem* 262(21):9942-9944

Lennartz MR, Cole FS,, Stahl PD (1989) Biosynthesis and processing of the mannose receptor in human macrophages. *J Biol Chem* 264(4):2385-2390

Lew DB, Rattazzi MC (1991) Mitogenic effect of lysosomal hydrolases on bovine tracheal myocytes in culture. *J Clin Invest* 88:1969-1975

Lew DB, Songu-Mize E, Pontow SE, Stahl PD, Rattazzi MC (1994) A mannose receptor mediates mannosyl-rich glycoprotein-induced mitogenesis in bovine airway smooth muscle cells. *J Clin Invest* (in press)

Magnusson S, Berg T (1993) Endocytosis of ricin by rat liver cells *in vivo* and *in vitro* is mainly mediated by mannose receptors on sinusoidal endothelial cells. *Biochem J* 291:749-755

Magnusson S, Berg T, Turpin E, Frenoy JP (1991) Interaction of ricin with sinusoidal endothelial rat liver cells. Different involvement of two distinct carbohydrate-specific mechanisms in surface binding and internalization. *Biochem J* 277-855

Marodi L, Schrieber S, Anderson DC, MacDermott RP, Korchak HM, Johnston RB Jr. (1993) Enhancement of macrophage candidacidal activity by interferon-γ. Increased phagocytosis, killing, and calcium signal mediated by a decreased number of mannose receptors. *J Clin Invest* 91(6):2596-2601

Mokoena T, Gordon S (1985) Modulation of mannosyl, fucosyl receptor activity *in vitro* by lymphokines, γ and α interferons and dexamethasone. *J Clin Invest* 75:624-631

Nathan C (1992) Interferon and inflammation. *In* "Inflammation: Basic principles and clinical correlates (Gallin JI, Goldstein IM, Snyderman R, eds.) Raven Press, New York p. 265-290

Ohsumi Y, Lee YC (1987) Mannose-receptor ligands stimulate secretion of lysosomal enzymes from rabbit alveolar macrophages. *J Biol Chem* 262:7955-7962

Otter M, Zockova P, Kuiper J, Van-Berkel TJC, Barrett- Bergshoeff MM, Rijken DC (1992) Isolation and charaterization of the mannose receptor from human liver potentialy involved in the plasma clearence of tissue-type plasminogen activator. *Hepatol* 16(1):54-59

Paul WE (1991) Interleukin-4: a prototypic immunoregulatory lymphokine. *Blood* 77(9):1859-1870

Petersen TE, Thogersen HC, Skorstengaard K, Vibe-Pedersen K, Sahl P, Sottrup-Jensen L, Magnusson S (1983) Partial primary structure of bovine plasma fibronectin: Three types of internal homology. *Proc NatAcad Sci USA* 80:137-141

Pitt A, Mayorga LS, Schwartz AL, Stahl PD (1992a) Transport of phagosomal components to an endosomal compartment. *J Biol Chem* 267(1):126-32

Pitt A, Mayorga LS, Stahl PD, Schwartz AL (1992b) Alterations in the protein composition of maturing phagosomes. *J Clin Invest* 90:1978-1983

Pontow SE, Kery V, Stahl PD (1992) Mannose Receptor. *Int Rev Cytol* 137B:221-244

Reis e Sousa C, Stahl PD, Austyn JM (1993) Phagocytosis of antigens by Langerhans cells *in vitro*. *J Exp Med* 178(2):509-519

Sato Y, Beuteler E (1993) Binding, internalization, and degradation of mannose-terminated glucocerebrosidase by macrophages. *J Clin Invest* 91(5):1909-1917

Schlesinger PH, Doebber TW, Mandell BF, White R, DeSchryver C, Rodman JS, Miller MJ, Stahl PD (1978) Plasma clearance of glycoproteins with terminal mannose and N-acetylglucosamine by liver non-parenchymal cells. *Biochem J* 176:103-109

Schreiber S, Blum JS, Chappel JC, Stenson WF, Stahl PD, Teitelbaum SL, Perkins SL (1990) Prostaglandin E specifically upregulates the expression of the mannose receptor on bone marrow-derived macrophages. *Cell Reg* 1:403-413

Schreiber S, Perkins SL, Teitelbaum SL, Chappel J, Stahl PD, Blum JS (1993) Regulation of mouse bone marrow macrophage mannose receptor expression and activation by prostoglandin E and IFN-γ. *J Immunol* 151(9):4973-4981

Schreiber S, Stensen WF, McDermitt RP, Stahl PD, Teitelbaum SL, Perkins SL (1991) Monomeric IgG2a promotes maturaton of bone-marrow macrophages and expression of the mannose receptor. *Proc Nat Acad Sci USA* 88:1616-1620

Sett R, Sarkar K, Das PK (1993) Macrophage-directed delivery of doxorubicin conjugated to neoglycoprotein using Leishmaniasis as the model disease. *J infect Disease* 168:994-999

Shepherd VL, Abdolnasuria R, Garrett M, Cowan HB (1990) Downregulation of mannose receptor activity in macrophages after treatment with lipopolysaccharide and phorbol esters. *J Immunol* 145:1530-1536

Shepherd VL, Campbell EJ, Senior RM, Stahl PD (1982) Characterization of the mannose/fucose receptor on human mononuclear phagocytes. *J Reticulo Soc* 32:423-431

Shepherd VL, Hoidal JR (1991) Clearance of neutrophil-derived myeloperoxidase by the macrophage mannose receptor. *Am J Resp Cell Mol Biol* 2:335-340

Shepherd VL, Lee YC, Schlesinger PH, Stahl PD (1981) L-Fucose-terminated glycoconjugates are recognized by pinocytosis receptors on macrophages. *Pro Nat Acad Sci USA*2:1019-1022

Shepherd VL, Tarnowski BI, McLaughlin BJ (1991) Isolation and characterization of a mannose receptor from human pigment epithelium. *Invest Ophthamol Vis Sci* 32:1779-1784

Simmons BM, Stahl PD, Russell J (1986) Mannose receptor-mediated uptake of ricin toxin and ricin A chain by macrophages. *J Biol Chem* 261(17):7912-7920

Simmons BM, Stahl PD, Russell J (1987) *In vivo* depletion of mannose receptor bearing cells from rat liver by ricin A chain: effects on clearance of β-glucuronidase. *Biochem Biophys Res Commun* 146:849-854

Stahl PD (1990) The macrophage mannose receptor. *Am J Respir Cell Mol Biol* 2:317-318

Stahl PD (1992) The mannose receptor and other macrophage lectins. *Curr Op Immunol* 4:49-52

Stahl PD, Rodman JS, Miller MJ, Schlesinger PH (1978) Evidence for receptor mediated binding of glycoproteins, glycoconjugates, and lysosomal glycosidases by alveolar macrophages. *Proc Nat Acad Sci USA* 75(3):1399-1403

Stahl PD, Rodman JS, Schlesinger, P (1976a) Clearance of lysosomal hydrolases following intravenous infusion. *Arch Biochem Biophys* 177:594-605

Stahl PD, Schlesinger PH, Sigardson E, Rodman JS, and Lee YC (1980) Receptor-mediated pinocytosis of mannose glycoconjugates by macrophages: characterization and evidence for receptor recycling. *Cell* 19:207-215

Stahl PD, Six H, Rodman JS, Schlesinger P, Tulsiani DRP, Touster O (1976b) Evidence for specific recognition sites mediating clearance of lysosomal enzymes *in vivo*. *Proc Nat Acad Sci USA* 73(11):4045-4049

Stein M, Keshav S, Harris N, Gordon S (1992) Interleukin 4 potently enhances murine macrophage mannose receptor activity: a marker of alternative immunologic macrophage activation *J Exp Med* 176:287-292

Taylor ME (1993) Recognition of complex carbohydrates by the macrophage mannose receptor. *Biochem Soc Trans* 21(2):468-73

Taylor ME, Bezouska K, Drickamer K (1992) Contribution to ligand binding by multiple carbohydrate-recognition domains in the macrophage mannose receptor. *J Biol Chem* 267(3):1719-1726

Taylor ME, Conary JT, Lennartz MR, Stahl PD, Drickamer K (1990) Primary structure of the mannose receptor contains multiple motifs resembling carbohydrate-recognition domains. *J Biol Chem* 265:12156-12162
Taylor ME, Drickamer K (1993) Structural requirements for high affinity binding of complex ligands by the macrophage mannose receptor. *J Biol Chem* 268(1):399-404
Tietze C, Schlesinger P, Stahl PD (1980) Chloroquine and ammonium ion inhibit receptor-mediated endocytosis of mannose-glycoconjugates by macrophages: apparent inhibition of receptor recycling. *Biochem Biophys Res Comm* 93:1-8
Tietze C, Schlesinger P, Stahl PD (1982) Mannose-specific endocytosis receptor of alveolar macrophages: demonstration of two functionally distinct intracellular pools of receptor and their roles in receptor recycling. *J Cell Biol* 92:417-424
Tietze C, Schlesinger PH, Wileman T, Boshans RL, Schlesinger PH, Stahl PD (1984) Monensin inhibits recycling of macrophage mannose-glycoprotein receptors and ligand delivery to lysosomes. *Biochem J* 220:665-675
Weis WI, Crichlow GV, Murthy HMK, Hendrickson WA, Drickamer K (1991a) Physical characterization and crystallization of the carbohydrate-recognition domain of a mannose-binding protein from rat. *J Biol Chem* 266:20678-20686
Weis WI, Drickamer K, Hendrickson WA (1992) Structure of a C-type mannose-binding protein complexed with an oligosaccharide. *Nature* 360:127-134
Weis WI, Kahn R, Fourme R, Drickamer K, Hendrickson WA (1991b) Structure of the calcium-dependent lectin domain from a rat mannose-binding protein determined by MAD phasing. *Science* 254:1608-1615
Wileman T, Boshans RL, Schlesinger P, Stahl PD (1984) Monensin inhibits recycling of macrophage mannose-glycoprotein receptors and ligand delivery to lysosomes. *Biochem J* 220:665-675
Wilson ME, Pearson RD (1988) Role of CR3 and mannose receptors in the attachment and ingestion of *Leishmania donovani* by human mononuclear phagocytes. *Infect Immun* 56:363-369
Wright SD (1992) Receptors for complement and the biology of phagocytosis. *In* "Inflammation: Basic principles and clinical correlates (Gallin JI, Goldstein IM, Snyderman R, eds.) Raven Press, New York p. 477-495

Cell Cycle Changes to the Golgi Apparatus in Animal Cells

Graham Warren
Imperial Cancer Research Fund
PO Box 123
44 Lincoln's Inn Fields
London WC2A 3PX

Morphology

Dramatic changes occur to the morphology of the Golgi apparatus at the onset of mitosis in animal cells (Warren, 1985, 1993) and figure 1 provides a schematic view of this process. The compact, juxta-nuclear reticulum found in interphase cells is converted, during prophase, to several hundred discrete Golgi stacks. This is thought to occur by scission of the tubules that connect equivalent cisternae in adjacent stacks (Rambourg and Clermont, 1990; Rothman and Warren, 1994). During the middle phases of mitosis (prometaphase, metaphase and anaphase), each stack undergoes complete vesiculation to yield Golgi clusters (Lucocq et al., 1987; Lucocq and Warren, 1987). These clusters then shed vesicles which become dispersed throughout the mitotic cell cytoplasm (Lucocq et al., 1989). During telophase, these processes are reversed; clusters grow by accretion of vesicles which then fuse to re-form Golgi stacks. The dispersed stacks then move to the peri-centriolar region, probably by movement along microtubules (Ho et al., 1989; Corthésy-Theulaz et al., 1992), where they undergo homotypic fusion to re-form the interphase Golgi apparatus (Lucocq et al., 1989). The end result of this stochastic process is that the original mother Golgi apparatus is equally distributed between the two daughter cells (Birky, 1983).

Membrane Traffic

These striking morphological changes to the Golgi apparatus are accompanied by an inhibition of membrane traffic (Warren, 1993). All vesicle-mediated processes on the endocytic and exocytic pathways are inhibited (with one exception; Kreiner and Moore, 1990) at the onset of mitosis and do not resume until telophase. Transport to the Golgi stack from the endoplasmic reticulum is inhibited (Featherstone et al., 1985) as is intra-Golgi transport (Stuart et al., 1993; Collins and Warren, 1992; Mackay et al., 1993). Both resume during telophase and assembly of the Golgi stack

NATO ASI Series, Vol. H 91
Trafficking of Intracellular Membranes
Edited by M.C. Pedroso de Lima N. Düzgüneş and D. Hoekstra

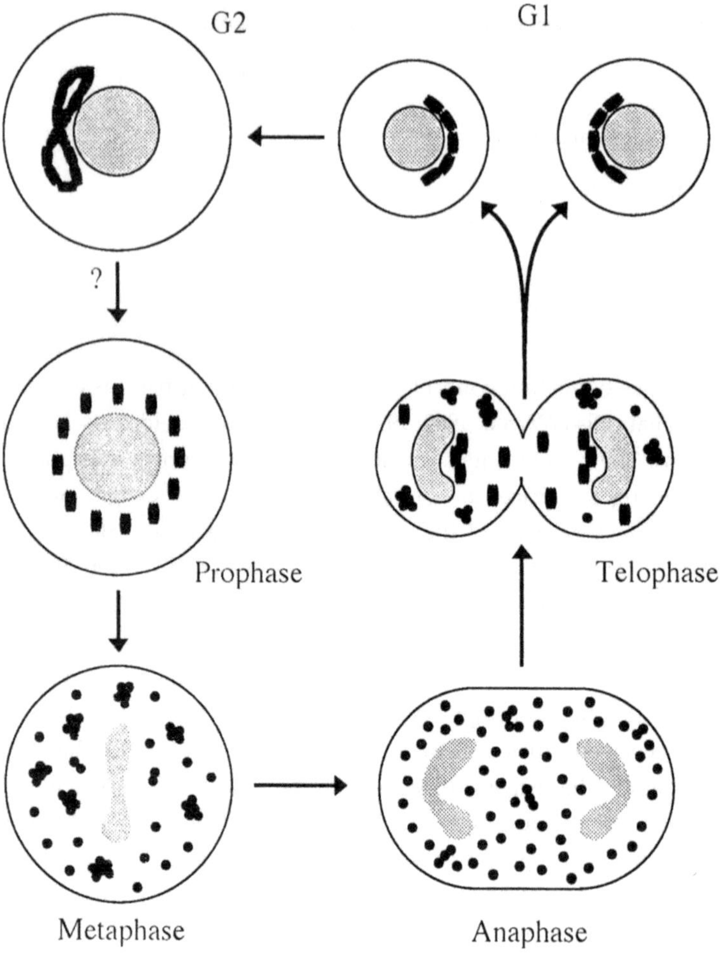

Figure 1. Schematic outline of the cell cycle changes to the Golgi apparatus in HeLa cells.

The juxta-nuclear reticulum fragments during prophase and the resulting stacks vesiculate during pro-metaphase and metaphase forming Golgi clusters. These shed Golgi vesicles which become dispersed throughout the mitotic cell cytoplasm by anaphase. During telophase the clusters re-grow by accretion of vesicles which fuse to form Golgi stacks. These congregate in the peri-centriolar region and fuse to re-form a single copy Golgi in each daughter cell.

is so rapid that it is complete before newly-synthesised proteins begin to arrive from the endoplasmic reticulum (Souter et al., 1993).

The fact that inhibition of intra-Golgi transport is accompanied by vesiculation of the Golgi apparatus has given rise to a simple idea to explain the vesiculation process (Warren, 1985, 1993). During interphase, transport vesicles bud from the dilated cisternal rims and fuse with the next cisterna in the stack towards the trans side (Figure 2, top). If, at the onset of mitosis, budding of transport vesicles were to continue, but fusion of the transport vesicles with the next cisterna in the stack were to be inhibited, then vesiculation of the Golgi stack would occur (Figure 2, bottom). Resumption of fusion during telophase would reverse this process.

Fragmentation in a cell-free system

In order to test this hypothesis and understand the molecular mechanism underlying Golgi vesiculation, we have reconstituted the mitotic disassembly of Golgi stacks in a cell-free system (Misteli and Warren, 1994). A schematic outline of this process is shown in figure 3. Rat liver Golgi stacks, when incubated with high levels of mitotic cytosol, underwent vesiculation in a manner dependent upon time, temperature, energy (added in the form of ATP) and the mitotic kinase, $p34^{cdc2}$. The dependence on the mitotic kinase was demonstrated by using a cell line with a temperature-sensitive mutation in this protein (Th'ng et al., 1990). Vesiculation was monitored either directly, using electron microscopy, or by using a biochemical assay. Low speed centrifugation was used to separate mitotic Golgi vesicles from Golgi remnants and the amount was quantitated by assaying for the trans Golgi enzyme, β1, 4-galactosyltransferase.

Stereological analysis showed that the membrane lost from stacked and single cisternae appeared in both small (50-100nm in diameter) and large (100-200nm in diameter) vesicles, the smaller vesicles constituting more than 50% of the total membrane. The half-time for this process was about 20mins., reasonably similar to that observed in vivo (Zieve et al., 1980). The production of these small vesicles depended upon the COP-mediated budding mechanism that ensures budding of transport vesicles at the level of both the endoplasmic reticulum and the Golgi apparatus (Rothman and Orci, 1992). This was shown by immunodepletion of one of the subunits of COP coats (the coatomer) from mitotic cytosol. Vesicles were no longer formed but highly fenestrated networks appeared, an effect reversed by the

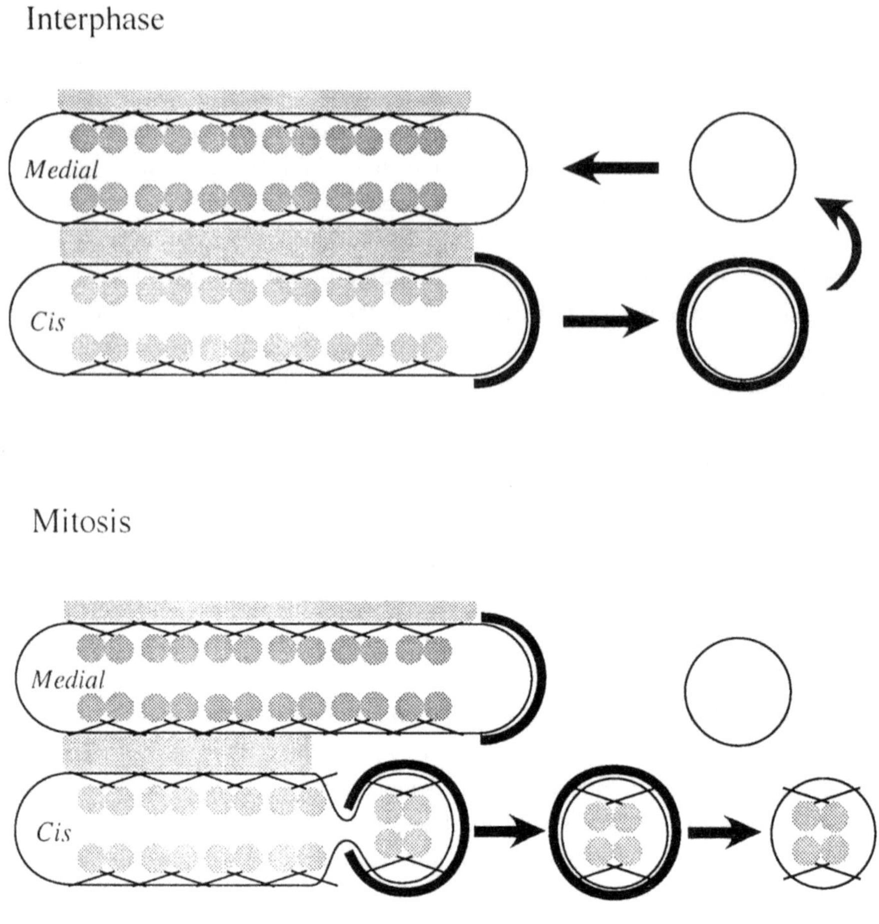

Figure 2. Model for the vesiculation of the Golgi stack.

Top: During interphase, transport vesicles bud from the dilated cisternal rims and fuse with the next cisterna in the stack towards the trans side. Bottom: During mitosis the budding of transport vesicles continues but they can no longer fuse with the next cisterna in the stack. Relaxation of the process that retains Golgi enzymes in the stack coupled with the unstacking of cisternae ensures that complete vesiculation occurs.

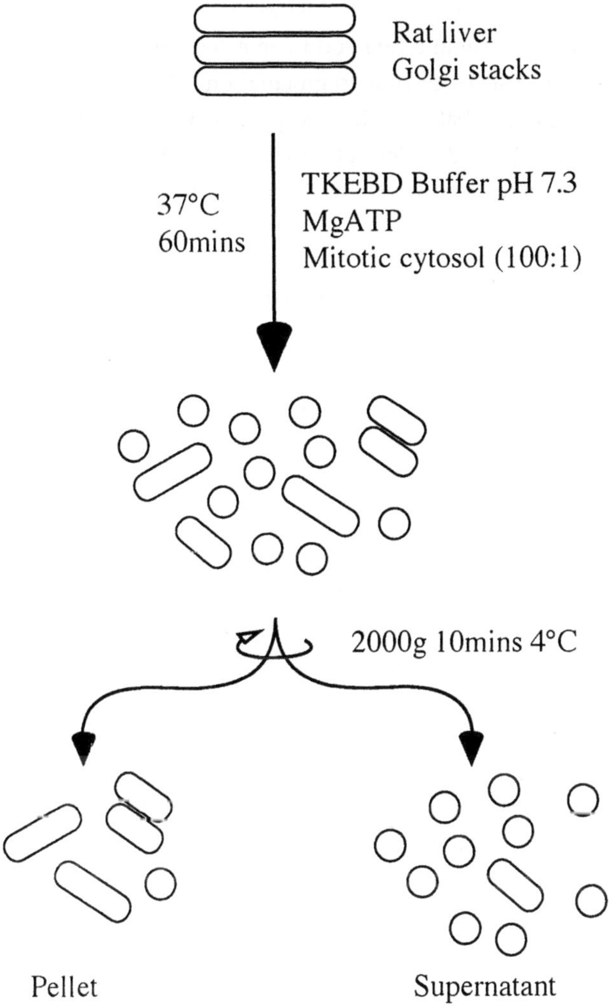

Figure 3. Fragmentation Assay.
Rat liver Golgi stacks are incubated with a 100-fold excess of mitotic cytosol in TKEBD Buffer (50mM Tris - HCl, pH 7.3, 50mM KCl, 10mM MgCl2, 20mM β-glycerophosphate, 15mM EGTA, 1mM DTT) containing 2mM ATP and incubated for 60 mins. at 37°C. Samples were then fixed directly for electron microscopy or first centrifuged at low speed to separate the mitotic Golgi vesicles (supernatant) from the remnants (pellet).

re-addition of purified coatomer. Using GTPγS to prevent uncoating, COP-coated vesicles were shown to form at a rate and to an extent sufficient to account for the production of small vesicles during fragmentation. The rate of production was similar to that found in incubations with interphase cytosol showing that budding is neither activated nor inhibited under mitotic conditions. Together these experiments provide strong support for our hypothesis (Warren, 1985 and 1993).

Retention of Golgi enzymes and cisternal stacking

Though an inhibition of fusion is the primary requirement of our hypothesis, it is clear that this alone would not be sufficient to explain the observed vesiculation of Golgi stacks. Transport vesicles could only continue to bud if the cisternae are separated from each other and if there is a relaxation of the mechanism that normally retains Golgi enzymes in the stack. It is now clear that the membrane-spanning domain of Golgi enzymes is responsible for their retention in the Golgi stack (Machamer, 1993) but the mechanism is still unclear. On the one hand it has been suggested that Golgi enzymes form specific hetero-oligomers that are too large to enter budding transport vesicles (Nilsson et al., 1993). These oligomers would depend in part upon the membrane-spanning domain for their formation (Nilsson et al., 1994). On the other hand it has been suggested that the length of the membrane-spanning domain is the primary determinant (Bretscher and Munro, 1993). This is based on earlier observations by Orci (1981) suggesting that there is a gradient of cholesterol along the exocytic pathway. Increasing levels of cholesterol increase the thickness of lipid bilayers so that Golgi enzymes would move along the exocytic pathway until the increasing thickness of the membrane prevented further movement. Whatever the mechanism, it must be capable of regulation during mitosis such that relaxation can occur allowing the resident enzymes to enter the budding transport vesicles.

The mechanism that stacks the cisternae is also unclear. We have isolated an intercisternal matrix that binds medial Golgi enzymes probably by their cytoplasmic tails (Slusarewicz et al., 1994). Separation of this matrix from the cytoplasmic tails of Golgi enzymes following phosphorylation would certainly lead to cisternal unstacking at the onset of mitosis. Experiments have still to be carried out to test this possibility.

Acknowledgements

I would like to thank the many people in my laboratory who, over the years, have contributed to the work described in this review. I also thank Louise Dewhurst for help in preparing the manuscript.

References

Birky C W (1983) The partitioning of cytoplasmic organelles at cell division. Int Rev Cytology 15: 49-89

Bretscher MS , Munro S (1993) Cholesterol and the Golgi apparatus. Science 261: 1-3

Collins R, Warren G (1992) Sphingolipid transport in mitotic HeLa cells. J biol Chem 267: 24906-24911

Corthésy-Theulaz I, Pauloin A, Pfeffer SR (1992) Cytoplasmic dynein participates in the centrosomal localization of the Golgi complex. J Cell Biol 118: 1333-1346

Featherstone C, Griffiths G, Warren G (1985) Newly synthesized G protein of vesicular stomatitis virus is not transported to the Golgi complex in mitotic cells. J Cell Biol 101: 2036-2046

Ho WC, Allan VJ, van Meer G, Berger E G, Kreis T E (1989) Reclustering of scattered Golgi elements occurs along microtubules. Eur J Cell Biol 48: 250-263

Kreiner T, Moore H-P (1990) Membrane traffic between secretory compartments is differentially affected during mitosis. Cell Regulation 1: 415-424

Lucocq JM, Berger EG, Warren G (1989) Mitotic Golgi fragments in HeLa Cells and their role in the reassembly pathway. J Cell Biol 109: 463-474

Lucocq JM, Pryde JG, Berger EG, Warren G (1987) A mitotic form of the Golgi apparatus in HeLa cells. J Cell Biol 104: 865-874

Lucocq JM, Warren G (1987) Fragmentation and partitioning of the Golgi apparatus during mitosis in HeLa cells. EMBO J 6: 3239-3246

Machamer CE (1993) Targeting and retention of Golgi membrane proteins. Curr Op Cell Biol 5: 606-612

Mackay DM, Kieckbusch R, Adamczewski JP, Warren G (1994) Cyclin A-mediated inhibition of intra-Golgi transport requires p34*cdc2* . FEBS Lett 336: 549-554

Misteli T, Warren G (1994) Transport vesicles are involved in the mitotic fragmentation of Golgi stacks in a cell-free system. J Cell Biol 125: 269-282

Nilsson T, Hoe MH, Slusarewicz P, Rabouille C, Watson R, Hunte F, Watzele G, Berger EG, Warren G (1994) Kin recognition between *medial* Golgi enzymes in HeLa cells. EMBO J 13: 562-574

Nilsson T, Slusarewicz P, Hoe M, Warren G (1993) Kin Recognition: A Model for the Retention of Golgi Enzymes. FEBS Lett 330: 1-4

Orci L, Montesano R, Meda P, Malaisse-Lagae F, Brown D, Perrelet A, Vassalli P (1981) Hetergeneous distribution of filipin-cholesterol complexes across the cisternae of the Golgi apparatus. Proc Natl Acad Sci (USA) 78: 293-297

Rambourg A, Clermont Y (1990) Three-dimensional electron microscopy: structure of the Golgi apparatus. Eur J Cell Biol 51: 189-200

RothmanJE, Orci L (1992) Molecular dissection of the secretory pathway. Nature 355: 409-416

Rothman JE, Warren G (1994) Implications of the SNARE hypothesis for the specificity dynamics topology of intracellular membranes. Curr Biol 4: 220-233

Slusarewicz P, Nilsson T, Hui N, Watson R, Warren G (1994) Isolation of a matrix that binds *medial* Golgi enzymes. J Cell Biol 124: 405-414

Souter E, Pypaert M, Warren G (1993) The Golgi stack reassembles during telophase before arrival of proteins transported from the endoplasmic reticulum. J Cell Biol 122: 533-540

Stuart R, Mackay D, Adamczewski J, Warren G (1993) Inhibition of intra-Golgi transport *in vitro* by mitotic kinase. J biol Chem 268: 4050-4054

Th'ng JP, Wright PS, Hamaguchi J, Lee MG, Norbury CJ, Nurse P, Bradbury EM (1990) The FT210 cell line is a mouse G2 phase mutant with a temperature-sensitive CDC2 gene product. Cell 63: 313-24

Warren G (1985) Membrane Traffic Organelle Division. Tr Biochem Sci 10: 439-443

Warren G (1993) Membrane partitioning during cell division. Ann Rev Biochem 62: 323-348

Zieve GW, Turnbull D, Mullins JM, McIntosh JR (1980) Production of large numbers of mitotic mammalian cells by use of the reversible microtubule inhibitor nocodazole. Exp Cell Res 126: 397-405

Index

NATO ASI Series H

Vol. 39: **Chemosensory Information Processing.**
Edited by D. Schild. 403 pages. 1990.

Vol. 40: **Dynamics and Biogenesis of Membranes.**
Edited by J. A. F. Op den Kamp. 367 pages. 1990.

Vol. 41: **Recognition and Response in Plant-Virus Interactions.**
Edited by R. S. S. Fraser. 467 pages. 1990.

Vol. 42: **Biomechanics of Active Movement and Deformation of Cells.**
Edited by N. Akkas. 524 pages. 1990.

Vol. 43: **Cellular and Molecular Biology of Myelination.**
Edited by G. Jeserich, H. H. Althaus, and T. V. Waehneldt. 565 pages. 1990.

Vol. 44: **Activation and Desensitization of Transducing Pathways.**
Edited by T. M. Konijn, M. D. Houslay, and P. J. M. Van Haastert. 336 pages. 1990..

Vol. 45: **Mechanism of Fertilization: Plants to Humans.**
Edited by B. Dale. 710 pages. 1990.

Vol .46: **Parallels in Cell to Cell Junctions in Plants and Animals.**
Edited by A. W Robards, W. J . Lucas, J . D. Pitts, H . J . Jongsma, and D. C. Spray. 296 pages. 1990.

Vol. 47: **Signal Perception and Transduction in Higher Plants.**
Edited by R. Ranjeva and A. M. Boudet. 357 pages. 1990.

Vol. 48: **Calcium Transport and Intracellular Calcium Homeostasis.**
Edited by D. Pansu and F. Bronner. 456 pages. 1990.

Vol. 49: **Post-Transcriptional Control of Gene Expression.**
Edited by J. E. G. McCarthy and M. F. Tuite. 671 pages. 1990.

Vol. 50: **Phytochrome Properties and Biological Action.**
Edited by B. Thomas and C. B. Johnson. 337 pages. 1991.

Vol. 51: **Cell to Cell Signals in Plants and Animals.**
Edited by V. Neuhoff and J. Friend. 404 pages. 1991.

Vol. 52: **Biological Signal Transduction.**
Edited by E. M . Ross and K . W. A. Wirtz. 560 pages. 1991.

Vol. 53: **Fungal Cell Wall and Immune Response.**
Edited by J. P. Latge and D. Boucias. 472 pages. 1991.

Vol. 54: **The Early Effects of Radiation on DNA.**
Edited by E. M. Fielden and P. O'Neill. 448 pages. 1991.

Vol. 55: **The Translational Apparatus of Photosynthetic Organelles.**
Edited by R. Mache, E. Stutz, and A. R. Subramanian. 260 pages. 1991.

Vol. 56: **Cellular Regulation by Protein Phosphorylation.**
Edited by L. M. G. Heilmeyer, Jr. 520 pages. 1991.

NATO ASI Series H

Vol. 57: **Molecular Techniques in Taxonomy.**
Edited by G . M . Hewitt, A. W. B. Johnston, and J. P. W. Young . 420 pages. 1991.

Vol. 58: **Neurocytochemical Methods.**
Edited by A. Calas and D. Eugene. 352 pages. 1991.

Vol. 59: **Molecular Evolution of the Major Histocompatibility Complex.**
Edited by J. Klein and D. Klein. 522 pages. 1991.

Vol. 60: **Intracellular Regulation of Ion Channels.**
Edited by M. Morad and Z. Agus. 261 pages. 1992.

Vol. 61: **Prader-Willi Syndrome and Other Chromosome 15q Deletion Disorders.**
Edited by S. B. Cassidy. 277 pages. 1992.

Vol. 62: **Endocytosis. From Cell Biology to Health, Disease and Therapie.**
Edited by P. J. Courtoy. 547 pages. 1992.

Vol. 63: **Dynamics of Membrane Assembly.**
Edited by J. A. F. Op den Kamp. 402 pages. 1992.

Vol. 64: **Mechanics of Swelling. From Clays to Living Cells and Tissues.**
Edited by T. K. Karalis. 802 pages. 1992.

Vol. 65: **Bacteriocins, Microcins and Lantibiotics.**
Edited by R. James, C. Lazdunski, and F. Pattus. 530 pages. 1992.

Vol. 66: **Theoretical and Experimental Insights into Immunology.**
Edited by A. S. Perelson and G. Weisbuch. 497 pages. 1992.

Vol. 67: **Flow Cytometry. New Developments.**
Edited by A. Jacquemin-Sablon. 1993.

Vol. 68: **Biomarkers. Research and Application in the Assessment of Environmental Health.** Edited by D. B. Peakall and L. R. Shugart. 138 pages. 1993.

Vol. 69: **Molecular Biology and its Application to Medical Mycology.**
Edited by B. Maresca, G. S. Kobayashi, and H. Yamaguchi. 271 pages. 1993.

Vol. 70: **Phospholipids and Signal Transmission.**
Edited by R. Massarelli, L. A. Horrocks, J. N. Kanfer, and K. Löffelholz. 448 pages. 1993.

Vol. 71: **Protein Synthesis and Targeting in Yeast.**
Edited by A. J. P. Brown, M. F. Tuite, and J. E. G. McCarthy. 425 pages. 1993.

Vol. 72: **Chromosome Segregation and Aneuploidy.**
Edited by B. K. Vig. 425 pages. 1993.

Vol. 73: **Human Apolipoprotein Mutants III. In Diagnosis and Treatment.**
Edited by C. R. Sirtori, G. Franceschini, B. H. Brewer Jr. 302 pages. 1993.

NATO ASI Series H

Vol. 74: **Molecular Mechanisms of Membrane Traffic.**
Edited by D. J. Morré, K. E. Howell, and J. J. M. Bergeron. 429 pages. 1993.

Vol. 75: **Cancer Therapy. Differentiation, Immunomodulation and Angiogenesis.**
Edited by N. D'Alessandro, E. Mihich, L. Rausa, H. Tapiero, and T. R. Tritton. 299 pages. 1993.

Vol. 76: **Tyrosine Phosphorylation/Dephosphorylation and Downstream Signalling.**
Edited by L. M. G. Heilmeyer Jr. 388 pages. 1993.

Vol. 77: **Ataxia-Telangiectasia.** Edited by R. A. Gatti, R. B. Painter. 306 pages. 1993.

Vol. 78: **Toxoplasmosis.** Edited by J. E. Smith. 272 pages. 1993.

Vol. 79: **Cellular Mechanisms of Sensory Processing. The Somatosensory System.**
Edited by L. Urban. 514 pages. 1994.

Vol. 80: **Autoimmunity: Experimental Aspects.**
Edited by M. Zouali. 318 pages. 1994.

Vol. 81: **Plant Molecular Biology. Molecular Genetic Analysis of Plant Development and Metabolism.**
Edited by G. Coruzzi, P. Puigdomènech. 579 pages. 1994.

Vol. 82: **Biological Membranes: Structure, Biogenesis and Dynamics.**
Edited by Jos A. F. Op den Kamp. 367 pages. 1994.

Vol. 83: **Molecular Biology of Mitochondrial Transport Systems.**
Edited by M. Forte, M. Colombini. 420 pages. 1994.

Vol. 84: **Biomechanics of Active Movement and Division of Cells.**
Edited by N. Akkaş. 587 pages. 1994.

Vol. 85: **Cellular and Molecular Effects of Mineral and Synthetic Dusts and Fibres.**
Edited by J. M. G. Davis, M.-C. Jaurand. 448 pages. 1994.

Vol. 86: **Biochemical and Cellular Mechanisms of Stress Tolerance in Plants.**
Edited by J. H. Cherry. 616 pages. 1994.

Vol. 87: **NMR of Biological Macromolecules.**
Edited by C. I. Stassinopoulou. 616 pages. 1994.

Vol. 88: **Liver Carcinogenesis. The Molecular Pathways.**
Edited by G. G. Skouteris. 502 pages. 1994.

Vol. 89: **Molecular and Cellular Mechanisms of H^+ Transport.**
Edited by B. H. Hirst. 504 pages. 1994.

Vol. 90: **Molecular Aspects of Oxidative Drug Metabolizing Enzymes: Their Significance in Environmental Toxicology, Chemical Carcinogenesis and Health.**
Edited by E. Arınç, J. B. Schenkman, E. Hodgson. 623 pages. 1994.

NATO ASI Series H

Vol. 91: **Trafficking of Intracellular Membranes: From Molecular Sorting to Membrane Fusion.** Edited by M. C. Pedroso de Lima, N. Düzgüneş, D. Hoekstra. 371 pages. 1995.

MIX
Papier aus verantwortungsvollen Quellen
Paper from responsible sources
FSC® C105338

If you have any concerns about our products,
you can contact us on
ProductSafety@springernature.com

In case Publisher is established outside the EU,
the EU authorized representative is:
Springer Nature Customer Service Center GmbH
Europaplatz 3, 69115 Heidelberg, Germany

Printed by Libri Plureos GmbH
in Hamburg, Germany